CHEMICAL AND STRUCTURE MODIFICATION OF POLYMERS

CHEMICAL AND STRUCTURE MODIFICATION OF POLYMERS

Edited by
Kajetan Pyrzynski, PhD
Grzegorz Nyszko, PhD
Gennady E. Zaikov, DSc

AAP APPLE
ACADEMIC
PRESS

Apple Academic Press Inc. | Apple Academic Press Inc.
3333 Mistwell Crescent | 9 Spinnaker Way
Oakville, ON L6L 0A2 | Waretown, NJ 08758
Canada | USA

First issued in paperback 2021

Exclusive worldwide distribution by CRC Press, a member of Taylor & Francis Group
No claim to original U.S. Government works

ISBN 13: 978-1-77463-543-8 (pbk)
ISBN 13: 978-1-77188-122-7 (hbk)

Typeset by Accent Premedia Services (www.accentpremedia.com)

Library and Archives Canada Cataloguing in Publication

Chemical and structure modification of polymers/edited by Kajetan Pyrzynski, Grzegorz Nyszko, Gennady E. Zaikov.

Includes bibliographical references and index.
Issued in print and electronic formats.
ISBN 978-1-77188-122-7 (hardcover).--ISBN 978-1-4987-2744-0 (pdf)
1. Polymers. 2. Polymerization. 3. Chemistry, Physical and theoretical. I. Zaikov, G. E. (Gennadiĭ Efremovich), 1935-, author, editor II. Pyrzynski, Kajetan, editor III. Nyszko, Grzegorz, editor

QD381.C54 2015   547'.7   C2015-905771-X   C2015-905772-8

Library of Congress Cataloging-in-Publication Data

Chemical and structure modification of polymers / [edited by] Kajetan Pyrzynski, Grzegorz Nyszko, Gennady E. Zaikov

pages cm
Includes bibliographical references and index.
ISBN 978-1-77188-122-7 (alk. paper)
1. Polymers. 2. Polymerization. I. Pyrzynski, Kajetan. II. Nyszko, Grzegorz. III. Zaikov, G. E. (Gennadii Efremovich), 1935-

QD381.C476 2015   547'.7--dc23   2015031587

Apple Academic Press also publishes its books in a variety of electronic formats. Some content that appears in print may not be available in electronic format. For information about Apple Academic Press products, visit our website at **www.appleacademicpress.com** and the CRC Press website at **www.crcpress.com**

CONTENTS

LIST OF CONTRIBUTORS

Shahab Hasan oglu Akhyari
Institute of Petrochemical Processes of National Academy of Sciences of Azerbaijan, Baku, Azerbaijan

E. Aleksandrov
N.M. Emanuel Institute of Biochemical Physics of Russian Academy of Sciences, 4 Kosygin St., Moscow 119334, Russian Federation; Phone: +7 (495) 939 7318; Mobile: +7-916-953-76-19; E-mail: 28en1937@mail.ru

Olga V. Alekseeva
G.A. Krestov Institute of Solution Chemistry, Russian Academy of Sciences, Akademicheskaya St., 1; Ivanovo, 153045, Russian Federation

B. S. Alikhadzhieva
Chechen State Pedagogical Institute, 33, Kievskaj St., 364037, Grozny, Russian Federation, E-mail: belkaas52@liSt.ru

Jimsher Aneli
Institute of Macromolecular Chemistry and Polymeric Materials, I. Javakhishvili Tbilisi State University, I. Chavchavadze Ave. 13, 0179 Tbilisi, Georgia

Esen Arkis
Specialist, Izmir Institute of Technology, Department of Chemical Engineering, Gulbahce Urla 35430 Urla Izmir Turkey

Arda Aytac
Master Student, Izmir Institute of Technology, Department of Chemical Engineering, Gulbahce Urla 35430 Urla Izmir Turkey, E-mail: ardaaytac@iyte.edu.tr

Nadezhda A. Bagrovskaya
G.A. Krestov Institute of Solution Chemistry, Russian Academy of Sciences, Akademicheskaya St., 1; Ivanovo, 153045, Russian Federation

E. Bakuradze
Department of Biology Faculty of Exact and Natural Sciences, Iv. Javakhishvili Tbilisi State University, Tbilisi, Georgia

Beste Balci
Master Student, Izmir Institute of Technology, Department of Chemical Engineering, Gulbahce Urla 35430 Urla Izmir Turkey, E-mail: beste.blc@hotmail.com

Devrim Balkose
Professor, Izmir Institute of Technology, Department of Chemical Engineering, Gulbahce Urla 35430 Urla Izmir Turkey, E-mail: devrimbalkose@gmail.com

Galina A. Batalova
North-East Agricultural Research Institute, 166-a Lenin St., Kirov, 610007, Russian Federation

Marina V. Bazunova
Docent of the Department of High-Molecular Connections and General Chemical Technology of the Chemistry Faculty of the Bashkir State University, 450076, Ufa, Zaks Validi Street, 32, Russian Federation, Tel.: (347) 229–96–86; Mobile: 89276388192; E-mail: mbazunova@mail.ru

T. I. Borodina
Joint Institute for High Temperatures, Russian Academy of Science, 13(2) St. Izhorskaya, Moscow, 125412, Russian Federation

Hayrullah Cetinkaya
Student, Izmir Institute of Technology, Department of Chemical Engineering, Gulbahce Urla 35430 Urla Izmir Turkey, E-mail: hayrullahcetinkaya@iyte.edu.tr

A. E. Chalykh
Frumkin Institute of Physical Chemistry and Electrochemistry, Russian Academy of Sciences, Leninskii pr. 31, Moscow, 119991, Russian Federation

Valentina Chernova
Bashkir State University, 450074, Zaki Validi St., 32, Ufa, Russian Federation

M. Chertenkov
N.M. Emanuel Institute of Biochemical Physics, Russian Academy of Sciences, 4, Kosygin St., Moscow 119334, Russian Federation

Funda Colak
Master Student, Izmir Institute of Technology, Department of Chemical Engineering, Gulbahce Urla 35430 Urla Izmir Turkey, E-mail: fundacolak@yahoo.com

Begum Can Dilhan
Undergraduate Senior Year Student, Izmir Institute of Technology, Department of Chemical Engineering, Gulbahce Urla 35430 Urla Izmir Turkey, E-mail: bgmdhncn@gmail.com

D. Dzidzigiri
Department of Biology Faculty of Exact and Natural Sciences, Iv. Javakhishvili Tbilisi State University, Tbilisi, Georgia, E-mail: d_dzidziguri@yahoo.com

Erol Erbay
Petkim Petrokimya Holding, Izmir, Turkey, E-mail: fnasirov@petkim.com.tr

Izabela Esartia
Iv. Javakhishvili Tbilisi State University, Department of Chemistry, I. Chavchavadze Ave., 3, Tbilisi 0179, Georgia

Alfiya Galina
Bashkir State University, 450074, Zaki Validi St., 32, Ufa, Russian Federation

V. K. Gerasimov
Frumkin Institute of Physical Chemistry and Electrochemistry, Russian Academy of Sciences, Leninskii pr. 31, Moscow, 119991, Russian Federation, E-mail: vladger@mail.ru

Ece Topagac Germen
Master Student, Izmir Institute of Technology, Department of Chemical Engineering, Gulbahce Urla 35430 Urla Izmir Turkey, E-mail: ecetopagac@iyte.edu.tr

N. Giorgobiani
Department of Biology Faculty of Exact and Natural Sciences, Iv. Javakhishvili Tbilisi State University, Tbilisi, Georgia

M. D. Goldfein
Saratov State University Named After N.G. Chernyshevsky, Russian Federation, E-mail: goldfeinmd@mail.ru

O. A. Gololobova
Joint Institute for High Temperatures, Russian Academy of Science, 13(2) St. Izhorskaya, Moscow, 125412, Russian Federation

Nemat Akif oglu Guliyev
Institute of Petrochemical Processes of National Academy of Sciences of Azerbaijan, Baku, Azerbaijan

A. K. Haghi
University of Guilan, Rasht, Iran

B. A. Howell
Central Michigan University, Chemical Faculty, Mount Pleasant, MI, USA, E-mail: bob.a.howell@cmich.edu

Marina Iskakova
Ak. Tsereteli Kutaisi State University, Department of Chemical Technology,

Alaz Izer
Izmir Institute of Technology Department of Chemical Engineering Gulbahce Urla Izmir Turkey

Natia Jalagonia
Iv. Javakhishvili Tbilisi State University, Department of Chemistry, I. Chavchavadze Ave., 3, Tbilisi 0179, Georgia

Nazil Fazil oglu Janibayov
Institute of Petrochemical Processes of National Academy of Sciences of Azerbaijan, Baku, Azerbaijan, E-mail: j.nazil@yahoo.com

Tugce Nefise Kahyaoglu
Izmir Institute of Technology Department of Chemical Engineering Gulbahce Urla Izmir Turkey

Mohammad Kanafchian
University of Guilan, Rasht, Iran

Motahareh Kanafchian
University of Guilan, Rasht, Iran

S. G. Karpova
Institute of the Russian Academy of Sciences N.M. Emanuel Institute of Biochemical Physics, Russian Academy of Sciences, Moscow, Russian Federation

V. T. Karpukhin
Joint Institute for High Temperatures, Russian Academy of Science, 13(2) St. Izhorskaya, Moscow, 125412, Russian Federation, E-mail: vtkarp@gmail.com

Z. S. Khasbulatova
Chechen State Teacher Institute, 33, Kievskaj St., 364037, Grozny, Russian Federation; E-mail: hasbulatova@liSt.ru

A. V. Khvatov
N.M. Emanuel Institute of Biochemical Physics of Russian Academy of Sciences, 4 Kosygin St., Moscow 119334, Russian Federation

N. N. Kolesnikova
N.M. Emanuel Institute of Biochemical Physics of Russian Academy of Sciences, 4 Kosygin St., Moscow 119334, Russian Federation

Sergei V. Kolesov
Professor of the Department of High-Molecular Connections and General Chemical Technology of the Chemistry Faculty of the Bashkir State University, 450076, Ufa, Zaks Validi Street, 32, Russian Federation, Tel.: (347) 229–96–86; E-mail: kolesovservic@mail.ru

N. N. Komova
Moscow State University of Fine Chemical Technology, 86 Vernadskii prospekt, Moscow 119571, Russian Federation, E-mail: Komova_@mail.ru

G. A. Korablev
Doctor of Chemical Sciences, Professor, Head of Department of Physics,Izhevsk State Agricultural Academy, Studencheskaya St., 11, 426069 Izhevsk, Russian Federation, E-mail: korablevga@mail.ru

Alexander L. Kovarski
N.M. Emanuel Institute of Biochemical Physics RAS, 119334, Kosygin St., 4, Moscow, Russian Federation

N. V. Kozhevnikov
Saratov State University Named After N.G. Chernyshevsky, Russian Federation

Georgiy V. Kozlov
Kh.M. Berbekov Kabardino-Balkarian State University, Chernyshevsky St., 173, Nal'chik-360004, Russian Federation

T. N. Kukhta
Scientific-research Institute BelNIIS RUE, 15b Frantsisk Skorina St., 220114, Minsk, Belarus, E-mail: kuhta_tatiana@mail.ru

Elena I. Kulish
Professor of the Department of High-Molecular Connections and General Chemical Technology of the Chemistry Faculty of the Bashkir State University, 450076, Ufa, Zaks Validi Street, 32, Russian Federation, Tel.: (347) 229–96–86; E-mail: onlyalena@mail.ru

Isil Kurtulus
PhD Student, Izmir Institute of Technology, Department of Chemical Engineering, Gulbahce Urla 35430 Urla Izmir Turkey, E-mail: isilkurtulus@iyte.edu.tr

Gulistan Kutluay
Master Student, Izmir Institute of Technology, Department of Chemical Engineering, Gulbahce Urla 35430 Urla Izmir Turkey, E-mail: glstnchml@gmail.com

V. Lidgi-Goryaev
N.M. Emanuel Institute of Biochemical Physics, Russian Academy of Sciences, 4, Kosygin St., Moscow 119334, Russian Federation

Eugene M. Lisitsyn
North-East Agricultural Research Institute, 166-a Lenin St., Kirov, 610007, Russian Federation; E-mail: edaphic@mail.ru

Yu. K. Lukanina
N.M. Emanuel Institute of Biochemical Physics of Russian Academy of Sciences, 4 Kosygin St., Moscow 119334, Russian Federation

M. M. Malikov
Joint Institute for High Temperatures, Russian Academy of Science, 13(2) St. Izhorskaya, Moscow, 125412, Russian Federation

Eliza Markarashvili
Iv. Javakhishvili Tbilisi State University, Department of Chemistry, I. Chavchavadze Ave. 1, 0179
Tbilisi, Georgia, E-mail: marinaiskakova@gmail.com

Abdulakh K. Mikitaev
Kh.M. Berbekov Kabardino-Balkarian State University, Chernyshevsky St., 173, Nal'chik-360004,
Russian Federation

I. Modebadze
Department of Biology Faculty of Exact and Natural Sciences, Iv. Javakhishvili Tbilisi State
University, Tbilisi, Georgia

T. V. Monakhova
Institute of the Russian Academy of Sciences N.M. Emanuel Institute of Biochemical Physics,
Russian Academy of Sciences, Moscow, Russian Federation

G. Mosidze
Department of Biology Faculty of Exact and Natural Sciences, Iv. Javakhishvili Tbilisi State
University, Tbilisi, Georgia

Omar Mukbaniani
Iv. Javakhishvili Tbilisi State University, Faculty of Exact and Natural Sciences, Institute of
Macromolecular Chemistry and Polymeric Materials, I. Chavchavadze Ave., 13, Tbilisi 0179,
Georgia; E-mail: omarimu@yahoo.com

Fuzuli Akber oglu Nasirov
Petkim Petrokimya Holding, Izmir, Turkiye, E-mail: fnasirov@petkim.com.tr

Lev N. Nikitin
Nesmeyanov Institute of Elementoorganic Compounds RAS, 119991, Vavilov St., 28, Moscow,
Russian Federation

Andrew V. Noskov
G.A. Krestov Institute of Solution Chemistry, Russian Academy of Sciences, Akademicheskaya St.,
1; Ivanovo, 153045, Russian Federation

Donari Otiashvili
Iv. Javakhishvili Tbilisi State University, Faculty of Exact and Natural Sciences, Institute of
Macromolecular Chemistry and Polymeric Materials, I. Chavchavadze Ave., 13, Tbilisi 0179,
Georgia

A. Petrov
N.M. Emanuel Institute of Biochemical Physics, Russian Academy of Sciences, 4, Kosygin St.,
Moscow 119334, Russian Federation, E-mail: 28en1937@mail.ru

N. G. Petrova
Specialist-Expert, Department of Information Security and Communications, Ministry of
Informatization and Communications, Udmurt Republic, Vadima Sivkova St., 186, 426057 Izhevsk,
Russian Federation, E-mail: biakaa@mail.ru

A. A. Popov
N.M. Emanuel Institute of Biochemical Physics of Russian Academy of Sciences, 4 Kosygin St.,
Moscow 119334, Russian Federation, E-mail: julialkn@gmail.com

N. R. Prokopchuk
Corresponding Member of Belarusian National Academy of Sciences, Doctor of Chemical Sciences, Professor, Head of Department (BSTU), Belarusian State Technological University, 13a Sverdlov St., 220006, Minsk, Belarus, E-mail: prok_nr@mail.by

E. G. Rozantsev
Saratov State University Named After N.G. Chernyshevsky, Russian Federation

M. Rukhadze
Department of Biology Faculty of Exact and Natural Sciences, Iv. Javakhishvili Tbilisi State University, Tbilisi, Georgia

S. N. Rusanova
Kazan National Research Technological University, K. Marx St., 68, Kazan, 420015, Tatarstan, Russian Federation

L. Rusishvili
Department of Biology Faculty of Exact and Natural Sciences, Iv. Javakhishvili Tbilisi State University, Tbilisi, Georgia

A. I. Sergeev
Institute of the Russian Academy of Sciences N.N. Semenov Institute of Chemical Physics, Russian Academy of Sciences, Moscow, Russian Federation

Anatolii B. Shapiro
N.M. Emanuel Institute of Biochemical Physics RAS, 119334, Kosygin St., 4, Moscow, Russian Federation

L. S. Shibryaeva
Institute of the Russian Academy of Sciences N.M. Emanuel Institute of Biochemical Physics, Russian Academy of Sciences, Moscow, Russian Federation

Lyudmila N. Shikhova
Vyatka State Agricultural Academy, 133 Oktyabrsky Ave., Kirov, 610010, Russian Federation

A. L. Shutova
Candidate of Technical Sciences, Assistant professor (BSTU), Belarusian State Technological University, 13a Sverdlova St., Minsk, 220006, Belarus, E-mail: VPSh_bstu@mail.ru, prok_nr@mail.by, a.l.shutova@mail.ru

Olga G. Sitnikova
V.N. Gorodkov Research Institute of Maternity and Childhood, Pobedy St., 20, Ivanovo, 153045, Russian Federation

S. Yu. Sofina
Kazan National Research Technological University, K. Marx St., 68, Kazan, 420015, Tatarstan, Russian Federation

Olga N. Sorokina
N.M. Emanuel Institute of Biochemical Physics RAS, 119334, Kosygin St., 4, Moscow, Russian Federation

O. V. Stoyanov
Kazan National Research Technological University, K. Marx St., 68, Kazan, 420015, Tatarstan, Russian Federation, E-mail: ov_stoyanov@mail.ru

D. A. Strikanov
Joint Institute for High Temperatures, Russian Academy of Science, 13(2) St. Izhorskaya, Moscow, 125412, Russian Federation

Tamar Tatrishvili
Iv. Javakhishvili Tbilisi State University, Department of Chemistry, I. Chavchavadze Ave., 3, Tbilisi 0179, Georgia

E. Tavdishvili
Department of Biology Faculty of Exact and Natural Sciences, Iv. Javakhishvili Tbilisi State University, Tbilisi, Georgia

N. E. Temnikova
Kazan National Research Technological University, K. Marx St., 68, Kazan, 420015, Tatarstan, Russian Federation

Irina Tuktarova
Bashkir State University, 450074, Zaki Validi St., 32, Ufa, Russian Federation

Utku Ulucan
PhD Student, Izmir Institute of Technology, Department of Chemical Engineering, Gulbahce Urla 35430 Urla Izmir Turkey, E-mail: utkuulucan@iyte.edu.tr

Denis R. Valiev
Student of the Department of High-Molecular Connections and General Chemical Technology of the Chemistry Faculty of the Bashkir State University, 450076, Ufa, Zaks Validi Street, 32, Russian Federation, Tel.: (347) 229–96–86; E-mail: valief@mail.ru

G. E. Valyano
Joint Institute for High Temperatures, Russian Academy of Science, 13(2) St. Izhorskaya, Moscow, 125412, Russian Federation

S. Varfolomeev
N.M. Emanuel Institute of Biochemical Physics of Russian Academy of Sciences, 4 Kosygin St., Moscow 119334, Russian Federation

K. I. Vinhlinskaya
PhD Student, Junior Scientific Researcher (BSTU), Belarusian State Technological University, 13a Sverdlova St., Minsk, 220006, Belarus

Gennady E. Zaikov
N.M. Emanuel Institute of Biochemical Physics of Russian Academy of Sciences, Moscow – 119334, Kosygin St., 4, Russian Federation, E-mail: chembio@sky.chph.ras.ru

V. Zavolzhsky
N.M. Emanuel Institute of Biochemical Physics, Russian Academy of Sciences, 4, Kosygin St., Moscow 119334, Russian Federation

LIST OF ABBREVIATIONS

AA	acetic acid
AC	activated carbon
AC	aerocellulose
Al	aluminum
ANOVA	analysis of variance
BD	beginning of the decay measurement
BM	binary mixtures
BOPP	biaxially oriented poly propylene
CB	carbon black
CNF	multiwalled carbon nanofilaments
CNT	carbon nanotubes
CO2	carbon dioxide
CPMG	Carr-Purcell-Meiboom-Gill
CS	cellulose
CS	chitosan
DCM	dichloromethane
DDA	degree of deacetylation
DEAC	diethylaluminumchloride
DMSO	dimethylsulfoxid
DSC	differential scanning calorimetry
EADC	ethylaluminumdichloride
ED	end of the decay measurement
FID	free induction decay
FTIR	Fourier-transform infrared spectroscopy
GPC	gel permeation chromatograph
IR	infrared spectroscopy
K	potassium
LAR	leaf area ratio
LDPE	low density polyethylene
LWR	leaf weight ratio
MAO	methylaluminoxane

MD	machine direction
MDA	malonic dialdehyde
MV	Mooney viscosity
MW	molecular weight
N	nitrogen
NA	nutrient agar
NB	nutrient broth
NC	nanocarbon
NS	number of scans
OGB	oxide tungsten bronzes
OI	oxygen index
OTB	oxide-tungsten bronzes
P	phosphorus
PEO	polyethylene oxide
PMHS	polymethylhydrosiloxanes
PP	polypropylene
PS	polystyrene
PVA	polyvinyl acetate
PWM	paintwork materials
RD	recycle delay between scans
SEM	scanning electron microscopy
SEP	spatial-energy parameter
SLA	specific leaf area
SV	solution viscosity
TAA	trialkylaluminum
TEA	triethylaluminum
TGA	thermo gravimetric analysis
TGIC	triglycidyl isocyanurate
THF	tetrahydrofuran
TSPC	thermostable protein complex
TWL	typical weight loss
UV	ultraviolet

LIST OF SYMBOLS

a	preexponential factor
A_{hi}	hydrodynamic invariant
c	concentration of the polymer in the solution (g/dL)
C	Curie's constant
c_0	aggregating particles initial concentration
D	absorbance
d	density of the adsorbate, g/cm^3
d	dimension of Euclidean space
D_{CNT}	carbon nanotube (nanofilament) diameter
dx/dt	rate of settling (cm/s)
E	the activation energy of charge transfer
E	system total energy
E_g^{bulk}	band gap of the solid
E_m	elasticity moduli of matrix polymer
E_n	elasticity moduli of nanocomposite
E_{nf}	elasticity moduli of nanofiller
h	medium heterogeneity exponent
h	Plank constant
k	Boltzmann's constant
k	rate constant
k_1	constant reflecting the reaction of the polymer with a solvent
L	grain size in nanomaterials
l	large of the cuvette
L_{CNT}	carbon nanotube (nanofilament) length
m	mass of the adsorbed benzene (acetone, n-heptane), g
M	mass of the dried sample, g
m	weight of the solution in the pycnometer
m_0	weight of the solvent in the pycnometer
m_1	number of hours of impact at particular values of service temperature

n	kinetic parameter
N	number of particles or realized interactions depending on the process type
N_0	number of particles in the sphere volume of the radius R
N_1 and N_2	number of homogeneous atoms in subsystems
N_A	Avogadro's number
n_i	number of electrons of the given orbital
η_{sp}	specific viscosity of polymer solution
P	ones in ground state
$P*$	radical decay products in excited state
P_E	effective p-parameter (effective SEP)
P_c	structural parameter
$P°$	spatial-energy parameter
r	bond dimensional characteristics
R	particle radius
r	radius of particle (cm)
R_{CNT}	ring-like structures CNT (CNF) radius
r_i	orbital radius of i orbital
s	boundary region width
S_0	sedimentation constant
S_M	entropy
T	absolute temperature
t	process duration
t	time expansion
T_c	crystallization temperature
U	mutual potential energy of material points
U_2 and U_1	potential energies of the system in final and initial states
W	number of available states of the system or degree of the degradation of microstates
W_i	orbital energy of electrons
W_n	nanofiller mass contents
$Z*$ and $n*$	nucleus effective charge and effective main quantum number

Greek Symbols

α	degree of crystallinity of the polymer matrix
β	constant value for the given class of structures

γ	surface energy of a gas bubble
ΔG	total free energy of activation
ΔG_η	free enthalpy of activation of molecular diffusion across the phase boundary
ΔU	resulting (mutual) potential energy of these interactions
ΔU_1 and ΔU_2	potential energies of material points on the elementary region of interactions
ε	molar extinction coefficient
η	viscosity of medium (g/cm.s)
η_0	dynamic viscosity of the solvent
$[\eta]$	intrinsic viscosity, dL/g
θ	contact angle of the particle with the bubble surface
v	partial specific volume, cm^3/g
v_0	volume of the pycnometer
ρ	density of particle (g/cm^3)
ρ_0	density of the solvent g/cm^3
ρ_{20}	resistivity at room temperature
ρ_o	density of medium (g/cm^3)
ρ_T	electrical resistivity at temperature T
Σm_i	total number of hours of impact at variable values of service temperature
τ_{Ts}	durability of a polymer article in years at particular value of article service temperature
$(1-v\rho_0)$	Archimedes factor or buoyancy factor

PREFACE

This is a timely book in many respects: the field of polymer science is expanding at the moment due to the continuing interest in and development of polymeric materials, which are expected to play a crucial role in this arena. A complete understanding of the structure and physical properties of these materials is essential. The book is well written in its coverage.

In this book, a considerable section of each chapter is devoted to the description of actual experiments and the applications of these various techniques, which is an extremely useful approach and which successfully demonstrates the advantages and disadvantages of each physical technique.

This book is highly recommended as a reading and advanced teaching tool to a wide range of researchers in the general field of polymer science. It is particularly well suited to the nonexperts in these various fields and serves as a practical guide to characterization. However, successful characterization of polymer systems is one of the most important objectives of today's experimental research of polymers. Considering the tremendous scientific, technological, and economic importance of polymeric materials, not only for today's applications but for the industry of the twenty-first century, it is impossible to overestimate the usefulness of experimental techniques in this field. Since the chemical, pharmaceutical, medical, and agricultural industries, as well as many others, depend on this progress to an enormous degree, it is critical to be as efficient, precise, and cost-effective in our empirical understanding of the performance of polymer systems as possible. This presupposes our proficiency with, and understanding of, the most widely used experimental methods and techniques.

This book is designed to fulfill the requirements of scientists and engineers who wish to be able to carry out experimental research in polymers using modern methods. Each chapter describes the principle of the respective method, as well as the detailed procedures of experiments with examples of actual applications. Thus, readers will be able to apply the concepts as described in the book to their own experiments.

Experts in each of the areas covered have reviewed the state of the art, thus creating a book that will be useful to readers at all levels in academic, industry, and research institutions.

This volume provides an overview of polymer characterization test methods. The methods and instrumentation described represent modern analytical techniques useful to researchers, product development specialists, and quality control experts in polymer synthesis and manufacturing. Engineers, polymer scientists and technicians will find this volume useful in selecting approaches and techniques applicable to characterizing molecular, compositional, rheological, and thermodynamic properties of elastomers and plastics.

ABOUT THE EDITORS

Kajetan Pyrzynski, PhD
President, P. I. W. Delta-Company, 5 Krupczyn St., 63-140 Dolsk, Poland
Tel.: (+48) 507 171 007; Fax: (+48) 61 28 30 718;
E-mail: biuro@delta-dolsk.pl

Dr. Kajetan Pyrzynski is a specialist in the fields of organic chemistry, physical chemistry and chemical technology. He is president of P.I.W. Delta Company in Poland. He established the company for realization of chemical ideas for practices: research, development, and production. He published about 100 scientific papers and one book. Some years ago he was Consul of Poland in Republic of Peru.

Grzegorz Nyszko, PhD
Director, Military Institute of Chemistry and Radiometry,
105 Allee of General A. Chrusciela, 00-910 Warsaw, Poland
Tel. (+48) 22 516 99 09; Fax (+48) 22 673 51 80;
E-mail: grzegorz.nyszko@wichir.waw.pl

Dr. Grzegorz Nyszko is the director of Military Institute of Chemistry and Radiometry in Warsaw, Poland. He has published several papers in International journals. He completed his Master of Science (technical physics) in 1994 at the Military University of Technology, Warsaw, Poland; and PhD (defense studies) in 2007 at the National Defense University, Warsaw, Poland. He started his career in 1994 as a research worker in Naval Academy (1994–1996); 1996–to date in Military Institute of Chemistry and Radiometry; 1996–2007 as research worker; 2007–2010 as head of the Department of NBC Skin Protection; 2010–2014 as Research and Development Deputy Director; 2014 as Director of the Institute; and as CBRN Defense Equipment Standardization Committee Leader.

Gennady E. Zaikov, DSc

Head, Polymer Division, N. M. Emanuel Institute of Biochemical Physics, Russian Academy of Sciences; Professor, Moscow State Academy of Fine Chemical Technology; Professor, Kazan National Research Technological University, Russia

Gennady E. Zaikov, DSc, is Head of the Polymer Division at the N. M. Emanuel Institute of Biochemical Physics, Russian Academy of Sciences, Moscow, Russia, and professor at Moscow State Academy of Fine Chemical Technology, Russia, as well as professor at Kazan National Research Technological University, Kazan, Russia. He is also a prolific author, researcher, and lecturer. He has received several awards for his work, including the Russian Federation Scholarship for Outstanding Scientists. He has been a member of many professional organizations and on the editorial boards of many international science journals.

CHAPTER 1

RECOVERY TECHNOLOGY OF DEPLETED OIL FIELDS

E. ALEKSANDROV, S. VARFOLOMEEV, G. ZAIKOV, V. LIDGI-GORYAEV, and A. PETROV

N.M. Emanuel Institute of Biochemical Physics of Russian Academy of Sciences, 4 Kosygin str., Moscow 119334, Russia; Phone: +7 (495) 939 7318; Mobile: +7-916-953-76-19; E-mail: 28en1937@mail.ru

CONTENTS

1.1 INTRODUCTION

The cornerstone problem of modern oil production is water intrusion in the productive layers resulted from water flood displacement (water introduction from water injecting to oil producing wells aimed at oil displacement and subsequent oil recovery). Water is typically introduced not earlier than the end of active and most cost effective stage of development (10–20 years). The water introduction is aimed at the increase of layer pressure. In Russia after pumping millions of tons of water and extracting about 40% of oil deposits the process gradually turns from oil extraction into water extraction. This process is accompanied by the formation of skin layer consisting of viscous heavy oil, which more easily adheres to the surface of pores and cracks as compared to light oil. Because of oil hydrophobic properties, the water driven from water injection to oil recovery wells moves along the washouts thus displacing oil on a very limited scale. As a result of this about 60% of Russian known oil reserves stay underground labeled difficult or impossible to extract. These difficulties are further aggravated by the ever-increasing cost of geological survey of new oil fields, almost all deposits at small depth (up to 5 km) being already known. Still during the recent years there has evolved a way to solve the above-described problems by means of thermochemical technology of Binary Mixtures (BM) [1–6].

1.2 TECHNOLOGY DEVELOPMENT MILESTONES

Binary mixtures are water solutions of petersalts (ammoniac and organic ones) with the reaction initiator of petersalt decomposition (metal hydrides or sodium nitrite) [1, 2]. The two components are injected into the oil well via two separate channels and the reagents react upon contact opposite or inside the productive layer emitting heat and gas. The binary mixtures found their application in the oil extraction domain starting from 1982. It should be noted that for the reason of considerable explosion hazard the Russian Technical Supervision Service used to restrict the injection amount to 1 ton of saltpeter per oil well. Up to the year of 2010 the BM reaction was maintained in an uncontrollable mode with its efficiency approximately amounting to 0.4 [3–5].

The year 2010 saw the development of the system of optimization and control over the heat emission inside the well, the BM reaction efficiency having increased from 0.4 to 0.8. The BM technology architects obtained the permission of the Russian Technical Supervision Service (№ 25–BL–19542–2010) allowing them to inject the unlimited quantities of petersalt into oil wells. The system of reaction control has greatly contributed to spot and study the exothermic reaction of saltpeter decomposition in a heated layer with the emission of heat (Q1) and oxygen oxidizing a small part of layer oil accompanied by heat emission (Q2). As Q2 ≈ 2Q1 the progress of heat front along the layer containing petersalt water solution in its cracks and pores is as a rule a self-maintaining process.

The whole process is explosion-proof as once in the productive layer petersalt emits heat produced during the reaction which is further absorbed by the rock, the petersalt and rock mass ratio amounting to 1:20.

By increasing the BM reagent mass injected into the oil well and the productive layer by tens of times the authors mastered the heating process as well as stage-by-stage removal of the skin layer accumulated near the wells as a result of their long performance.

In 2011, the partial removal of skin layer blocking the Usinsk oil field No. 1242 and No. 3003 by means of BM [6] led to the increase in oil extraction by 4.95 and 8.44 t/day, respectively. The amount of additional oil extracted from the mentioned oil wells in 2012 amounted to 3.4 thousand of tons, that is, approximately 1.7 thousand per oil well.

In 2012, in the course of treatment of the Usinsk oil fields No. 6010, No. 600, No. 1283, No. 7169 and No. 8198 managed by E. Alexandrov and V. Zavolzhsky the technology was further improved. The reaction catalyst of petersalt decomposition (sodium nitrite) was abandoned for thermal catalyst. The water vapor heat preliminary injected into the layer at the temperature of 250°C promoted saltpeter decomposition and oxidation of small part of oil by the emitted oxygen. During the year following the treatment the extraction rate from the five described oil wells amounted to 13232 tons of additional oil. It is likely in this case that the treatment process promoted more thorough cleaning of the skin layer than the one described in oil wells No. 1242 and No. 3003. The increase of the average annual growth of oil extraction per one oil well from 1700 tons (2011) to 2646 tons (2012) is a safe indicator.

In July 2013 oil wells No. 8 and No. 10 of the Eastland oil field (Texas, USA) were treated by the BM reaction products. The mentioned oil field was abandoned in 1994 labeled unprofitable. Starting from July 1 to July 7 just before treatment, the layer fluid was pumped from the well. The fluid consisted of water (99.999%) and oil film (<0.001%). On July 8 after injecting saltpeter solutions (about 35 tons) and sodium nitrite (about 12 tons) into the oil well both oil wells produced oil with the industrial ratio of 30% of oil and 70% of water. Although the production performance proved profitable solely from oil well No. 8 the customer company Viscos Energy put an advertisement mentioning the successful oil recovery at an abandoned field with the application of the Russian technology. The difference in the oil recovery rate at oil well No. 8 and oil well No. 10 is explained by the difference in their location. Oil well No. 8 is located in the center of the field while oil well No. 10 is situated on its border.

1.3   MAIN RESULTS AND PROSPECTS OF THE BM TECHNOLOGY

1. The modern BM technology developed by the authors since 1997 at present differs from all the rest of similar technologies in the heat emission optimization in the course of reagent injection. For this reason the technology ensures a much more thorough elimination of the skin layer around the producing oil wells. Accumulated during tens of years the skin layer restricts the profitable extraction to less than a half of the deposited oil.

2. Defined by the market competition if compared to other technologies in respect to their cost the modern BM technology holds the second place and as related to the rate of oil extraction increase trails only the hydro fracturing technology (USA and Canada). Starting from the first patent (RU 2126084 97111229/031997.06.30) and up to the latest ones of 2010–2012 (Patent WO 2010/043239, April 22, 2010 и Patent WO 2012/025150, March 1, 2012) the efficiency increase of the technology application during the last 16 years is clearly seen from the increase in treatment cost of one oil well:

    * $\approx$ \$5,000 (1997, Customer – the owner of the Vostochno-Poltavskoye oil field, Ukraine);
    * $\approx$ \$40,000 (2010–2011, Customer – LLC LUKOIL-KOMI);
    * $\approx$ \$60,000 (2013, Customer – Viscos Energy Holding ltd. USA).

3. We tend to interpret the facts described in points 1–2 as a turning point from over the century accumulation of nonrecoverable oil deposits to their profitable extraction. It should be noted that at present the amount of hydrocarbons still deposited underground is far greater than the deposits at the oil fields under current treatment. For this reason

discovering the efficient technology for nonrecoverable oil extraction equals discovering a new large-scale oil field.

4. The oil recovery revival of the previously unprofitable Russian oil fields by means of the BM technology can be regarded a new direction of field thermochemistry development, which can promote a radical improvement of Russian economy as a great energy extracting power.

KEYWORDS

- **binary mixtures**
- **water flood displacement**
- **heat emission**
- **BM technology**

REFERENCES

1. Alexandrov, E., Kuznetsov, N., Karotazhnik Science and Research Magazine, 2007, № 4, pp. 113–127.
2. Aleksandrov, E., Koller, Z. Technology of oil and bitumen output stimulation by heat from reactions of binary mixtures (BM), TCTM limited, 2008, 76 p.
3. Merzhanov, A., Lunin, V., Aleksandrov, E., Lemenovsky, D., Petrov, A., Lidgi-Goryaev, V. Science and technology applied to industry, 2010. vol. 2. pp. 1–6.
4. Aleksandrov, E., Varfolomeev, S., Lidgi-Goryaev, V., Petrov, A. Tochka Opory Magazine, November 2012 № 158 pp. 14–15.
5. Aleksandrov, E., Aleksandrov, P., Kuznetsov, N., Lunin, V., Lemenovsky, D., Rafikov, R., Chertenkov, M., Shiryaev, P., Petrov, A., Lidgi-Goryaev, V. Neftekhimiya, 2013, vol. 53, № 4, pp. 312–320.
6. Patent. Gas evolving oil viscosity diminishing compositions for stimulating the productive layer of an oil reservoir, Aleksandrov, E., Lemenovsky, D., Koller, Z. Patent WO2010/025150 A1.

CHAPTER 2

EFFECT OF POLYSTYRENE/ FULLERENE COMPOSITES ON THE LIPID PEROXIDATION IN BLOOD SERUM

OLGA V. ALEKSEEVA,[1] OLGA G. SITNIKOVA,[2]
NADEZHDA A. BAGROVSKAYA,[1] and ANDREW V. NOSKOV[1]

[1]*G.A. Krestov Institute of Solution Chemistry, Russian Academy of Sciences, Akademicheskaya str., 1; Ivanovo, 153045, Russia*

[2]*V.N. Gorodkov Research Institute of Maternity and Childhood, Pobedy str., 20, Ivanovo, 153045, Russia*

CONTENTS

ABSTRACT

Effect of polystyrene/fullerene composite on free-radical processes in blood serum has been investigated in vitro. The parameters of lipid peroxidation in native serum after adding of the composite films were determined by chemiluminescent analysis and spectrophotometry. It was revealed that fullerene-containing nanocomposites can manifest antioxidant activity in blood serum.

2.1   INTRODUCTION

Nanocomposite materials are promising field of advanced material sciences with scope of using in biology, medicine, pharmacology, etc. Nanomaterials are based on nanoparticles with unique characteristics resulting from their microscopic size [1].

The discovery of soccer-ball shaped Buckminster fullerene in 1985 [2] was an exciting and unexpected discovery that established an entirely new branch of chemistry. Fullerene is molecular compound belonging to the class of carbon allotropes. Fullerene molecule is a closed convex polyhedron made of carbon atoms that arrange at the vertices of regular hexagons and pentagons.

Fullerene is possessed the unusual physical and chemical properties. Based on these properties the new materials were obtained, such as superconducting "fullerenes/alkali metal atoms" compounds, thin films and solutions of fullerenes with nonlinear optical properties, new super-hard materials based on fullerenes [3].

Currently, various derivatives of fullerenes exhibit a broad spectrum of biological activity: anticancer, antiviral, antibacterial, neuroprotective, and antioxidant activities [4]. Biological abilities of fullerenes are due to lipophilic properties, which facilitate cell penetration and lack of electrons, helping to react with free radicals and generate active oxygen species [5].

Interaction mechanisms of fullerenes with cell are still poorly understood. To research these processes, various methods have been used, such as molecular dynamics simulations [6], radioactivity method [7], etc. It was demonstrated in Ref. [7] that fullerene derivative $C_{61}(CO_2H)_2$ could cross the external cellular membrane, and it localized preferentially to the mitochondria. Formation of oxygen free radicals occurs due to the electrons leakage in the mitochondrial electron transport chain. Therefore, the localization of fullerenes near derivatives mitochondrions may contribute to their antioxidant action.

The mechanism of biological role of fullerene is shown to depend on its aggregate form: crystalline, colloid or soluble organic complex [8, 9]. Soluble organic fullerene complex has highest bioactivity [8]. The authors of Ref. [8] explained this effect by low association of nanocarbonic particles.

A new direction of fullerene science has been developed, and the related studies in this direction have concerned the preparation of fullerene-containing polymers, which would combine the unique characteristics of fullerene with the useful properties of matrix polymers [10].

One of polymers able to complex with nanoparticles is polystyrene (PS), which is widely spread in industry. Also polystyrene is used, for example, for the manufacture of surgical instruments, as well as an arterial embolization to stop bleeding. Replacement of pure polymer by fullerene-containing one will impart it antibacterial and antimicrobial properties. Therefore, polystyrene/fullerene composites are the subject of numerous studies [11–13]. It is considered that integration of fullerenes into polymer matrix can produce biocomposites which have medical potential as drug transporters, antiseptics, and antioxidants. Polystyrene/fullerene composite materials may be used to manufacture containers for blood storing too.

Early in Refs. [14, 15] we reported on studies of the polystyrene/fullerene composites by scanning electron microscopy, infrared spectroscopy, and X-ray diffraction. In scanning electron microscopy images (Figure 2.1) it can be observed, the surface is composed of rounded particles with irregular relief, the size of irregularities being comparable with the polystyrene molecule size (10 nm).

The above listed methods and mathematical modeling have been used to study the influence of the fullerene additives on structure of polystyrene [14, 15]. We have concluded that during the solvent evaporation and film formation the individual polymer molecules collide and stick together into aggregates. In films without fullerenes, packing of straightened chains

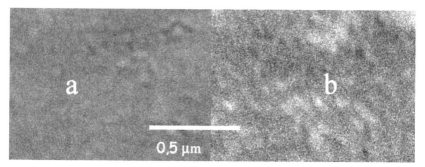

FIGURE 2.1   Electron microscopic image of the surface of the film: a) polystyrene film; b) polystyrene film filled with fullerene (0.035 wt % of $C_{60}+C_{70}$). (This figure was presented in Ref. [14]).

parallel to each other predominated. This shows that a polystyrene molecule, which had the shape of a coil in solution, straightened and stretched itself along the aggregate surface when attached to it. In polystyrene/fullerene composite, the intermolecular interactions between polystyrene and fullerene are appreciable, and under the influence of fullerene, polystyrene molecules straightened with the formation of ordering elements in the arrangement of chains.

So the fullerene additives effect on structure of polystyrene. Such modification may result to occurrence of new properties of polymer, for example, biological activity.

Bioactivity of material may be evaluated by its effect on the free radical processes in biologic fluid. Currently regulation of free-radical processes is adjusted by both natural and synthetic pharmaceutical compositions [16]. As any other medicine some antioxidants may produce adverse events. Thus, finding of safe preparations with high antioxidant activity is still actual.

We have analyzed the literature on studies of the polystyrene/fullerene composites, and have concluded that the antioxidant activity of these materials was not researched enough, although for fullerenes it was known. For our opinion, it is necessary to fill this gap. Therefore the goal of present research was to investigate the influence of polystyrene/fullerene nanocomposites on free-radical processes in biologic fluid (blood serum) in vitro.

2.2 EXPERIMENTAL PART

2.2.1 MATERIALS AND SUBJECTS

We have chosen polystyrene (Aldrich, Germany, M_n=1.4×10^5, M_w/M_n=1.64) as a matrix for fabrication of fullerene-containing nanocomposites, because it has high solubility in aromatic hydrocarbons like fullerene itself. C_{60}+C_{70} fullerene mix (Fullerene Technologies Ltd., Russia) was preliminary purified [17].

Fullerene-polystyrene composition films were produced as follows. Batches of polymer and C_{60}+C_{70} were solved separately in o-xylene (or toluene). Concentrations of fullerenes in o-xylene solutions were equal to 0.018, 0.054, 0.18, and 1.8 g/L. Concentrations of fullerenes in toluene solutions were equal to 0.18, and 0.9 g/L. Polystyrene batches were dissolved in respective solvents (179.38 g/L of PS) too. Then, fullerene/o-xylene solutions and polystyrene/o-xylene solutions were mixed together, so that weight fractions of fullerene were equal to 0.01, 0.03, 0.1, and 1 wt %. The mixed solutions were stirred for about 1 day before being cast into thin films. Similar actions were carried out with toluene solutions. After casting the solvent was slowly evaporated over several days to produce the polystyrene/fullerene composite films.

Subject of research was native blood serum of 10 patients managed in V.N. Gorodkov Research Institute of Maternity and Childhood (Ivanovo, Russia). Specimen of pure PS or composite film (size 1.5 cm^2, weight 5 mg) was put into blood serum (1 mL). System has been incubated for 1 h at 4°C.

2.2.2 METHODS

After contact of blood with pure polystyrene films or fullerene-containing polystyrene films, we have studied the free radical processes in blood. Also we have studied the free radical processes in native blood serum without preliminary contact, choosing as control. The parameters of lipid peroxidation have been determined by chemiluminescent analysis and spectrophotometry.

2.2.3 *CHEMILUMINESCENCE*

Induction of chemiluminescence (ChL) by hydrogen peroxide and iron sulfate is based on Fenton reaction: at mixing the components the catalytic decomposition of hydrogen peroxide takes place by divalent iron ions. Thus formed free radicals oxidize the lipoproteins of blood serum in the test samples, leading to the formation of new free radicals. At the recombination of radicals, unstable products are formed and decomposed with the release of photons.

Generally, a chain reaction in which formation of radicals leads to chemiluminescence, may be represented by the scheme [18]:

$$\longrightarrow R \bullet \xrightarrow{\ k\ } P^* \xrightarrow{\ k_e\ } P + photon$$

where – the free radicals; P^* – the radical decay products in excited state; P – ones in ground state.

The intensity of chemiluminescence, I, is proportional to the rate reaction, v:

$$I = \eta v, , \tag{1}$$

where η – quantum efficiency of the chemiluminescent reaction.

For the process, the ChL curve dips because of antioxidant agents. The decay rate constant of free radicals, k, is defined by the dip rate of ChL curve. Therefore, the main indicator of the antioxidant activity of the system is tangent of maximum slope angle of ChL curve towards time axis, $\tan\alpha$.

Also, we used following parameters:

- I_m is maximum intensity of ChL during the experiment. Value of I_m quantifies the level of free radicals, that is, gives an idea of the potential ability of the blood serum to free radical lipid peroxidation;
- S is an area covered by intensity curve or total light sum. Value of S is inversely proportional to the antioxidant activity of the sample;
- $S_n = S/I_m$ is normalized light sum. The value of S_n evaluates antioxidant activity more correctly than the value of S, because the total area covered by ChL curve depends on value of I_m.

The induced chemiluminescence tests have been performed by BChL-07 luminometer (Medozons, Russia). We have used hydrogen peroxide and ferric sulfate as inductors of ChL. 0.1 mL of serum, 0.4 mL of phos-phate buffer (pH 7.5), 0.4 mL of 0.01 M ferric sulfate and 0.2 mL of 2% hydrogen peroxide have been put into cuvette. Luminescence has been registered for 40 s.

Free radical processes in serum have been studied after contact with original polystyrene films and fullerene-containing polystyrene films. The mean values of ChL parameters in native serum without addition of the film have been used as controls. The results have been expressed as per-centages relative to controls and have been given as mean values ± stan-dard deviations. Level of significance, p, was 0.05.

2.2.4   SPECTROPHOTOMETRY

Also we have defined the malonic dialdehyde (MDA) concentration in blood samples after contact with PS and PS/C_{60}+C_{70} films. This indicator reflects the amount of the lipid peroxidation products. MDA concentration has been determined by the SF-46 spectrophotometer (Russia) at wave-length of 532 nm. This method based on the formation of complex MDA with 2-thiobarbituric acid [19]. The concentration of MDA has been cal-culated as follows:

$$C_{MDA} = \frac{D}{l\varepsilon}, \tag{2}$$

where D is absorbance; l is the large of the cuvette; ε is the molar extinc-tion coefficient.

Value of total antioxidant reactivity, TAR, has been evaluated by measuring the absorbance before and after incubation of samples by equation [20]:

$$TAR = 1 - \frac{D_t - D_0}{D_t^{st} - D_0^{st}}, \tag{3}$$

where D_0 and D_t are values of absorbance for blood samples before and after incubation, respectively; D_t^{st} and are the same quantities for solution of linolenic acid choosing as standard.

2.3   RESULTS AND DISCUSSION

Using above-described technique and o-xylene as solvent, we have prepared one sample of polystyrene film and four samples of polystyrene/ fullerene composite films with various $C_{60}+C_{70}$ percentages. Unmodified polystyrene samples were colorless, whereas the polystyrene/fullerene composite films were light purple. The intensity of color depended on the content of $C_{60}+C_{70}$ in the composite.

Figure 2.2 shows kinetics of chemiluminescence in blood serum after contact with original polystyrene and nanocomposite films. Peak of chemiluminescence due to free radical production was in 2 s of reaction. This can be explained by production of active oxygen species (HO_2^*, O_2^*, O_2^-, OH^-). The highest intensity, I_m, was registered in case of nanocomposites with 0.01 and 0.03 wt % of fullerenes.

In Table 2.1 we have represented the main ChL parameters for films prepared by casting of o-xylene solution. It can be seen in case of original polystyrene film the ChL parameters were approximate to controls. Value of I_m for the PS/$C_{60}+C_{70}$ composites is higher than for control serum samples. A light sum, S, was significantly increased only for films with

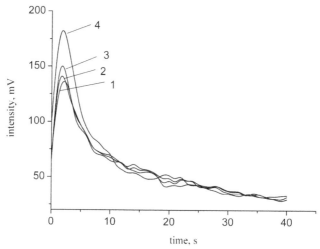

FIGURE 2.2   Kinetic chemiluminescence profiles of native blood serum (*1*) and after contact of it with studied materials: *2* – PS; *3* – PS/$C_{60}+C_{70}$ (1.0 wt %); *4* – PS/$C_{60}+C_{70}$ (0.03 wt %). Solvent: o-xylene.

TABLE 2.1 Chemiluminescence Parameters in Blood Serum After Contact With Original Polystyrene Film and Fullerene-Containing Nanocomposites (Solvent: o-xylene)

Fullerene content, wt %	S, %	I_m, %	tan α, %	S_n, %
Controls	100.0	100.0	100.0	100.0
0	96.0±9.0	97.0±7.0	97.5±10.5	99.0±11.5
0.01	121.5±10.5*	131.5±11.5 *	139.0±27.0*	92.0±6.5 *
0.03	115.0±8.0*	125.0±11.0 *	146.0±27.0*	92.0±7.0 *
0.1	105.0±3.0	110.0±6.0 *	110.0±8.0 *	95.5±5.0
1.0	96.0±12.0	111.0±6.0 *	121.0±7.0 *	86.5±7.0 *

*Significant differences compared to control ($p<0.05$).

0.01 and 0.03 wt % of fullerenes ($p<0.05$). For nanocomposite containing 1 wt % of $C_{60}+C_{70}$, no significant change in value of S was revealed. In addition we found both significant growth in value of tanα and reduction in value of S_n for all fullerene-containing films (Table 2.1). So regardless of the fullerenes content, the antioxidant activity of PS/$C_{60}+C_{70}$ composites is higher than for original polystyrene. It seems nanocomposites containing fullerenes were easy to react with oxygen species, preventing lipid peroxidation.

Also intensity of lipid peroxidation has been estimated by malodic dialdehyde concentration and total antioxidant reactivity assessed by spectrophotometry (Table 2.2). We have obtained that nanocomposite with 0.03 wt % of $C_{60}+C_{70}$ increased MDA level in blood serum ($p<0.05$).

TABLE 2.2 Lipid Peroxidation Parameters (MDA, TAR) in Blood Serum After Contact With Original Polystyrene Film and Fullerene-Containing Nanocomposites (solvent: o-xylene)

Fullerene content, wt %	MDA	TAR
Controls	100.0	100.0
0	103.0±23.0	108.0⊥4.0
0.03	113.0±5.0 *	116.0±5.0 *
0.1	94.0±8.0	110.0±4.5 *
1.0	96.5±10.5	107.0±4.0 *

*Significant differences compared to control ($p<0.05$).

The increasing of MDA level indicates the activation of lipid peroxidation and the accumulation of free radicals in the blood serum. But the film specimens with higher fullerene content decreased this parameter. Note the similar trend correlates with dependence of I_m value on fullerene concentration in film (Table 2.1).

It can be seen in Table 2.2 that total antioxidant reactivity was increased in serum samples after contact with nanocomposites containing 0.03, 0.10, and 1.00 wt % of fullerenes ($p<0.05$). This proved antioxidant effect of experimental materials. It appears the amount of active centers, which able to effectively capture and inactivate the free radicals, increases with concentrations of fullerenes in composite material.

It is interesting to reveal the effect of the medium in which the films have been fabricated. For this we have performed experiments for films prepared by casting of other aromatic compound – toluene.

The main ChL parameters for "toluene" films are given in Table 2.3. It can be seen that in case of original polystyrene film all values were approximate to controls. But using nanocomposites (0.5 wt % of $C_{60}+C_{70}$) we have found significant increase in value of tanα. It demonstrates the antioxidant activity of researched PS/$C_{60}+C_{70}$ composites.

In addition, it can be seen in Tables 2.1 and 2.3 that at the same concentration of fullerene (0.1 wt %), the value of tanα is higher for film formed of o-xylene than for film formed of toluene. This can be explained by that the solubility of fullerenes in o-xylene is higher than in toluene [21]. It appears that in toluene solution the fullerene molecules are in the form of clusters, which does not ensure uniform distribution of the nanoparticles in the composite film during its formation of solution.

TABLE 2.3   Chemiluminescence Parameters in Blood Serum After Contact With Original Polystyrene Film and Fullerene-Containing Nanocomposites (solvent: toluene)

Fullerene content, wt %	S, %	I_m, %	tanα, %	S_n, %
Controls	100.0	100.0	100.0	100.0
0	98.0±7.5	99.0±7.0	98.5±7.5	100.0±7.0
0.1	101.5±10.5	120.5±14.0 *	100.5±10.5	84.0±8.5 *
0.5	103.0±7.5	101.0±7.5	130.0±14.0 *	102.0±5.0

*Significant differences compared to control ($p<0.05$).

Note also we have performed the preliminary experiments with PS films containing fullerene that have been fabricated by casting of aliphatic compound – chloroform. The results of chemiluminescent analysis and spectrophotometry for serum samples after contact with both original polystyrene film and composite films regardless of the fullerenes content were approximate to controls. This again emphasizes the significance of the medium in which the films were fabricated.

2.4   CONCLUSION

In conclusion, our investigation proved that polystyrene/fullerene nano-composites have ability to inhibit the lipid peroxidation in blood serum. Furthermore, possibility of such inhibition depends on conditions for forming composite.

ACKNOWLEDGMENTS

The study was supported by the Russian Foundation for Basic Research (project no. 12-03-97528-a).

KEYWORDS

- antioxidant activity
- chemiluminescence
- free-radical processes
- fullerene-containing nanocomposites
- malonic dialdehyde

REFERENCES

1. Chakraborty, M., Jain, S., Rani, V. (2011) *Appl. Biochem. Biotechnol. 165*, 1178–1187.
2. Kroto, H. W., Heat, J. R., O'Brien, S. C., Curl, R. F., Smalley, R. E. (1985) *Nature. 318*, 162–163.

3. Bezmel'nitsyn, V. N., Eletskiĭ, A. V., Okun' M. V. (1998) *Physics-uspekhi. 41*, 1091–1114.

4. Andreev, I., Petrukhina, A., Garmanova, A., Babakhin, A., Andreev, S., Romanova, V., Troshin, P., Troshina, O., DuBuske, L. (2008) *Fuller. Nanotub. Car. N. 16*, 89–102.

5. Da Ros, T. (2008), in: Carbon Materials: Chemistry and Physics, vol. 1: Medicinal Chemistry and Pharmacological Potential of Fullerenes and Carbon Nanotubes (Cataldo, F., Da Ros, T., eds.), Springer, pp. 1–21.

6. Wong-Ekkabut, J., Baoukina, S., Triampo, W., Tang, I-M., Tieleman, D. P., Monticelli, L. (2008) *Nat. Nano. 3*, 363–368.

7. Foley, S., Crowley, C., Smaihi, M., Bonfils, C., Erlanger, B. F., Seta, P., Larroque, C. (2002) Biochem. Biophys. Res. Commun. *294*, 116–119.

8. Piotrovskiy, L. B., Eropkin, M. Y., Eropkina, E. M., Dumpis, M. A., Kiselev, O. I. (2007) *Psychopharmacol. Biol. Narcol. 2*, 1548–1554 [in Russian].

9. Lyon, D. Y., Adams, L. K., Falkner, J. C., Alvarez, P. J. (2006) *Environ. Sci. Technol. 40*, 4360–4366.

10. Verner, R. F. and Benvegnu, C. (eds.) (2012) *Handbook on fullerene: synthesis, properties and applications.* Nova Science Publishers, Inc. New York.

11. Weng, D., Lee, H. K., Levon, K., Mao, J., Scrivens, W. A., Stephens, E. B., Tour, J. M. (1999) *Eur. Polym. J. 35*, 867–878.

12. Badamshina, E. R. and Gafurova, M. P. (2008) *Polym. Sci. B. 50*, 215–225.

13. Alekseeva, O. V., Barannikov, V. P., Bagrovskaya, N. A., Noskov, A. V. (2012) *J. Therm. Anal. Calor. 109*, 1033–1038.

14. Alekseeva, O. V., Bagrovskaya, N. A., Kuz'Min, S. M., Noskov, A. V., Melikhov, I. V., Rudin, V. N. (2009) Russ. J. Phys. Chem. A. *83*, 1170–1175.

15. Alekseeva, O. V., Bagrovskaya, N. A., Kuz'min, S. M., Noskov, A. V., Rudin, V. N., Melikhov, I. V. (2008) Dokl. Phys. Chem. *422*, 275–278.

16. Okovitiy, S. V. (2003) *FARMindex: Praktik. 2003*, 85–111 [in Russian].

17. Evlampieva, N. P., Zaitseva, I. I., Ryumtsev, E. I., Dmitrieva, T. S., Melenevskaya, E. Yu. (2007) Polymer Sci. A. *49*, 284–291.

18. Vladimirov, Yu. A. and Proskurnina, E. V. (2009) *Biochemistry (Moscow). 2009*, 1545–1566.

19. Ishihara, M. (1978) *Clin. Chim. Acta. 84*, 1–9.

20. Promyslov, Sh. M., Demchuk, M. L. (1990) *Voprosy Meditsinskoi Khimii. 36*, 90–92 [in Russian].

21. Zhou, X., Liu, J., Jin, Z., Gu, Z., Wu, Y., Sun, Y. (1997) *Fullerene Sci. Techn. 5*, 285–290.

CHAPTER 3

CHARACTERIZATION OF A PEARLESCENT BIAXIALLY ORIENTED MULTILAYER POLYPROPYLENE FILM

ESEN ARKIS,[1] HAYRULLAH CETINKAYA,[2] ISIL KURTULUS,[3] UTKU ULUCAN,[4] ARDA AYTAC,[5] BESTE BALCI,[6] FUNDA COLAK,[7] ECE TOPAGAC GERMEN,[8] GULISTAN KUTLUAY,[9] BEGUM CAN DILHAN,[10] and DEVRIM BALKOSE[11]

[1]*Specialist, Izmir Institute of Technology, Department of Chemical Engineering, Gulbahce Urla 35430 Urla, Izmir, Turkey*

[2]*Student, Izmir Institute of Technology, Department of Chemical Engineering, Gulbahce Urla 35430 Urla, Izmir, Turkey, E-mail: hayrullahcetinkaya@iyte.edu.tr*

[3]*PhD Student, Izmir Institute of Technology, Department of Chemical Engineering, Gulbahce Urla 35430 Urla, Izmir, Turkey, E-mail: isilkurtulus@iyte.edu.tr*

[4]*PhD Student, Izmir Institute of Technology, Department of Chemical Engineering, Gulbahce Urla 35430 Urla, Izmir, Turkey, E-mail: utkuulucan@iyte.edu.tr*

[5]*Master Student, Izmir Institute of Technology, Department of Chemical Engineering, Gulbahce Urla 35430 Urla, Izmir, Turkey, E-mail: ardaaytac@iyte.edu.tr*

[6]*Master Student, Izmir Institute of Technology, Department of Chemical Engineering, Gulbahce Urla 35430 Urla, Izmir, Turkey, E-mail: beste.blc@hotmail.com*

[7]*Master Student, Izmir Institute of Technology, Department of Chemical Engineering, Gulbahce Urla 35430 Urla, Izmir, Turkey, E-mail: fundacolak@yahoo.com*

[8]*Master Student, Izmir Institute of Technology, Department of Chemical Engineering, Gulbahce Urla 35430 Urla, Izmir, Turkey, E-mail: ecetopagac@iyte.edu.tr*

[9]*Master Student, Izmir Institute of Technology, Department of Chemical Engineering, Gulbahce Urla 35430 Urla, Izmir, Turkey, E-mail: glstnchml@gmail.com*

[10]*Undergraduate Senior Year Student, Izmir Institute of Technology, Department of Chemical Engineering, Gulbahce Urla 35430 Urla, Izmir, Turkey, E-mail: bgmdhncn@gmail.com*

[11]*Professor, Izmir Institute of Technology, Department of Chemical Engineering, Gulbahce Urla 35430 Urla, Izmir, Turkey, E-mail: devrimbalkose@gmail.com*

CONTENTS

ABSTRACT

The morphology, composition, optical, thermal and mechanical properties of a commercial pearlescent and multilayer BOPP film were determined in the present study. The film was polypropylene and it was biaxially oriented as shown by FTIR spectroscopy and X-ray diffraction. FTIR spectroscopy indicated carbonate ions, EDX analysis indicated the presence of Ca element, X-ray diffraction showed the presence of calcite and thermal gravimetric analysis indicated 11.2% calcite was present in the film. The 30 μm film consisted of a core layer filled with calcite and 4 μm thick upper and lower layers without any filler and from different polymers. There were long air cavities in the core layer with aspect ratios of 23 and 19 in machine and transverse directions making the film pearlescent. The surfaces of the film were very smooth and had surface roughness in the range of 3.052 nm and 11.261 nm as determined by AFM. The film melted at 163.6°C had 51% crystallinity and had 6.3.nm polymer crystals when heated at 10°C/min rate. The film thermally degraded in two steps. The first and the second steps were for the polymer fraction and decomposition of calcite, respectively. For 10°C/min heating rate the onset of polypropylene degradation was 250°C and calcite decomposition was 670°C. The activation energies for polypropylene degradation and calcite decomposition were 64.8 kJ/mol and 204.8 kJ/mol. The tensile strength of the film in machine and transverse directions were 97.7 and 35.9 MPa, respectively.

3.1 INTRODUCTION

Polypropylene (PP) is one of the most preferred polymer in food packing, protective coating and printing applications with its high stiffness, high temperature resistance, good chemical resistance, lower moisture transmission rate and high mechanical stress properties [1 4]. The polypropylene film that is stretched in both machine direction (MD) and across machine (transvers) direction to improve mechanical properties is called biaxially oriented polypropylene (BOPP). BOPP is widely used in packaging and in a variety of other applications due to their great potential in terms of barrier properties, brilliance, dimensional stability and processability [2]. Different fillers such as talc and calcium carbonate and pigments may be added to BOPP films in order to improve its optical properties and provide a pearly esthetic look [5, 6]. Thus flexible packaging companies are willing to use pearl films for their inexpensive prices, good decoration, and excellent performance. Generally, because they have a certain pearl effect, they are often used in cold drink packaging such as: ice cream, heat seal label, sweet food, biscuits, and local flavor snack packaging [1].

Mineral particles, such as calcium carbonate and talc powders, are widely used in biaxially oriented films, which are also called cavitated and pearlized structures. Pearled film is based on orientation process, where the interface around the particles is stretched forming small cavities in the polymer structure. The foam extent of the film is low but the film becomes highly opaque because of inter scratches [4, 7, 8]. Pearl film is a kind of BOPP film by adding pearl pigments into plastic particles and through biaxial stretch heat setting. A typical pearl film is BOPP pearl film produced by A/B/A layer coextruded biaxial stretch [9]. Three layer films are coextruded where the surface is optimized in order to attain good printability. In fact, the more pigment is in the system the more light is scattered outward, making the system appear opaque and white [10]. Calcium carbonate particles having 0.7–3 μm size are often used in producing micro porous films [11].

The surface morphology of BOPP film could be investigated by atomic force microscopy. The polymer film is characterized by a nanometer-scale, fiber-like network structure, which reflects the drawing process used during the fabrication of the film. The residual effects of the first stretching of the film surface can provide information on the way in which morphological development of the BOPP occurs [12, 13].

The aim of the present study is characterization of a commercial pearlescent BOPP film by advanced analytical techniques. The functional groups, crystal structure, morphology, surface roughness, light transmission and reflection, melting and thermal degradation of the film and mechanical properties were investigated.

3.2   EXPERIMENTAL PART

3.2.1   MATERIALS

The pearlescent films that were kindly supplied by BAK Ambalaj Turkey, which was produced in their plant in Izmir. They were kindly supplied in form of A4 sized sheets with 30 μm thickness.

3.2.2 METHODS

The functional groups in pearlescent film were determined by infrared (IR spectroscopy). IRPrestige-21 FT-IR 8400S by Shimadzu was used to obtain FTIR spectrum of the film by transmission technique. The DRIFT FTIR spectra of the both surfaces of the film were obtained in Digilab Excalibur FTIR spectrophotometer using Harricks Praying Mantis attachment.

Crystal structure of the films was determined by x-ray diffraction using Phillips X'Pert Pro diffractometer system. Cu Kα radiation was used and a scan rate of 2° θ/min was applied.

SEM micrographs of upper and lower surfaces and cross section of gold-coated Pearlescent BOPP films taken by a FEI Quanta 250 FEG type scanning electron microscope. Chemical composition of the film surface was determined by EDX analysis using the same instrument.

AFM (Nanoscope IV) and silicon tip was used to obtain surface morphology and roughness of the film. 1 Ohm Silicon tip has coating: front side: none, back side: 50 +/– 10 nm Al. Cantilever properties are T:3.6–5.6 µm, L:140–180 µm, k:12–103 N/m, fo:330–359 kHz, W:48–52 µm. To achieve surface properties of pearlescent BOPP film, it was cut in 1×1 cm size, then it was put in sample holder in AFM. USRS 99–010, AS 01158–060 serial no OD57C-3930 standard was used in reflection mode. For the reflection spectrum, a black CD was placed at back of the film.

The film was thermally treated under pressure to eliminate its pores. Thus the transparency of pearlescent film and heat-treated film were tested. The pearlescent film's thickness was reduced from 30 µm to nearly 23 µm by in a compression-molding machine (Shinto) in two stage. Film is exposed in the hot press with preheated for 3 min under 0 kg/cm^2 pressure, then heated for 3 min at 60 kg/cm^2. After this stage, film was placed in a cold press for 3 min at 150 kg/cm^2. The light transmission from the films was tested by covering the surface of a paper with our Institute's logo.

The stress strain diagrams of the film in machine direction and transverse direction were obtained with Texture Analyzer TA-XT2 (Stable microsystem, Godalming, UK) having Exponent stable Micro System software. The test is done in ASTM D882. The strips with 5 mm width and 10 mm length were strained with 5 mm/min rate.

3.3   RESULTS AND DISCUSSION

3.3.1   FTIR SPECTROSCOPY

Figure 3.1a shows the pearlescent BOPP film FTIR spectrum taken by the transmission method. The peaks between 2950 and 2800 cm^{-1} correspond to the various aliphatic CH stretching modes. The peaks near 1450 cm^{-1} and 1380 cm^{-1} are the CH$_2$ and CH$_3$ deformation bands, respectively [14]. The other peaks below 1400 cm^{-1} are the well-known "fingerprint" of iso-tactic PP. The peak at around 1500 cm^{-1} of pearlescent BOPP film is wide which is caused existence of calcite. The reason of the increase of the peaks around this region is calcite. The bands at ~1420, ~874 and ~712 cm^{-1} could be attributed to vibrations of CO$_3$ group of calcite [15].

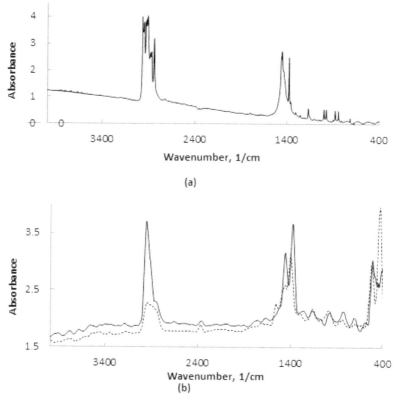

FIGURE 3.1   (a) DRIFT FTIR spectra of front (dotted line) and back (continuous line) surfaces of the film, (b) transmission spectrum.

The DRIFT FTIR spectra of the both surfaces of the film are seen in Figure 3.1b. The peak around 3000 cm^{-1} for back surface is similar to previous result but for front surface, peak is very small. The spectrum of the surfaces of the pearlescent film was very different from the transmission spectrum, indicating they were made out of a different polymer. PVdC and acrylic coatings were used for making the pearlescent film heat sealable and printable. However, without further characterizations it was not possible to identify the polymer surfaces of the film.

3.3.2 X-RAY DIFFRACTION

In Figure 3.2, x-ray diffraction diagram of the film in 5–35 ° 2 theta range is seen. The maximum reflection points of biaxially orientes isotactic polypropylene were observed at 14.2 (110); 17 (040); 18.85 (130); (111) 21.4; (–131) 21.8 2 theta values in the figure [16].

The sharp peak at 29.4° 2 theta value can be attributed to 104 planes of calcite. The x-ray diffraction diagram of the film in 35–65° 2theta values is seen in Figure 3.3 Observed peaks at 36.03, 39.4, 43.2, 47.2, 47.4, 47.6, 48.5° 2theta values are very close to peaks of calcite reported in JCPDS Card Index File, Card 5–5868 [17], which are two theta values of 36.03, 39.4, 43.2, 47.2, 47.5, 48.6. Thus the presence of calcite was also confirmed by x-ray diffraction.

3.3.3 SEM AND EDX

The SEM micrographs of the cross sections of the film in machine and transverse direction are seen in Figure 3.4. The film has a there layer structure. The top and bottom surface layers which had 4 μm thickness do not have any solid particles. FTIR analysis had indicated that the two surfaces were made out of two different polymers other than the core layer. The SEM micrographs of the surfaces indicated that they were very smooth. The core layer with 22 μm thickness had a stratified structure. There were holes having very high aspect ratio created by the 0.8–3 μm sized particles and the orientation process. The dimensions of the pores in machine direction are length 16.4±6.2 μm and width 0.7±0.3 μm, in transverse direction are length 9.14± 3.99 μm and width 0.47± 0.5 μm. Mean aspect ratios

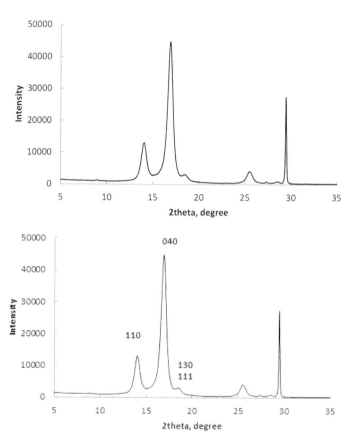

FIGURE 3.2   *x*-ray diffraction diagram of the film in 5–35 ° 2 theta range

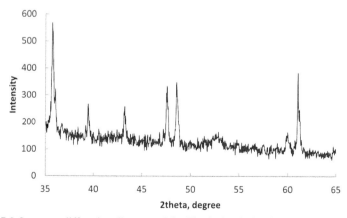

FIGURE 3.3   *x*-ray diffraction diagram of the film in 35–65 ° 2theta range

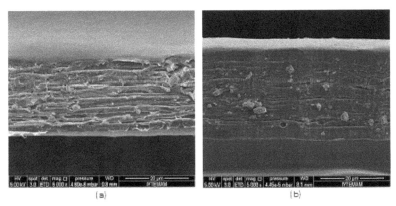

FIGURE 3.4 SEM micrographs of the cross-sections of the film in: (a) machine direction (b) transverse direction.

(length/width) of pores observed in Figures 3.4a and 3.4b are 23 and 19, respectively.

The EDX analysis of the surface of the filler particles indicated that they consisted of Ca, C and O elements. They had a composition similar to $CaCO_3$, which had 40% Ca, 12% C and 48% O. EDX analysis of the particles showed that the particles had 42.8±1.6% Ca, 22.3±3.73% C and 34.9 ±5.2% O. The particles were calcite and they were coated by a compound, which was rich in C.

3.3.4 AFM STUDY

Typical images of the surface of the pearlescent film in two and three dimensions are seen in Figure 3.5. The surface consists of spherical particles. No network structure was observed as indicated by previous studies for 8:1 draw ratio, indicating the draw ratio of the pearlescent film in machine and transverse direction were close to each other. The surface roughness of the films were determined in three different regions and reported in Table 3.1. The root mean square roughness (Rms) was between 3.052 and 11.261 nm and average roughness(Ra) was in the range 2.330–7.326 nm. This low roughness values indicated that the surface of the pearlescent films was very smooth.

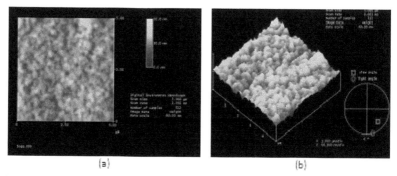

FIGURE 3.5 AFM micrographs of the surface of the pearlescent films: (a) two dimensional, (b) three dimensional appearance.

TABLE 3.1   Image Statistics of Pearlescent BOPP Films at Three Different Regions

Scan size, μm x μm	5×5	5×5	1×1
Z range, nm	38.155	113.93	21.031
Raw mean, nm	25.591	53.191	−37.563
Rms (Rq), nm	4.861	11.261	3.052
Ra, nm	3.821	7.326	2.330
Srf. Area, μm²	25.057	25.121	1.003

3.3.5   DSC ANALYSIS

DSC analysis was used to determine melting point, melting heat and crystallinity, the crystallite size and activation energy of the melting process. The DSC curves of the sample heated at different rates are seen in Figure 3.6. A shoulder corresponding to the melting of small crystallites was observed at all heating rates. This shoulder was also observed for biaxial oriented polypropylene by previous investigators [18]. The melting temperature shifts to higher temperatures as the rate of heating was increased. Table 3.2 shows enthalpy of melting, melting temperature and crystallinity determined at different rates of heating of the film.

The degree of crystallinity (Xc) of the samples from DSC melting peaks were determined using Equation 1.

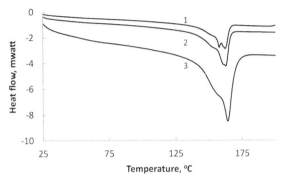

FIGURE 3.6   DSC curves of the film at: (a) 5°C/min, (b) 10°C/min, (c) 15°C/min heating rates.

TABLE 3.2   Enthalpy of Melting, Melting Temperature and Crystallinity Determined by DSC

B, °C/min)	ΔHm, J/g	Tm, °C	Crystallinity, %	Lc, nm
5	87.63	162.9	48	6.1
10	93.75	163.6	51	6.3
15	129.07	164.2	60	6.5

$$X_c (\%) = \frac{\Delta H_m}{w \Delta H_f^0} \times 100 \qquad (1)$$

where ΔH_m is the melting enthalpy of the samples (J/g) and ΔH_f^0 is the heat of the fusion of PP at 100% crystallinity, correspondent to 207 J.g^{-1}[18]. The crystallinity of the film also increases with the rate of heating.

The Thompson–Gibbs equation predicts a linear relationship between T_m and the reciprocal of crystal thickness.

$$T_m = T_m^0 \left(1 - \frac{2\sigma}{L_c \rho_c \Delta H_f^0} \right) \qquad (2)$$

where σ is the fold surface free energy, T_m^0 is the equilibrium melting temperature, ρ_c is the crystal phase density of pp, ΔH_f^0 is the heat of the fusion of PP at 100% crystallinity, correspondent to 207 J.g^{-1} [19], and L_c is the thickness of the lamellar crystals. $T°_m$ is 459.1 K [20], ρ_c is 946 kg/m^3 [21] and σ is 30.1 mN/m [22]. The crystal thickness values determined by using the melting temperature for different heating rates are reported in Table 3.2. They were in the range of 6.1 to 6.5 nm.

3.3.6 TG ANALYSIS

Thermo gravimetric analysis (TGA) method was also employed to understand thermal degradation behavior of the BOPP film. Typical weight loss (TGA) curves of BOPP film at heating rate 5, 10 and 15°C/min under nitrogen is seen in Figure 3.7. It is observed that thermal degradation process of BOPP film proceeds in two stages. The first stage corresponds in to the degradation of polymer. The second stage is related to the decomposition of calcium carbonate. The degradation of the BOPP film started at 235°C, 258°C and 265°C at heating rate of 5°C, 10°C and 15°C/min, respectively. The maximum rate of degradation of BOPP film was 358, 404.6 and 412°C at 5°C, 10°C and 15°C/min heating rates, respectively. The second step of the mass loss observed in Figure 3.7 was for the decomposition of calcite. Figure 3.7 display that the degradation of the calcium carbonate started at 648°C, 670°C and 675°C at the heating rate 5, 10 and 15°C/min. and its rate was maximum at 690°C, 721.5°C and 714.7°C. for 5°C, 10°C, 15°C heating rates.

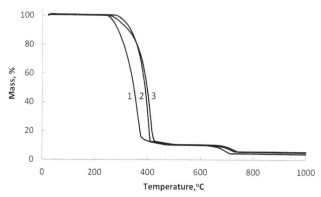

FIGURE 3.7   TG curves of the film for: (a) 5°C/min, (b) 10°C/min, (c) 15°C/min heating rates.

The second stage of the mass loss belongs to decomposition of calcium carbonate. The calcium carbonate decomposes calcium carbonate and carbon dioxide;

$$CaCO_{3(s)} \longrightarrow CaO_{(s)} + CO_{2(g)} \tag{3}$$

If one mole of calcium carbonate decomposes, one mole of calcium oxide and one mole of carbon dioxide would form. Thus the second step is for the evolution of CO_2 from $CaCO_3$. From the mass loss of the second step of the degradation curve it was found that the film contained 11.2% $CaCO_3$.

In this study decomposition activation energy was determined by using Flynn and Wall equation. Flynn and Wall is derived a convenient method to determine the activation energy from weight loss curves measured at several heating rates. The following relationship is used to calculate the activation energy [23].

$$E = \frac{-R}{b}[\frac{dlog(\beta)}{d\left(\frac{1}{T}\right)}] \tag{4}$$

where E = activation energy (j/mol), R = gas constant (8.314 j/mol K) and b = constant (0.457) [23].

The values of 1, 2 and 5% decomposition level were chosen to determine the activation energy for degradation of the polypropylene and temperatures for these conversions were red from Figure 3.7. The activation energy was determined directly by plotting the logarithm of the heating rate versus 1000/T at constant conversion. The plotted data produced straight lines with r2 values higher than 0.93. From the slopes the activation energy values were found and they are reported in Table 3.3. The average activation energy was 64.8 kJ/mol.

TABLE 3.3   Activation Energy For Degradation of Polypropylene and Calcite

Polypropylene			Calcite		
Mass loss, %	R^2 value	Ea, kJ/mol	Mass loss, %	R^2 value	Ea, kJ/mol
1	0.99	64.8	931	0.98	-195.6
2	0.96	66.4	92	0.99	-209.6
5	0.93	63.4	93	0.99	-209.1

The activation energy for the decomposition of the calcite in the film was determined in the same manner and reported in Table 3.3. The average activation energy for the decomposition of calcite was found as 204.8 kJ/mol.

3.3.7 OPTICAL PROPERTIES

The unique luster of pearls depends upon the reflection, refraction, and diffraction of light from the translucent layers. The thinner and more numerous the layers in the pearl, the finer the luster. The iridescence that pearls display is caused by the overlapping of successive layers, which breaks up light falling on the surface [24]. The film under study had polypropylene layers separated by long and thin air pockets formed by orientation process and calcite particles as seen electron micrograph in Figure 3.4. Thus it shows pearlescent behavior.

Polypropylene polymer can reflect only a very small percentage of incoming light. Using Fresnel equation

$$R = \frac{(n_1 - n_2)^2}{(n_1 + n_2)^2} \tag{5}$$

In the above equation, 'n1' and 'n2' indicate the reflection indices of polypropylene and air, respectively. Reflection values were calculated considering Frensel's equation. Reflection index of polypropylene is 1.49 and 1.0 for air [25]. Then, reflection (R%) was calculated as 3.87%. However, the film under study reflects 85% of light at 400 nm and 65% at 700 nm as seen in its reflection spectrum in Figure 3.8. The thin layers and the air gaps between them is the cause of this pearlescent effect.

It was reported that BOPP films were transparent to light and the their smoother surface they had, the more transparent they were [26]. It followed that the clearest films were obtained from sheets with the most homogeneous texture, such as obtained by quenching from the melt, and by orienting at the lowest temperature, which minimized the amount of melting [26]. The film in the present study was a sandwich type BOPP film having a core layer with calcite. The film reflects light but it does not transmit it. The transmission spectrum of the film in Figure 3.9 indicated that 0.36% of incident light was transmitted at 400 nm and 0.5% was transmitted at 700 nm.

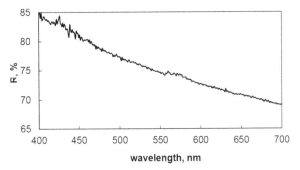

FIGURE 3.8   Reflection spectrum of the film under study.

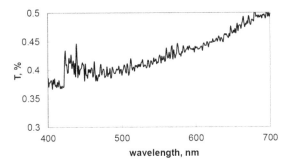

FIGURE 3.9   Transmission spectrum of the film in visible region.

The light was not transmitted from the film because of the air holes in the film. When these holes were removed by hot pressing the film, it became transparent as seen in Figure 3.10. The logo of our Institute covered by the pressed film was visible, but when the logo was covered by the pearlescent film it was not visible. The pearlescent film was opaque and when the air holes was removed it was transparent even if it contained 12% calcite. It was the air gaps not the calcite making the film opaque and pearlescent.

3.3.8   MECHANICAL PROPERTIES

The mechanical properties of BOPP in machine and transverse direction are different. In Figures 3.11 and 3.12 stress strain diagrams of the film in transverse and machine directions are seen. The tensile strength in transverse direction is lower and strain at break is higher than those of machine direction as reported in Table 3.4. Tensile stress 35.9 MPa and 97.7 MPa,

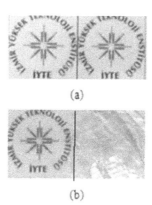

(a)

(b)

FIGURE 3.10 (a) Paper surface without sample films, (b) paper surface covered with sample films: Left side was covered with nearly 23 micron film pressed film. Right side was covered with 30 micron original white BOPP film sample.

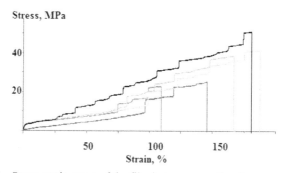

FIGURE 3.11 Stress strain curve of the film in transverse direction.

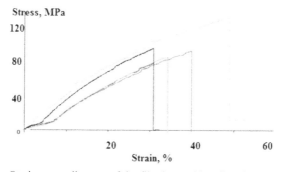

FIGURE 3.12 Strain stress diagram of the film in machine direction.

TABLE 3.4   Mechanical Properties of the Film in Machine and Transverse Directions

Direction	Stress, MPa	Strain, %	Young Modulus, MPa/ %
Transverse	35.9 ± 11.62	157.7 ± 31	0.129 ± 0.065
Machine	97.2 ± 20.6	37 ± 7.7	2.93 ± 0.51

elongation at break 157% and 37% for transverse and machine directions., respectively. No yield point was observed in the BOPP film with calcite. BOPP film without calcite was characterized by Yuksekkalayci et al. [27] and it had the yield point (34.2 MPa and 42.2 Mpa) machine direction and transverse direction. The film without calcite had higher values of tensile stress (151 MPa and 270 MPa) [27] in machine and transverse directions than the film with calcite. However, the elongation at break values (150%, 32%) were closed to the values for the film with calcite. The presence of pores lowers the tensile strength, however the elongation values were closer. The modulus of elasticity of the film under study also changed with direction. It was lower (0.129 MPa/%) in transverse direction than in machine direction (2.93 MPa/%). However, the film without calcite had much higher elastic modulus values 2.8 to 5.9 GPa. Thus the calcite filled BOPP film was much more flexible than the film without calcite.

3.4   CONCLUSION

A pearlescent packing material supplied by BAK Ambalaj Turkey was characterized for obtaining information about its properties for its application fields and its recycle in industry. The advanced characterization techniques such as FTIR spectroscopy, X-ray diffraction, SEM, EDX, AFM, DSC, TG analysis, visible spectroscopy and tensile testing were used for this purpose. The bulk film was polypropylene and it was biaxially oriented as shown by FTIR spectroscopy and X-ray diffraction, respectively. FTIR spectroscopy indicated presence of carbonate ions, the presence of Ca element was indicated EDX analysis. X-ray diffraction showed the presence of calcite and 11.2% calcite was present in the film as indicated by TG analysis. The 30 μm film consisted of a core layer of polypropylene filled with calcite and 4 μm thick upper and lower layers without any filler were from different polymers. There were long air cavities in the core layer with aspect ratios of 23 and 19 in machine and transverse directions making the film pearlescent. The surfaces of the film were very smooth and had a surface roughness in the range of 3.052 and 11.261 nm as determined by AFM. The film melted at 163.6°C had 51% crystallinity and had 6.3 nm polymer crystals for 10°C/min heating rate. The film thermally degraded in two steps. The first step was for the polymer fraction and the second step was for decomposition of calcite. For 10°C/min heating rate the onset of polypropylene degradation was 250°C and calcite decomposition was 670°C. The activation energies for polypropylene degradation and calcite decomposition were 64.8 kJ/mol and 204.8 kJ/mol. The film reflected but not transmitted visible light. The tensile strength of the film in machine and transverse directions were different and it was 97.7 and 35.9 MPa, respectively.

ACKNOWLEDGEMENT

The authors thanks to Bak Ambalaj Turkey for providing the pearlescent films for this study.

KEYWORDS

- AFM
- BOPP
- pearlescent film
- SEM
- tensile strength
- X-ray diffraction

REFERENCES

1. Nie, H. Y., Walzak, M. J., McIntire, N. S., Atomic force Microscope Study of Biaxially Oriented Films, *Journal of Materials Engineering and Performance*,2009, 13 (4) 4511–460.

2. Introduction of BOPP Film http://decrobopp.wordpress.com (2013).

3. Longo, C., Savaris, M., Zeni, M., Brandalise, R. N., Grisa, A. M. C., Degradation Study of Polypropylene and Bioriented Polyproylene in the Environment. *Materials Research*, 2011, 14 (4) 442–48.

4. Raukula, J. I., A New Technology to Manufacture Polypropylene Foam Sheet and Biaxially Oriented Foam Film; Thesis, Technical Research Centre of Finland (1998).

5. Ulku, S., Balkose, D., Arkis, E., Sipahioglu, M., A study of chemical and physical changes during biaxially oriented polypropylene film production, *Journal of Polymer Engineering*, 2003, 23, 437–456.

6. Kalapat, N., Amornsakchai, T., Surface modification of biaxially oriented poly propylene (BOPP) film using acrylic acid-corona treatment: Part, I. Properties and characterization of treated films, *Surface and Coating Technol*ogy, 2012, 207, 594–601.

7. Nago, S., Mizutani, Y., Microporous Polypropylene Sheets Containing $CaCO_3$ Filler: Effects of Stretching Ratio and Removing $CaCO_3$ Filler *Journal of Applied Polymer Science*, 1998, 68, 1543–1553.

8. Opaque polymeric films and processes making the same US patent 6183856B1,2001

9. Koleske, J. V., Paint and Coating Testing Manual, 14th ed., the Gardner-Sward Ed., Philadelphia, PA: ASTM, 1995.

10. http://www.specialchem4polymers.com (2013).

11. http://www.plastemart.com/upload/literature/246_art_bopp_in_foodpack.asp (2014).

12. Biswas, J., Kim, H., Lee, B. H., Choe, S., Air-hole properties of calcite-filled polypropylene copolymer films. *Journal of Applied Polymer Science*, 2008, 109, 1420–1430.

13. Nie, H.-Y, Walzak, M. J., McIntire, N. S. Atomic Force Microscopy Study of Biaxially Oriented Polypropylene Films *JMEPEG,* 2004, 13,451–460.

14. Nie, H.-Y., Walzak, M. J., McIntire, N. S. Draw-ratio-dependent morphology of biaxially oriented polypropylene films as determined by atomic force microscopy *Polymer*, 2000, 41, 2213–2218.

15. Izer, A., Kahyaoglu, T. N., Balkose, D., Calcium Soap Lubricants, Science Journal of Volgograd State University, 10, 1, 16–25.

16. Chen, J., Xiang L., Controllable synthesis of calcium carbonate polymorphs at different temperatures, *Powder Technology*, 2009, 189, 64–69.

17. Diez, F. J., Alvarino, C., Lopez, J., Ramirez, C. Abad, M. J., Cano, J., Garcia-Garabal, S., Barral, L., "Influence of the Stretching in the Crystallinity of Biaxially Oriented Polypropylene (BOPP) Films, " *Journal of Thermal Analysis and Calorimetry,* 2005, 81, pp. 21–25.

18. Data from JCPDS Card Index File, Card 5–5868.

19. Yang, W., Li Z-M., Xie, B.-H., Feng J-M., Shi, W., Yang M-B. Stress-Induced Crystallization of Biaxially Oriented Polypropylene *Journal of Applied Polymer Science*, 2003, 89, 686–690.

20. Bu, H. S., Cheng, S. Z. D., Wunderlich, B. Addendum to the thermal properties of polypropylene. *Die Makromolekulare Chemie Rapid Communications* 1988, 89(2), 75–77.

21. Yamada, K., Hikosaka, M., Toda, A., Yamazaki, S., Tagashira, K. Equilibrium melting temperature of isotactic polypropylene with high tacticity: 1. Determination by differential scanning calorimetry *Macromolecules,* 2003,36,4790–4801.
22. http://en.wikipedia.org/wiki/Polypropylene.
23. http://www.surface-tension.de/solid-surface-energy.htm.
24. http://www.tainstruments.co.jp/application/pdf/Thermal_Library/Applications_ Briefs/TA125.PDF.
25. http://perlas.com.mx/en/quality/luster.html, 2014.
26. Birley, A. W., Haworth, B., Batchelor, J., *Physics of Plastics: Processing Properties and Materials Engineering*, Hanser Publishers, Munich (1992).
27. Lin, Y. J., Dias, P., Chum, S., Hiltner, A., Baer, E., Surface roughness and light transmission of biaxially oriented polypropylene films, *Polymer Engineering and Science*, 2007,47, 1658–1665.
28. Yuksekkalayci, C., Yilmazer, U., Orbey, N., Effects of Nucleating Agent and Processing Conditions on the Mechanical, Thermal and Optical Properties of Biaxially Oriented Polypropylene Films, *Polymer Engineering and Science*, 1999, 39,1216–1222.

CHAPTER 4

POROUS STRUCTURE OF CELLULOSE TREATED BY SUPERCRITICAL CARBON DIOXIDE AS STUDIED BY SPIN PROBES TECHNIQUE

ALEXANDER L. KOVARSKI,[1] OLGA N. SOROKINA,[1] ANATOLII B. SHAPIRO,[1] and LEV N. NIKITIN[2]

[1]*Emanuel Institute of Biochemical Physics RAS, 119334, Kosygin str., 4, Moscow, Russia;*

[2]*Nesmeyanov Institute of Elementoorganic Compounds RAS, 119991, Vavilov str., 28, Moscow, Russia*

CONTENTS

ABSTRACT

Highly porous cellulose samples (aerocellulose), obtained by the processing of stitched and not stitched original gel with supercritical carbon dioxide have been studied by the spin probes technique of electron paramagnetic resonance. The width of the distribution in rotational correlation time of spin probe (τ) and its average value at room temperature have been determined by the comparison of experimental and calculated electron paramagnetic resonance spectra. These parameters characterize porous polymer structure. The data obtained in this work show that the width of the Gaussian distribution in τ value at 1/e part of it height (α) after processing of the original gels by supercritical CO_2 may be as much as 2–3 orders of magnitude. The greatest value of α is typical for the sample obtained from previously stitched original gel (5 orders of magnitude). The results show that the cellulose with different porous structure can be obtained by processing of the original gels with supercritical CO_2, varying the concentration of polymer in the gel, and introducing cross-linkers.

4.1 INTRODUCTION

Porous materials with pore sizes at micro – and nanoscale, as well as composite materials on their basis, is very popular as active media for medicine, pharmaceuticals, cosmetics, biology, materials science and other fields. This is especially important for the tasks that require biocompatible and biodegradable materials, such as three-dimensional supporting matrices for tissue medicine [1], containers for targeted drug delivery [2], safe biodegradable packaging and insulating materials [3]. These materials include porous solid polysaccharides, including cellulose, which is an ecofriendly renewable raw material.

The porous cellulose previously was obtained usually by freeze drying (lyophilization) of gels [4]. In recent years highly porous systems started to be formed by the treatment with supercritical carbon dioxide (SC-CO$_2$) with low critical parameters: the temperature of 31.1°C and pressure 7.38 MPa. It is known that at a temperature and pressure above the critical the substances combine the properties of liquids and gases. It is important that SC-CO$_2$ is quite safe and ecologically favorable reagent. This allows to eliminate the use of toxic solvents, so SC technologies are usually associated with environmentally friendly "green" chemistry. Varying the parameters (temperature, pressure, duration of exposure) and impact mode of SC-CO$_2$ (running, stationary), it is possible to influence the morphology of the formed porous structure in aerocellulose (AC). The significant results have been achieved when the creating porous cellulosic materials with pore sizes in nano – and microrange [5–10]. The porosity of AC exposed to SC-CO$_2$ can reach values higher than 95%. The specific surface of the pores may exceed 200 m^2/g, and density varies from hundredths to tenths parts of g/cm^3 depending on the formation conditions.

There are many methods for measuring the specific surface area and total pore volume. These include mercury porometry, low-temperature sorption of nitrogen vapors, BET method. In this chapter, the authors have analyzed the possibilities of electron paramagnetic resonance (EPR) of spin probes technique to study the distribution of the pores sizes in the samples of AC obtained at different conditions. Paramagnetic particles – stable nitroxide radicals were used as probes. The main parameter, which is measured from the EPR spectra, is the rotational correlation time of spin probes – τ [11–12]. The value of τ characterizes the period of particle reorientation at a sufficiently large angle ($\sim\pi/2$). The other parameters used here are rotation frequency $v=\tau^{-1}$ and coefficient of rotational diffusion $D = (6\ \tau)^{-1}$.

It was shown [12–17] that in polymers at temperatures below T_g the intensity of spin probes rotation is characterized by the values of τ in the range from 0.4 to 10 ns, and inversely proportional to the value of the specific surface S_{sp} determined by the method of low-temperature adsorption of nitrogen vapors. For example, the rotation of the probe 2,2, '6,6'-tetra-methylpiperidine-1-oxyl (TEMPO) in polyvinyl acetate (PVA) and polyvinyl chloride (PVC) at T_g is realized with correlation time of 10 ns, that is significantly higher than for the loose packed polymers such as polyvinyl-trimethylsilane (τ = 0.4 ns), polystyrene (τ = 1.7 ns), cellulose (τ =3.5 ns) [12, 17–19]. The spin probe mobility is proved to be connected with the pore sizes referring to [15]. The authors of [15, 19] investigated porous polymers based on PS, stitched by monochlordimethyl ether. The increase in the amount of cross-linking agent resulted in the formation of products of greater porosity (according to mercury porometry), but with a higher glass transition temperature. It turned out that with the increase in total pore volume the spin probe rotates faster despite the lower segmental mobility of the polymer, which is characterized by the increase in T_g. Another convincing proof of the connection of low molecular weight particles mobility with pore sizes are the data of the study of dehydrated synthetic zeolites, which represents a crystalline mineral sorbents with pores of certain and equal sizes [18]. It was shown that zeolites CaX with the size of the "input window" 0.4 nm, that is, less than the Van-der-Waals radius of the probe TEMPO (Figure 4.1), did not absorb this radical. In zeolites NaX with the size of the "input window" 10 nm the TEMPO-probe rotated with a high frequency ($\tau = 2$ ns) at room temperature. Similar results were obtained in the study of the mobility of TEMPO probes in the cavities of inclusion compounds (clathrates), such as thiourea, cyclohexane [20].

Thus, the main factor determining the rotational mobility of particles in rigid porous systems is not their molecular mobility but the parameters of the porous structure. In the present work we used the spin probes method to study the cellulose samples with high porosity.

4.2 EXPERIMENTAL PART

4.2.1 PREPARATION OF AEROCELLULOSE SAMPLES

In this chapter, we used natural microcrystalline cellulose "Avicel PH 101" (FMC Corp.) trademark, NaOH of 97% purity ("Aldrich") acetone of 99.98% and ethanol of 99.9% purity was purchased in "Bioblock." CO_2 with a purity of 99.9% was acquired in "Air-Liquide."

The samples of aerocellulose AC-1, AC-2 and AC-4 were obtained from a 5% solution of cellulose in a mixture of 8% NaOH and water (gel 1). The sample AC-3 was obtained from a 7% solution of cellulose in a mixture of the same composition (gel 2). Regeneration of all cellulose samples was performed in a 25% solution of ethanol-water at room temperature. The water from the gels was replaced with acetone to improve

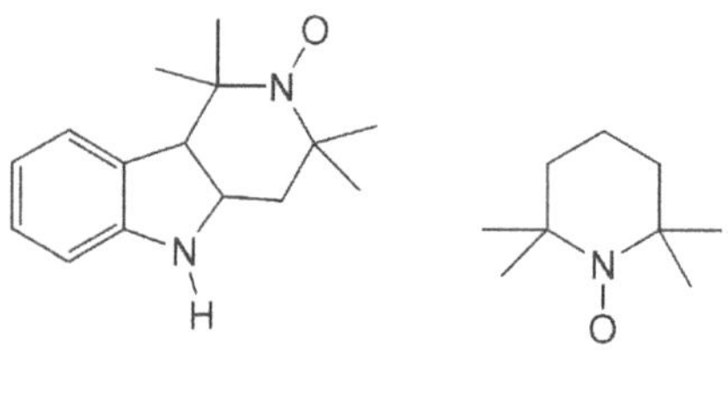

I   II

FIGURE 4.1   Probes I (carbolic) and II (TEMPO).

drying in SC-CO$_2$. The sample AC-1 was obtained by drying the gel 1 in air for 48 h at room temperature.

The sample AC-2 was obtained by drying the gel 1 in SC-CO$_2$. For drying the coagulated gel was placed in an autoclave with a volume of 1000 cm3, then the pressure of CO$_2$ was raised up to 5 MPa and the temperature was run up to 37°C and. After sustaining the sample for 2 h the pressure was raised up to 8 MPa, exposed for 1 h and subjected to the flowing mode of pumping of SC-CO$_2$ (5 kg CO$_2$ per hour) within 7 h. After that, the CO$_2$ slowly (over 12 h) released from autoclave (0.4 MPa/hour at 37°C) to avoid condensation of moisture and the samples cooled to room temperature. The dried samples of AC were the solid porous cylinders of white color.

The sample AC-3 was obtained by the drying of the gel 2 in SC-CO$_2$ by the same method as for the samples AC-2. The sample AC-4 was obtained by cross-linking of gel 1 in the presence of epichlorohydrin and dried as well as samples AC-2 and AC-3 using supercritical carbon dioxide.

The main parameters of the porous structure of the AC-2 sample are known. They were previously defined [10] from the adsorption isotherms of nitrogen vapors at a temperature of −196°C, measured on high-vacuum installation ASAP-2020 MP "Micromeritics" (USA) in the range of pressures from 10^{-6} to 0.99 ATM. The total specific surface area was calculated from the isotherms of adsorption of nitrogen vapors according to BET equation. They were: S_{BET} = 196 m^2/g, pore volume – 0.57 cm^3/g.

4.2.2 SPIN PROBES SELECTION AND INTRODUCTION INTO POLYMERS

The following nitroxide radicals were used as spin probes: 2,2', 6,6'-tetramethylpiperidine-1-oxyl (TEMPO) and 2,2',4,4'-tetramethyl-1,2,3,4-tetrahydro-γ-carbolin-3-oxyl (carbolic probe), the structural formula of which are shown in Figure 4.1. Van-der-Waals volumes of radicals were ~169 A^3 and ~257 A^3, respectively [12].

EPR spectra were recorded on the X-band spectrometer "Bruker EMX" at a temperature of 373 K. The parameters of the device were: amplitude modulation – 2 G, power of the microwave field – 10 mW.

First of all, it should be noted that the volume of probe molecules is not always the main characteristic parameter in the study of the porous structure. So for example, in the radical I, having an elongated shape, an important structural element, limiting the penetration of this radical into the pores, is a radical fragment (oxopiperidine ring with methyl groups), the diameter of which according to X-ray analysis is approximately 1 nm. However, the friction coefficient of the particles that determine their rotational mobility, sharply depend on the dimensional parameters [12, 13, 21]. Note also that during introduction of the radical into the polymer from the joint solution with subsequent drying of the sample, the probe itself forms the cavity which is comparable to its size. Studies performed with technology of hydrostatic compression of the sample [21–23], showed that the activation volume V^* for probes rotation (minimum free space required for reorientation of the particles) at $T \leq T_g$ is about 10% of their volume. However, with the introduction of probes in the system, the pore size of which is much greater than the dimensions of the probe, they can accumulate inside the pores in high concentrations, which lead to intermolecular dipole-dipole or spin-exchange broadening of EPR lines. Increasing of the local concentrations of probes is accompanied by a shift of the side components of the triplet to the center ("blur" of hyperfine structure, HS) and the formation of one line – singlet (Figure 4.2).

Such spectra are observed, for example, after exposure the samples in vapors of the radical II, even at short time (1–3 min). From these spectra it is impossible to obtain information on the rotational dynamics of the probes. For this reason, we have refused from using probe II.

In this paper we have introduced the probes I in the samples from the weak solution of radicals in toluene (~10^{16} spin/ml). Cellulose is not soluble in toluene, and, consequently, the exposure time in solution did not

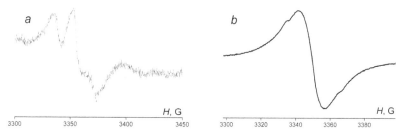

FIGURE 4.2 EPR spectra of probe II, introduced into the cellulose samples from a solution in toluene with a concentration of radical 10^{16} spin/ml (*a*), and from a gas phase at exposure of 10 min (*b*).

lead to structural changes of the sample. The exposure time of the samples in solution was 3–5 min. Then the samples were dried overnight in air, and then placed in a vacuum box for 3 h. Drying was carried out at room temperature.

4.3 RESULTS

Experimental EPR spectra of probe I in the samples of AC were compared with simulated ones in Figure 4.3.

The analysis of the spectra shows that beside three lines of nitrogen HFS the additional lines have appeared. These lines indicate the existence of a set of correlation times. To determine the width (and shape) of the distribution in τ we used "Atlas of EPR spectra of spin labels and probes" [24]. This atlas collects spectra simulated with the use of the stochastic Liouville equation (method of random trajectories [25–29]) for uncorrelated jump-like reorientations of particles, including Gaussian distribution in τ, which is described by the equation:

$$n(lg\tau) = \frac{1}{\alpha\pi^{1/2}} \exp\left[-\left(\frac{lg\tau - lg\tau_{av}}{\alpha}\right)^2\right]$$

Graphic depiction with Gaussian distribution by correlation times is shown in Figure 4.4.

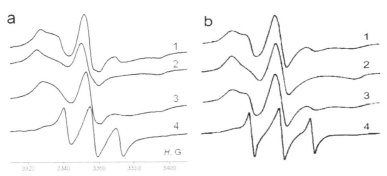

FIGURE 4.3 Experimental (a) and simulated (b) EPR spectra of probe I in the samples AC-1 – AC-4 at 373 K.

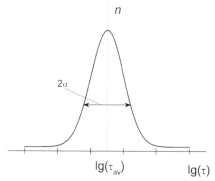

FIGURE 4.4   Gaussian distribution over τ [24].

Simulated spectra are most similar in shape to the experimental ones shown in Figure 4.3b. It should be noted that the values of electron-spin parameters for radicals used in theoretical calculations of the spectra in [24] and experimental values for the radical I, vary slightly and may not be the cause of noticeable errors. The distribution parameters are given in Table 4.1.

The error in the determination of τ_{av} and α does not exceed 30%.

4.4   DISCUSSION AND CONCLUSION

Despite the fact that the spin probes do not allow to determine accurately the sizes of the pores in which they are located, they can be used to determine the minimum pore size, and the width of the size distribution. The

TABLE 4.1   Parameters of the Distributions Over τ for Probe I at 373K: Average Allocation Value (τ_{av}), Logarithm of the Half-Width of Gaussian Distribution at 1/e Part of it Height (α)

Sample number	τ_{cp} (c)	α
AC-1	2×10^{-8}	2
AC-2	1×10^{-7}	2
AC-3	1×10^{-7}	3
AC-4	7×10^{-9}	5

minimum pore volume equals to the volume of the molecule probe (257 Å^3 for probe I) plus the activation volume V^* for reorientation of the probe at $T \leq T_g$ equals as indicated above, $\sim 10\%$ of the probe volume. Thus, the minimum pore volume according to the probe I is $\sim 280 \text{ Å}^3$. Unfortunately, we do not have data on the dependence between the values of the correlation time of the probe rotation τ and a pore sizes, but it's easy to assume that these parameters antibatic. In Refs. [12, 21, 30] it was shown that the magnitude of lgτ is directly proportional to the volume of probe molecules. The same conclusion can be drawn for the correlation time of the same probe in pores of different sizes. Adhering to this assumption, the distribution over lgτ reflects the distribution by volume of pores in the specimens examined (in the framework of the Gaussian distribution).

Now consider the parameters of the distributions obtained in this work (Table 4.1). From these data it follows that the width of the distribution depends on the conditions of preparation of dry samples of cellulose from the original gel. Recall that all samples except AC-1 were subjected to treatment with carbon dioxide under supercritical conditions, but differed in the percentage of cellulose in the original gel, and the original gel, which was used for obtaining a sample AC-4, was subjected to the stitching. This sample was characterized by the most wide distribution over lgτ (half-width α was 5 orders of magnitude) and the lowest value of the average correlation time ($\tau_{av} = 7 \times 10^{-9}$ c). There is sufficient data to indicate that cross-linked glassy polymers have high porosity, because the cross-links prevent the dense packing of macromolecules (see, e.g., [15]). For the non-stitching samples AC-3 and AC-2 the half-width is drastically lower (2–3 orders of magnitude), and the average correlation time is longer (10^{-7} s). A difference in the half-widths of the distributions between the two samples is due to the difference in concentrations of cellulose in gels used for dry samples production.

The original sample AC-1, obtained by drying the gel in air for 48 h at room temperature was not subjected to the carbon dioxide treatment or stitching, and was prepared by drying of the initial gel in the air. This sample is characterized by the same full width at half maximum of the distribution over lgτ as the sample AC-2. However, the average value τ_{av} for it is higher than for the samples AC-2 and AC-3. From these data it follows that the packing in the original polymer looser than in AC-2 and AC-3, but pores distribution in it is not wide.

Thus, it was shown that the spin probe method in contrast to other methods of investigation of the porous structure gives useful information about the width of the size distribution of pores, as well as the packing density of polymer glasses. Acting on the original gel by supercritical carbon dioxide, changing the concentration of polymer in the gel, and introducing of cross-links one can obtain cellulose with different characteristics of the porous structure.

The authors are grateful to Professor T.V. Budtova for preparing the samples of aerocellulose. Centre de Mise en Forme des Materiaux (CEMEF) Mines ParisTech, BP 207, 06904 Sophia-Antipolis, France (France).

The work was carried out with the financial support of Russian Foundation for Basic Research (project 12–03–00790) and the Russian Academy of Sciences (grant within the framework of the complex program OHNM-2).

KEYWORDS

- **aerocellulose**
- **supercritical carbon dioxide**
- **porosity**
- **spin probes**
- **EPR-spectroscopy**

REFERENCES

1. Sezer, A. D., Cevher, E. "Biopolymers as Wound Healing Materials: Challenges and New Strategies." In "Biomaterials Applications for Nanomedicine." Ed. Prof. Rosario Pignatello. Croatia. "In Tech." Publishing House. 2011, P. 383–414.
2. Mehling, T., Smirnova, I., Guenther, U., Neubert, R. H. H. J. Non-Crystalline Solids. 2009, v. 355, p. 2472.
3. Mohanty, A.K, Misra, M., Hinrichsen, G. Macromol. Mater. Eng. 2000, v. 276–277, p. 1–24.
4. Benoıt, J., Duchemin, C., Staiger, M. P., Tucker, N., Newman, R. H. J. Appl. Polymer Science. 2010, v. 115, p. 216.

5. Hao, J., Nishiyama, Y., Wada, M., Kuga, S. Colloids and Surfaces, Ser A: Physico-chem. Eng. Aspects, 2004, v. 240, p. 63.
6. Fischer, F., Rigacci, A., Pirard, R., Berthon-Fabry, S., P. Polymer, 2006. v. 47, p. 7636.
7. Gavillon, R.; Budtova, T. Biomacromolecules. 2008, v. 9, № 1, p. 269.
8. Sescousse, R., Gavillon, R., Budtova, T. J. Mat. Sci., 2011, v. 46, p. 759–765.
9. Liebner, F., Haimer, E., Wendland, M., Neouze, M.-A., Schlufter, K., Miethe, P., Heinze, T., Potthast, A., Rosenau, T. Macromol. Biosci., 2010, v. 10, p. 349.
10. Krasil'nikova, O. K., Grankina, T.Yu., Budtova, T. V., Solovtsova, O. V. The Reports of Russian Academy of Sciences (in Rus.) 2013, v. 451. № 5, p. 528–532.
11. Buchachenko, A. L., Kovarski, A. L., Wasserman, A. M. "Study of Polymers by Para-magnetic Probe Technique" in "Advances in Polymer Science." Ed. Z. A. Rogovin. New York, "Halsted press", Wiley, 1974. p. 26–42.
12. Wasserman, A. M., Kovarski, A. L., Spin probes and Labels in Physical Chemistry of Pjlymers, Moscow, "Nauka" ("Science", in Rus.) Publishing House. 1986.
13. Kovarski, A. L., Spin Probes and Labels. A Quarter of a Century of Application to Polymer Studies. Polymer Year Book/Ed. R. A. Pethric. Chur: Harwood Academic Publishers. 1996. v. 13, p. 113.
14. Kovarski, A. L., Saprygin, V. N., Rappoport, N.Ya. "Mechanics of Composite Mate-rials (in Rus.), 1979, № 2, p. 351.
15. Tsylipotkina, M. V., Tager, A. A., Davankov, V. A., Tsurupa, M. P. Macromolecular Compounds (in Rus.). 1976. v. 18 B, № 11, p. 874.
16. Kovarski, A. L., Wasserman, A. M., Buchachenko, A. L./"Spin Probe Studies in Polymer Solids", in "Molecular Motion in Polymers by ESR", Eds. R. F. Boyer and, S. E. Keinath. Chur, Switzerland. "Harwood Academic Press", 1980. p. 177.
17. Yampolskii Yu.P., Wasserman, A. M., Kovarski, A. L. Reports on Russian Academy of Sciences, (in Rus.). 1979. v. 249, № 1, p. 150.
18. Kovarskii, A. L., Placek, J., Szocs, F. Polymer. 1978. v. 19, № 10, p. 1137.
19. Wasserman, A. M., Kovarski, A. L., Alexandrova, T. A., Buchachenko A.L. "Spin Labels and Probes in Physical Chemistry of Polymers" in "Modern Methods of Poly-mer Research." Ed. G. L. Slonimskii. Moscow, "Khimia" (Chemistry, in Russian) Publishing House. 1982, p. 121.
20. Meierovitch, E. J. Phys. Chem. 1983, v. 87, № 17, p. 3310.
21. Kovarski, A. L. "Molecular Dynamics of Additives in Polymers." Utrecht, Nether-lands: VSp. 1997.
22. Kovarski, A.L "Molecular Dynamics of Polymers under High Pressure", in "High Pressure Chemistry and Physics of Polymers", Ed. A. L. Kovarski, Boca Raton, USA: CRC-Press. 1994. p. 117.
23. Kovarski, A. L. Macromolecular Compounds (in Rus.). 1986, v. 28 A, № 7, p. 1347.
24. Antsiferova, L. I., Wasserman, A. M., Ivanova, A. N., Published in: "Atlas of ESR Spectra of Spin Labels and Probes." Moscow. "Nauka" ("Science", in Rus.) Publish-ing House. 1977.
25. Korst, N. N., Khazanovich, T. N. J. Exp. Theor. Phys. (in Engl.)., V.18, № 4, P.1049
26. Silescu, H., Kivelson, D., J. Chem. Phys. 1968, V.48, № 8, p. 3493
27. Freed, J. H., Bruno, G. V., Polnaszek, C. F. J. Phys. Chem. 1971. v. 75, № 22, p. 3385.
28. S A. Goldman, G. V., Bruno, C. F., Polnaszek, Freed, J. H. J. Chem. Phys. 1972. v. 56, № 2, p. 718.

29. Antsiferova, L. I., Korst, N., Stryukov, V., Ivanova, A., Nazemets, N., Rabinkina, N. Mol. Phys. 1973. v. 25, № 4, p. 909.
30. Barashkova, I. I., Kovarski, A. L., Wasserman, A. M. Macromolecular Compounds (in Rus.). 1982. v. 24 A, № 1, pp. 91–95.

CHAPTER 5

MELTING BEHAVIOR OF POLYPROPYLENES OF DIFFERENT CHEMICAL STRUCTURE

YU. K. LUKANINA, A. V. KHVATOV, N. N. KOLESNIKOVA, and A. A. POPOV

N.M. Emanuel Institute of Biochemical Physics of Russian Academy of Sciences, 4 Kosygin str., Moscow 119334, Russia, E-mail: julialkn@gmail.com

CONTENTS

ABSTRACT

The melting behavior of polypropylenes of different chemical structure (isotactic homopolypropylene, propylene-based block and random copolymers and maleic anhydride grafted polypropylene) was studied by Differential Scanning Calorimeter (DSC) and Microscopy. Melting behavior and the crystal structure of polypropylene and its copolymers were observed depending on the crystallization rate, chemical nature of co-monomer unites and regularity of co-monomer units arrangement in the polypropylene main chain.

5.1 AIMS AND BACKGROUND

Polypropylene and its copolymers are the most important commodity thermoplastic polymers, which are found in a wide variety of applications due to its excellent strength, toughness, high chemical resistance and high melting point. Polypropylene versatility along with its low cost and technological potentials has rendered it as one of the most used materials worldwide [1–4].

Isotactic homopolypropylene (PP) is a crystalline polymer whose properties are dependent on the degree of crystallinity and it exhibits three crystalline forms: the monoclinic a-phase, the hexagonal b-phase, and the orthorhombic c-phase [5–8]. When PP is formed, its crystal formation changes according to the heat treatment temperature and the conditions of cooling process. These changes create differences in strength, heat resistance and pressure bonding properties. Accordingly, it interesting and important to find out how to manage the structure of the polymer by varying the crystallization conditions during processing.

Propylene-based block copolymer (bPP) – the chain of molecules of propylene gap in the chain of ethylene copolymer. Propylene block copolymers produced in the form of homogeneous color, of the granules with high impact strength (at low temperatures), and high flexibility; increased long-term thermal stability; resistance to oxidative degradation during production and processing of polypropylene as well as the operation of the product out of it. Owing to the crystalline structure of the polypropylene block copolymer is a thermoplastic structural sufficiently economical polymer [9, 10].

Propylene-based random copolymers (rPP) are made by copolymerizing propylene and small amounts of ethylene (usually up to 7 wt.%) and their structure is similar to isotactic polypropylene, but the regular repeating of propylene units along the macromolecular chains is randomly disrupted by the presence of the comonomer ones. The presence of ethylene units reduces the melting point and crystallinity by introducing irregularities into the main polymeric chain. The advantages of this class of polymers are improved transparency, relative softness, lower sealing temperature, and moderate low-temperature impact strength due to the lowered glass-transition temperature [11–14].

Maleic anhydride grafted PP (mPP) is the most common functional adhesion promoter. It was proved to be effective functional molecule for the reactive compatibilization [6, 15, 16].

It is generally known that the properties of semicrystalline polymers depend strongly on the size and shape of their supramolecular structure [1, 16, 17]. The purpose of this paper is to present new results concerning the melting behavior and the crystal structure of polypropylene and its copolymers in depends on the crystallization rates, chemical nature of co-monomer unites and regularity of co-monomer units arrangement in the polypropylene main chain.

5.2 EXPERIMENTAL PART

5.2.1 MATERIALS AND SAMPLE PREPARATIONS

All polymers used in this study are commercial available. PP – Caplen 01030, melt index 1.2 g/10 min; rPP – SEETEC R3400, content of ethylene groups – 3 wt%, melt index 8 g/10 min; bPP – SEETEC M1400, content of ethylene groups – 9 wt%, melt index 8 g/10 min, mPP – Polybond 3002, content of polar groups – 0.6 wt%, melt index 12 g/10 min.

5.2.2 DSC MEASUREMENTS

DSC measurements were made on DSC thermal system Microcalorimeter DSM-10 ma. Its temperature scale was calibrated from the melting characteristics of indium. The experiments were conducted in non-isothermally mode. The sample was first heated up to 200°C and maintained at this temperature for 10 min in order to erase any previous morphological history which the sample might be carrying. The sample then non-isothermally crystallized when it was cooled down to room temperature at different cooling rates (1, 2, 4, 8, 16, 32, 64°C/min). It was subsequently heated at a heating rate of 8°C/min. The sample then repeatedly non-isothermally crystallized with the same cooling rates to permit structure microphotographs. The samples were approx. 7 mg. All curves were normalized to the unit weight of the sample.

The percent crystallinity, X_c, of the samples is calculated by:

$$X_c = \Delta H_m / [W_{PP} \times \Delta H_{m0}] \tag{1}$$

where, ΔH_m and ΔH_{m0} are melting enthalpy of sample and 100% crystalline PP (147 J/g [12]), respectively, and W_{PP} is weight fraction of PP in the samples.

5.2.3 MICROSCOPY

The Axio Imager microscopes Z2 m (Collective use center "New materials and technologies," IBCP RAS) with Transmitted-light differential interference contrast (TL DIC) was used to obtained microphotographs.

5.3 RESULTS AND DISCUSSION

The influence of crystallization rate on melting behavior of PP and copolymers were investigated. The melting endotherms of PP and its copolymers after various crystallization rates are presented in Figure 5.1. Henceforth, for all investigated polymers, the peak at the lower temperature is called peak-1, and at the higher temperature is called peak-2. The values of melting temperature (T_m) and crystallinity (X_c) are listed in Table 5.1.

Figure 5.1a shows the corresponding DSC thermographs of the samples PP recorded after cooling with different rates. From Figure 5.1a, it can be seen that melting behavior depends remarkably on the cooling rate. At lower crystallization rates <8°/min, the general feature of the DSC curves is the single melting peak which localized mainly at a temperature about 163°C. With increase of cooling rate one melting peak transformed into double melting peaks. At crystallization rate 8°/min on the DSC melting curves appears shoulder peak at about 165°C, which grows

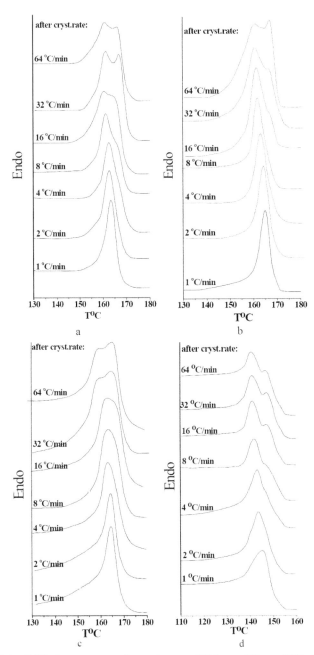

FIGURE 5.1   DSC thermograms of samples: a – PP, b – mPP, c – bPP, d – rPP after crystallization with different rates.

TABLE 5.1   Thermal Parameters of Polymers Samples During Melting Process After Different Crystallization Rates

Sample	Cryst. rate ($°C/min$)	T_m ($°C$)	X_c(%)	Sample	Cryst. rate ($°C/min$)	T_m ($°C$)	X_c(%)
PP	64	161, 166	50	mPP	64	160, 166	63
	32	161, 166	49		32	160, 166	62
	16	160, 165	46		16	161, 166	49
	8	161	47		8	162	54
	4	162	49		4	163	53
	2	163	54		2	164	55
	1	163	58		1	165	63
bPP	64	160, 166	62	rPP	64	140, 147	31
	32	160, 165	60		32	140, 147	29
	16	163, 166	59		16	141, 148	29
	8	163	60		8	142, 147	29
	4	163	61		4	143, 147	25
	2	164	63		2	144	25
	1	165	63		1	145	26

into a complete peak-2 at the crystallization temperature 32°C/min and 64°C/min, whereas the position of the peak-1 displaced at the location of lower temperature 161°C. Such dependence is explained by the formation of more advanced and stable crystal structures at low cooling rate, whereas with increasing cooling rate increases super-cooling and a large amount of defective crystals exposed to reorganization and recrystallization during heating process.

In Figure 5.1b, mPP samples shows melting endotherms after different crystallization rates. As well as pure PP, the DSC curve for mPP shows single melting peak under low cooling rate. It should be noted that appearance and increases of shoulder peak for such copolymer observed at higher cooling rates – 16°C/min. Another distinguishing feature of mPP melting after different crystallization rates is significant decrease of melting temperature of peak-1 (from 165°C after cooling rate 1°C/min to 160°C after cooling rate 64°C/min).

Figure 5.1c gives the heating DSC thermographs for the bPP samples prepared also at various cooling rates. Curves have the similar character described above by maintaining the main polypropylene chain under the block introduction of ethylene units. This later transition from a single peak to a double peak will explain in terms of polypropylene chemical structure change by the introduction of functional groups. Since the regular introduction of functional groups decreases mobility of the system and increases the viscosity of the system, at low cooling rates this leads to the rapid formation of crystallization centers around them to form more perfect crystals than in pure PP. This displayed the higher melting temperatures of mPP and bPP. In the transition to higher cooling rates of up to 16°/min, at which the viscosity increases, the rearrangement of molecules copolymers is difficult, which in turn reduced the rate of crystallization, resulting in the formation of qualitatively less perfect package. The melting temperature T_m decreases on 5°C. A further increase of the material crystallization rate contributes to recrystallization and formation of amid bulk imperfect crystals a high-melting crystalline structure.

The statistical introduction of ethylene units into the main polypropylene chain affects greatly the crystalline structure of rPP in Figure 5.1d. Firstly, a decrease in the melting temperature of the polymer matrix of about 20°C is observed. Secondary, the peak-2 as a shoulder peak appears at crystallization rate 4°C/min. For the propylene-based random copolymer the ethylene-monomers are often considered as a defect points in polypropylene matrix. This induces structure heterogeneity in the long chains of polypropylene namely short propylene sequences leading to a decrease in crystallizable sequence length. Even small co-monomer content (3%), as in our case results in short crystalline sequence.

From the Table 5.1 one can observe the relationship between the crystallization rate and crystalline of polymers samples. Comparing the crystalline degree of the samples at a rate of crystallization (for example 1°/min) one can note that the regular introduction of the co-monomer units of any chemical nature (mPP and bPP) leads to an increase in the polypropylene crystalline degree by 5%. Such incorporated co-units is not defect points, its lead to the facilitating folding crystalline structure. Due to described above-statistical distribution of ethylene co-units – defect points in the propylene random copolymer rPP crystallinity decrease on 30%. On the other hand, crystalline of PP, mPP and bPP characterized by decrease with increase crystallization rate, followed by crystallization increase at higher rates (32°C/min and 64°C/min). Samples of rPP characterized by the constancy of crystalline degree at low cooling rates and its growth of 4–5% even at a cooling rate of 8°/min. These dependences are as well as formation of the shoulder and peak-2 on the DSC thermograms are explainable in terms of reorganization and recrystallization of the polypropylene supramolecular structure.

Described above crystalline structures of PP, mPP, bPP and rPP obtained after different crystallization rates were visualized by microscopy. Figure 5.2 shows microphotographs of PP, mPP, bPP and rPP obtained after different cooling rates. For all polymer samples in the train of crystallization the line radial bundles emanating from a single point – polypropylene crystalline structure

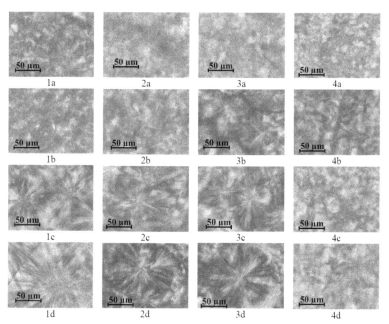

FIGURE 5.2   Microscope photographs of 1-PP, 2- mPP, 3 – bPP, 4 – rPP obtained after different crystallization rates: a – 64°C/min, b – 16°C/min, c – 4°C/min, d – 1°C/min.

are formed. The crystalline sizes of the samples vary widely depending on the polymer crystallization rate and its chemical structure. At higher cooling rates and structural inhomogeneous of the propylene long chains in random copolymer formed submicroscopic crystal structures – finely grain-like structure. Crystallization at a lower cooling rates leads to the formed crystals in diameter of about 100 microns. It is necessary to note that the size and shape of the supramolecular structures have great influence on the mechanical properties of the polymer. Samples with finely grain-like structure have high strength and have good elastic properties. Samples with large crystals destroyed fragile. Elasticity loss of crystalline polymers is manifested in the appearance of cracks and breaks at the crystals interface. The increase of its size leads to the fragility increased and decrease of strength.

5.4 CONCLUSION

The melting behavior of four polymers: polypropylene, propylene-based block copolymer, propylene-based random copolymer and maleic anhydride grafted polypropylene, was studied with DSC and Microscopy. Obtained results indicate that both chemical structure of polypropylene chain and crystallization rate show great influence on the crystalline structure of polypropylene. It was found that the regular introduction of the co-monomer units of any chemical nature (mPP and bPP) and higher crystallization rates leads to the formation of large crystals. On the other hand, structural inhomogeneous of polypropylene chain and increase of crystallization rates promote to the formation of polymer finely grain-like structure.

KEYWORDS

- crystal structure
- DSC
- melting
- microscopy
- polypropylene

REFERENCES

1. Papageorgiou, G. Z., Achilias, D. S., Bikiaris, D. N., Karayannidis, G. P., Crystallization kinetics and nucleation activity of filler in polypropylene/surface-treated SiO2 nanocomposites. Thermochim Acta 2005, 427(1–2), 117–28.
2. Seki, M., Nakano, H., Yamauchi, S., Suzuki, J., Matsushita, Y., Miscibility of isotactic polypropylene/ethylene-propylene random copolymer binary blends. Macromolecules 1999, 32(10), 3227–34.
3. Cai, H. J., Luo, X. L., Ma, D. Z., Wang, J. M., Tan, H. S., Structure and properties of impact copolymer polypropylene, I. Chain structure. J Appl Polym Sci 1999, 71(1), 93–101.
4. Feng, Y., Hay, J. N., The measurement of compositional heterogeneity in a propylene–ethylene block copolymer. Polymer 1998, 39(26), 6723–31.
5. IDE F., Hasegawa, A., J Appl Polym Sci 1974, 18, 963.
6. Hosoda, S., Kojima, K., Kanda, Y., Aoyagi, M., Polym Networks Blends 1991, 1, 51.
7. Cho, K.; Li, F. Macromolecules, 1998, 31, 7495.
8. Holsti-Miettinen, R., seppala, J., Ikkala, O. T., Polym Eng Sci 1992, 32, 868.
9. Jacoby, P., Bersted, B. H., Kissel, W. J., Smith, C. E., Studies on the b-crystalline form of isotactic polypropylene. J Polym Sci, Part B, Polym Phys 1986, 24(3), 461–91.
10. Lovinger, A. J., Microstructure and unit-cell orientation in apolypropylene. J Polym Sci, Polym Phys Ed 1983, 21(1), 97–110.
11. Purez, E., Zucchi, D., Sacchi, M. C., Forlinu, F., Bello, A., Obtaining the c phase in isotactic polypropylene, effect of catalyst system and crystallization conditions. Polymer 1999, 40(3), 675–81.
12. Ambrogi, I., Polypropylene, L., Chimia 1967, 316.
13. Qing Qiang, G., Hui, L., Zhen Qiang, Y., Er Qiang, C., Yu Dong, Z., Shou Ke, Y., Effect of crystallization temperature and propylene sequence length on the crystalline structure of propylene-ethylene random copolymers. Chinese Science Bulletin 2008, vol.53. N12. pp. 1804–1812.
14. Papageorgiou, D. G., Papageorgiou, G. Z., Bikiaris, D. N., Chrissafis, K., Crystallization and melting of propylene-ethylene random copolymers. Homogeneous nucleation and b-nucleating agents. European polymer journal 2013. V49, pp. 1577–1590
15. Nandi, S., Ghosh, A. K., Crystallization kinetics of impact modified polypropylene. Journal polymer research 2007. v.14, pp,387–396
16. Cho, K., Li, F., Choi, J., Crystallization and melting behavior of polypropylene and maleate polypropylene blends. Polymer. 1999. V 40. pp. 1719–1729
17. Lukanina Yu. K., Kolesnikova, N. N., Khvatov, A. V., Likhachev, A. N., Popov, A. A., Influence of polypropylene structure during micromycete growth on their composition. Journal of the Balkan tribological association. 2012, Vol.18, № 1, p. 142–148.

CHAPTER 6

RESEARCH OF POSSIBLE SYNTHESIS OF ALKYD-STYRENE RESINS

K. I. VINHLINSKAYA,[1] N. R. PROKOPCHUK,[2] and A. L. SHUTOVA[3]

[1]*PhD Student, Junior Scientific Researcher (BSTU);* [2]*Corresponding Member of Belarusian National Academy of Sciences, Doctor of Chemical Sciences, Professor, Head of Department (BSTU);* [3]*Candidate of Technical Sciences, Assistant professor (BSTU), Belarusian State Technological University, 13a Sverdlova Str., Minsk, 220006, Belarus, E-mail: VPSh_bstu@mail.ru, prok_nr@mail. by, a.l.shutova@mail.ru*

CONTENTS

ABSTRACT

This chapter covers possible methods of synthesis of alkyd-styrene resins, formulations and areas of application this film-formers in paint production. The article summarizes the main regularities that can help to get in the future alkyd-styrene resins of home production. Possibility of production alkyd-styrene resins at the chemical companies opens wide prospects for creation quick and naturally drying paint materials, which will reduce the dependence on imports, provide energy savings and correspondingly increase the competitive ability of products of companies by cheapening of coatings preparation processes.

6.1   INTRODUCTION

Currently conventional semifinished alkyd lacquers dominate in the CIS market among the modified film formers, but in recent years due to the constant rise in energy prices there appeared a demand for paint-and-lacquer materials that reduce energy costs in the preparation of varnish-and-paint coatings.

Energy intensity degradation of coating production can be achieved using natural hardening capable of quick forming of coating (drying time – no more than 30 min).

Alkyd-styrene resins are a particular class of film-forming materials, which are prepared based on these natural quick drying paints.

Styrenated alkyd resins in comparison with unmodified styrene resins have some advantages: faster drying, increased resistance to water and chemicals, their films are less susceptible to brightness reversion. The disadvantages of such resins include the occurrence of defects during drying and solvent resistance reduction that often results in "blistering" (delamination) during application of a secondary layer. The defect of the upper layers of alkyd-styrene resins is their low endurance, especially to scratching, so they are mainly used in undercoating [1].

This binding group is used in formulas of anticorrosion paints for painting ships, equipment, equalizing compositions and fillings, both one and multilayer coatings in natural and hot drying [2].

In this regard, there is a constant demand for the film former at the enterprises that they have to meet at the expense of purchase of raw materials in the international market. Therefore, the quality and competitiveness of domestic coatings are influenced by such factors as irregular supply, oxidation and loss of raw materials while transporting, as well as the high cost compared with other domestic film formers.

Thus, the creation of alkyd-styrene resins is one of the urgent problems that can both reduce the dependence on import of enterprises and provide energy saving, respectively, increasing the competitiveness of market at the expense of the cost reduction in coloring process.

6.2 EXPERIMENTAL PART

To produce domestic alkyd-styrene resins satisfying the requirements of world standards, we analyzed the scientific literature on the preparation, recipes, as well as the possibilities of application of alkyd-styrene resins in paint-and-varnish production.

Such properties of polystyrene as good coloring maintained unchangable for a long time, the exceptional water resistance, resistance to alkalis and high dielectric properties contributed to intensive search for methods of applying it as a filming. Unfortunately, polystyrene has a number of significant shortcomings along with the good properties.

One of the ways of the removal of shortcomings is to combine its positive properties with substance properties, which are film-forming themselves.

Getting styrenated drying oils and alkyd resins created the fourth important group of synthetic products for accelerated drying for decorative and protective coatings in addition to phenolic resins, alkyd resins and nitrocellulose [3].

According to the literature, styrene was polymerized with dehydrated oils in the early 1940 s for the first time. The first patent for alkyd resins, modified by styrene was obtained in the UK in 1942 [1].

Styrenated alkyds appeared in commercial quantities in 1948 and occupied a certain place in the fast-drying coatings and hot air drying. Large production capacity of styrene and its relatively low price stimulated its implementation in coating manufacture [4].

The basis of obtaining alkyd-styrene copolymer process is styrene copolymerization reaction with the double bonds of the fatty acid residues of oils [5].

The copolymerization reaction is characteristic for vinyl group. Styrene can be readily copolymerized by any conventional methods: in block or in solution. The process of copolymerization is carried out by heating for 20 h or more at about 140°C.

Xylene is usually used for copolymerization in the solution. The reaction rate and the amount of the resulting product are changed when xylene is replaced by other solvents. For example, the reaction time increases from 23 to 49 h when xylene is replaced by dipentene. The products obtained from dipentene solution are better aligned but they dry slower. Therefore it is recommended to replace only a part of xylene by dipentene [4].

Limiting styrene content is about 40%. Loading it in large quantities deteriorates compatibility with other film formers as well as it makes worse solubility in solvents; moreover, obtained coatings are brittle. Optimum performance is observed at a ratio of styrene: oil = 40:60.

As it is known [6], the molecular weight of the polymer depends on a number of factors – the concentration of monomers, temperature, initiator catalyst content and chain regulators in the reaction mixture, etc. As a means of reducing the concentration of monomer may be its slow introduction (droplet method) into heated to a certain temperature oil, fatty acid oils or alkyd resins. At the same time the decrease in molecular weight will be determined not only by a low concentration of monomer, but also high content of α-methylene groups per injected styrene particle that, in addition, will provide polystyrene chains, significantly enriched with oil component. High concentration of α-methylene groups eliminates the necessity for the introduction of specific growth regulator of circuit. The method of the slow introduction of the monomer allows to obtain homogeneous products with a higher styrene content (up to 70% in the reaction mixture) than conventional mortar method, although a high styrene content (50–70% in the mixture), significantly increases viscosity and reduces elasticity and shock resistance of the films.

A prerequisite to obtain a homogeneous reaction product is the initiator (3–4% by weight of monomer), which contributes not only to accelerate the reaction, but also to decrease molecular weight. As an initiator catalyst of copolymerization process it is better to use alkyl peroxides with relatively high decomposition temperature (tertiary-butyl peroxide). The use of peroxides with low decomposition temperature (lauryl peroxide, benzoyl peroxide, hydroperoxide of isopropylbenzene) gives a smaller effect [7].

Mixture consisting of styrene and α-methyl styrene enables to obtain homogeneous light products with a large number of oil. The ratio between α-methylstyrene and styrene is usually 3:7. α-Methylstyrene is very helpful to get compatible products, moderating reaction and slowing entry of styrene into copolymer.

Thus, one must regulate the quantitative ratios of the reactants in the copolymerization reaction, considering type and amount of initiator catalyst, temperature and solvent, if copolymerization is carried out in solution. Technically homogeneous products can be obtained when applied the same process conditions.

As the literary analysis showed, fundamentally different methods for alkyd-styrene copolymers are possible:

1. copolymerization of fatty acids of vegetable oils with styrene and further reacting of the obtained copolymer with other components of alkyd oligomers (phthalic anhydride, glycerol) by fatty acid method (method 1);
2. copolymerization of the styrene oil and the subsequent synthesis of the alkyd by the method of using the glyceride with oil copolymer (method 2);
3. copolymerization of monoglycerides with styrene and their further esterification (method 3);
4. copolymerization of preformed alkyd with styrene (method 4).

According to method 1 preparation of an alkyd-styrene resin can be represented as the simplified scheme shown in Figure 6.1.

A number of papers concern studying the process of copolymerization of styrene with fatty acids of vegetable oils. In particular, the copolymerization of styrene with α-eleostearic acid which has three conjugated bonds is under consideration. Tung oil is unique in content of eleostearic acid (80%). This research has made it possible to explain the mechanism of formation of styrene copolymers with fatty acids containing conjugated double bonds.

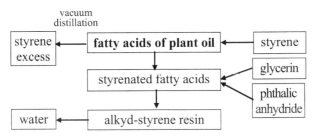

FIGURE 6.1   The scheme for producing alkyd-styrene resin according to method 1.

In this case, this reaction proceeds as in the copolymerization of styrene-butadiene: styrene-butadiene attachment occurs primarily in the 1,4-position, and chain growth of styrene via dienes [8].

Thus, the fatty acids with conjugated double bonds in the molecule (eleostearic, 9,11-linoleic) form two types of products with styrene – the true copolymers of high molecular weight (Figure 6.2), and adducts of the Diels-Alder (diene synthesis) formed in the addition of styrene to fatty acid in a molar ratio of 1:1 (Figure 6.3) [7].

This chapter of Shneyderova V. V. deals with copolymerization of styrene with linoleic acid containing two isolated double bonds (9,12-linoleic acid), which is one of the main components of most drying and semidrying oils. Polymerization was accomplished in sealed nitrogen-filled glass ampoules at 150°C with benzoyl peroxide (1% of styrene content) and the molar ratios in the mixture of styrene and linoleic acid 94:6, 90:10, 86:14 and 80:20. This paper shows that increasing the molar proportion of fatty acid in the initial reaction mixture dramatically increases the percentage of low molecular weight copolymers of styrene with linoleic acid [9].

It is known [4] that the tung oil acids react with styrene slower than acid of oiticica oil. Dehydrated castor oil acids react with styrene with the greatest reaction rate. It is hard to explain, because tung oil has the greatest number and the greatest degree of conjugacy of double bonds. Perhaps the reason for this phenomenon is almost complete copolymerization of styrene with tung oil acids. Because of this, the polymerization of the styrene with the greatest reaction rate occurs to a lesser extent with formation of a certain amount of polystyrene. Since the reaction rate of polymerization styrene is higher than the reaction of copolymerization the styrene flow rate at a particular time must be greater using dehydrated castor oil acids.

The formulation and technique [4] of obtaining styrene acids of dehydrated castor oil are known (Table 6.1).

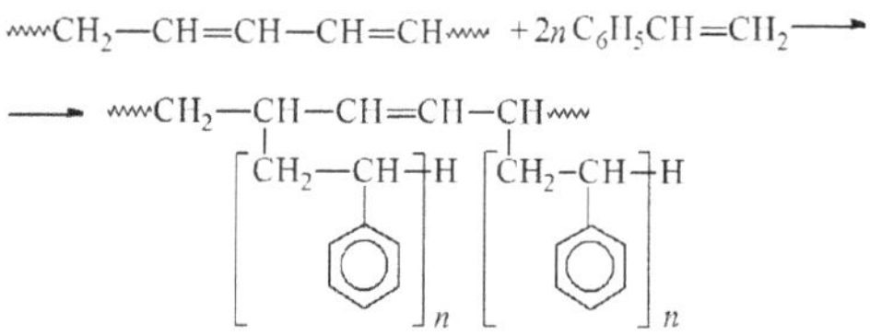

FIGURE 6.2   The scheme for producing alkyd-styrene copolymers.

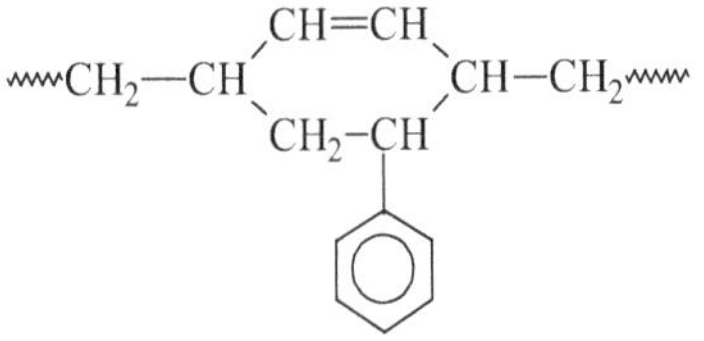

FIGURE 6.3 The adduct of the Diels-Alder.

TABLE 6.1 Formulation of Styrene Acids of Dehydrated Castor Oil

Components	Component content, wt %
Acids of dehydrated castor oil	40.9
Styrene	59.1
Total	100.0

Note. Initiator – 3% benzoyl peroxide (from styrene content).

According to this recipe, dehydrated castor oil acids are heated in a flask with stirrer and cooler to 115°C; then when stirred, a mixture of styrene with a catalyst is added for an hour and the temperature raises during this time to about 145°C. This temperature is maintained for 30 min, after that the unreacted styrene is subjected to vacuum distillation.

When acids of dehydrated castor oil were styrenated it was established that if all components are all heated, rapid exothermal reaction occurs at 120°C. The product obtained is thus turbid and inhomogeneous representing an incompatible mixture of unreacted fatty acids, a quantity of styrenated acids and a considerable amount of polystyrene. The product formed by slow addition of a mixture of styrene with the catalyst to fatty acids becomes transparent and homogeneous. The presence of free polystyrene is not determined, but it was established that with increasing amounts of loaded styrene the amount of styrene reacting with fatty acids increases. At the same time the molar ratio of styrene to fatty acid is 4:1, taken in formulation (see Table 6.1), is considered to be optimal.

The formulation of obtaining of alkyd resins, modified by styrene for synthesis according to method 1 is also known (Table 6.2) [10].

Deriving alkyd styrene resin according to method 2 (Figure 6.4) is under consideration.

It is known [8] that in 1946 a mechanism of copolymerization of oils with styrene was proposed which is different for oils with isolated and

TABLE 6.2   Formulation of Alkyd-Styrene Resin For Synthesis According to Method 1

Components	Component content, wt %
Acids of dehydrated castor oil	34.8
Styrene	24.4
α-Methylstyrene	10.4
Glycerine	13.1
Phthalic Anhydride	17.3
Total	100.0

Note. Initiator – 2% benzoyl peroxide (from styrene content).

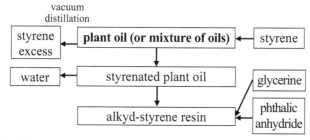

FIGURE 6.4   The scheme for producing alkyd-styrene resin according to method 2.

conjugated double bonds. In the first case, the oil serves as a chain transfer agent, playing the same role as the solvent. In the presence of oils with isolated double bonds chain stopping can occur due to unstable hydrogen of the methylene group located in the α-position to the double bond.

Drinberg et al. [11] have also investigated the copolymerization of styrene with vegetable oils. They found that the copolymerization reaction runs only with tung oil containing a conjugated bond system. Linseed oil is possible to introduce into the copolymerization reaction, being preoxidized by atmospheric oxygen. Crude sunflower oil as well as oxidized do not enter in the copolymerization reaction with styrene. Copolymerization reaction takes place only when oxidized mixtures of sunflower and tung oils are used together and it is sufficient to introduce 5% of tung oil. It is also established that the rate of conversion into a three-dimensional polymer increases with increasing styrene content in the copolymer [11].

Polystyrenes have relatively low melting, but at normal temperature they are incompatible with the oil. Homogeneous product is formed when oil styrene monomer is heated with reflux condenser and peroxidate catalyst as a result of copolymerization, which proceeds readily with oils having conjugated bonds, and only to a small extent with oils having non-conjugated bonds [4].

A high content of conjugated double bonds in oils, during their copolymerization with styrene can easily cause gelation as a result of the formation of spatial polymers due to crosslinking radicals of fatty acids [8].

Styrene and oil can be copolymerized either in solution or in block. Copolymerization in solution makes it possible to control the process well, but it is slow and, of course, if you need a product without solvent, it must then be removed. Usual formulation comprises 25 parts of oil, 25 parts of styrene and 50 parts of solvent. The mixture is heated until the desired degree of conversion. The residual monomer styrene can be distilled to remove, and it is accompanied by removal of most solvent, which requires subsequent addition of fresh solvent.

Block copolymerization proceeds much faster than in solution, but in this case styrene polymerizes itself readily as well as it copolymerizes. When copolymerized by this method, the products are incompatible and turbid due to the incompatibility of polystyrene with oil. Therefore, to obtain bright and homogeneous products it is recommended to use a mixture of styrene and α-methylstyrene [3].

Table 6.3 shows the possible formulations of styrenated oils and describes the technological copolymerization process in the block.

The table shows that almost all the copolymerization products dry quickly, except made by a mixture of soybean and tung oils. The films produced from a mixture of 50% linseed and 50% dehydrated castor possess the best drying ability. However, the least time-consuming process is the copolymerization of dehydrated castor oil with a mixture of styrene and α-methylstyrene.

The formulation [5] of obtaining alkyd-styrene resins with oxidized soybean oil according to method 2 is known (Table 6.4).

Alkyd-styrene resins obtained by this formulation have an acid number of not more than 16 mg KOH/g. The films of these alkyds dry at $(20 \pm 2)°C$ to degree 1 for 22 min, and up to degree 3 for not more than 8 h. They have good elasticity, and resistance to water, alkalis, and solvents.

TABLE 6.3   Features of Copolymerization Process in the Block According to Method 2

Components	Composition of the mixture of oils, wt %			
	100% dehydrated castor	90% linseed 10% tung	80% soybean 20% tung	50% linseed and 50% dehydrated castor
Formulations of styrenated oils				
Oil	45	45	45	50
Mixture of styrene and α-methylstyrene	55	55	55	50
Reaction time, h				
Addition of styrene	6	6	6	6
Temperature rise to 250°C	5	6	4	5
Exposure at 250°C	6	6	6	7
Heating 250–285°C	0.5	–	–	–
Heating 250–300°C	–	0.5	0.5	1
Exposure at 285°C	1.5	–	–	–
Exposure at 300°C	–	2.5	3.5	3
Overall process	19	21	20	22
Rates of obtained styrenated oils				
Amount of unreacted styrene, %	1.2	1.5	2.0	1.2
Drying time at temperature (20 ± 2)°C, h, not more:				
a) to degree 1	0.75	1	2.5	0.5
b) to degree 3	6.0	7.0	24	5

Notes. In all cases before styrene adding oil was heated to 160°C. Initiator – 3% benzoyl peroxide (from styrene content). Solidification was carried out with a mixture of desiccants: 0.5% Pb and 0.02% Co based on the weight of metal to oil. The dilutant was a mixture of 70% mineral spirits and 30% solvent resulted in obtaining of 60% solution.

Another known method of obtaining styrenated alkyds (method 3) consists in styrenating of monoglycerides and their subsequent esterification (Figure 6.5).

Table 6.5 shows the formulation of alkyd-styrene resins with a mixture of linseed and dehydrated castor oils to obtain alkyd-styrene resin [5].

TABLE 6.4   Formulation of Alkyd-Styrene Resin For the Synthesis According to Method 2

Components	Component content, wt %
Oxidized soya bean oil	37.7
Styrene	19.1
α-Methylstyrene	8.2
Glycerine	10.9
Phthalic anhydride	24.1
Total	100.0

Notes. Initiator – 3% benzoyl peroxide (from styrene content). The reaction of transesterification was carried out with 0.01% (based on the oil content) of calcium oxide.

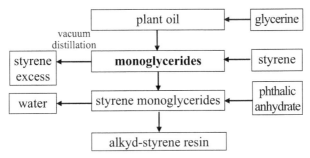

FIGURE 6.5   The scheme for obtaining of alkyd-styrene resins according to method 3.

TABLE 6.5   Formulation of Alkyd-Styrene Resin For the synthesis According to Method 3

Components	Component content, wt %
Linseed oil	19,0
Dehydrated castor oil	19.0
Glycerine	8.7
Styrene	35.0
Phthalic anhydride	18.3
Total	100.0

Notes. The reaction of transesterification was carried out with 0.06% (based on the oil content) of calcium oxide. 4.62% (by weight basis) of xylene was added for azeotropic distillation of water of reaction. Copolymerization reaction initiator – 0.85% (the content of the styrene) of cumene hydro-peroxide (73% solution in xylene).

According to the formulation given in Table 6.6, the process of preparation of styrenated alkyd consists of several stages:

1. alcoholysis of oils: linseed and dehydrated castor oil, glycerol and calcium oxide is heated to 230°C and kept at that temperature for about 1 hour to obtain a product which is soluble in methanol, 1:3;
2. styrenazing of monoglycerides: the half part of the styrene and the initiator solution are added to monoglycerides and the mixture is heated to 160°C, then a reflux condenser on and the residue of styrene is added for 3 h. Then in the next 4–5 hours the temperature is raised to 210°C;
3. esterification for obtaining of the styrenated alkyd: phthalic anhydride is added to styrenated monoglycerides and xylene is added for azeotropic solution. The mixture of these substances is heated at 215–230°C for about 3 h to achieve the required viscosity and acid number, after which the resin is dissolved and filtered.

Such a styrene-alkyd resin has an acid number of not more than 7 mg KOH/g. Films of alkyds dry at (20 ± 2)°C to 1 degree for 10 min.

The most common method of obtaining industrial alkyd-styrene resins is method 4 (Figure 6.6).

Low-viscosity alkyd is used for copolymerization with styrene. Homogeneous resins with good properties are obtained when applying medium fatty alkyd resins. Fatty acids with conjugated double bonds, are used, unsaturated dibasic acids or a mixture of saturated dibasic acids with a small amount of maleic anhydride are used in the synthesis of alkyd.

Maleic anhydride is involved in the reaction of a polyesterification and serves the source of both double bonds necessary for copolymerization

TABLE 6.6   Formulation of Alkyd-Styrene Resin For the Synthesis According to Method 3

Components	Component content, wt %
Glyptal resin (50%-concentration)	50.0
Styrene	25.0
Xylene	25.0
Total	100.0

Note. Initiator – 2% tertiary butyl peroxide (from styrene content).

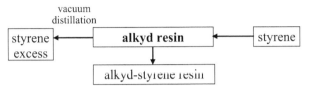

FIGURE 6.6 The scheme for obtaining alkyd-styrene resin according to method 4.

with styrene. The maleic anhydride content in alkyd should be precisely calculated. Optimal conditions for the copolymerization are provided for the introduction of maleic anhydride in the amount that one double bond of maleic anhydride is on three molecules of phthalic anhydride. In this case the final product is transparent and has a relatively stable viscosity [12].

Further copolymerization process is carried out at the temperature from 140 to 170°C in xylene with initiator. The main initiator is di-tret-butyl peroxide, the amount of which depends on the temperature and varies from 1 to 4% from weight of monomer. Pre-oxidation and polymerization of oils (linseed, sunflower) increases the yield of the copolymer. The copolymerization involves the method, in which styrene is gradually introduced into the mixture for several hours together with an initiator. After addition of monomers and initiators the reaction temperature is maintained; the degree of conversion is determined by measuring the mass of the nonvolatile part. If necessary, for the completeness of the reaction a small amount of initiator is additionally injected. The quantitative ratio of the alkyd resin and the copolymer is generally from 60:40 to 85:15. The reaction is usually carried out until it terminates at 95–97%. The residue of unreacted styrene is removed under suction at the end of the process.

To improve the odor and light-fastness it is desirable to remove small amounts of residual monomer styrene. For this, CO_2 is blown through the reaction product before it is left to cool. The content of volatile substances in the drying oils is negligible, and thus, the degree of conversion can be estimated by collecting matter removed by blowing [3].

The formulation of a styrene-alkyd resin for synthesis based on glyptal resin is known by method 4 (Table 6.6) [13]. Xylene solution of glyptal resin (50%), styrene and xylene are placed into flask. The resulting mixture is stirred for 30 min and is heated. When temperature runs up to temperature 140°C, 50% solution of t-butyl peroxide or other peroxide in xylene is given into the reaction mixture in four equal portions at intervals of 1.5 h. Aggregate exposure is carried out at 140°C for 20–25 h to obtain a dry residue 49.5–50.0%. The viscosity of the reaction solution should be 45–50 seconds. Then the solution is cooled to room temperature and filtered. Alkyd-styrene resin obtained by this formulation has an acid value of not more than 7 mg KOH/g, the drying time up to 3 degrees at (20 ± 2)°C – less than 8 h.

Depending on the styrene content styrene-alkyd resins can be divided into 3 groups [7]:

1. resins containing 30% of styrene and more; they dry faster, diluted with white spirit and are suitable after dilution for brushing. On the basis of these resins solid, water-resistant, fast drying coating are obtained (drying from dust for 20 min, the real drying 1.5 h). Lacquers and enamels, based on these resins have a high concentration of film-forming substance and they may be applied for painting the chassis of motor vehicles, heavy machinery and equipment, cables, decks, etc.;

2. resins containing 15–25% of styrene are used in the primers and in baking enamels. Particularly their use is appropriate with melamine-formaldehyde resins with which they conjunct well. The obtained coatings of drying possess high hardness, luster, resistance to water, alkalis and detergents, as well as weather resistance. They can be used for dyeing of washing machines, etc.;

3. resins containing 10% of styrene are well diluted with white spirit, and can be brushed. These resins can be used for coloring of the inside of a building.

Alkyd-styrene resins are available as solutions in white spirits or xylene. For quick hardening films of air drying xylene solution is more profitable, but for grinding pigments and improving coatings with high gloss the solution in slowly evaporating white spirit should be applied [4].

Film formation occurs primarily through physical drying (evaporation of solvent), as well as oxidative polymerization due to the remaining double bonds of the fatty acid residues of oils. Air curing is carried out with siccatives [14].

It is known [4] that usual combination of lead and cobalt driers are added to the coatings of air-drying based on styrenated alkyds. The amount of the metal input depends on the type of pigment, as well as in conventional oil and alkyd lacquers. In styrenated alkyds, drying at 120°C or below a small amount of cobalt is put, and in styrenated alkyds drying at a temperature above 120°C driers are not usually put.

6.3 CONCLUSION

This chapter deals with possible methods of obtaining, formulations, as well as the possibility of applying alkyd-styrene resins in paint and varnish production.

On the basis of this work one can distinguish four main methods of the synthesis of alkyd-styrene resins, which have their own characteristics, advantages and disadvantages. But the most common method is copolymerization of alkyd with styrene.

KEYWORDS

- alkyd resin
- alkyd-styrene resins
- areas of application
- copolymerization
- fatty acids of vegetable oils
- formulations
- initiator
- methods of synthesis
- plant oil
- styrene
- xylene

REFERENCES

1. Paints, coatings and solvents. Stoye, D., Freitag, W. (ed.). – 2. completely rev. ed. Weinheim; New York: Basel: Cambridge; Tokyo: Wiley-VCH. 1998.
2. Pot, W. Polyesters and alkyd resins/W. Pot; translated from German L.V. Kazakova. – M.: Paint-Media. 2009. 232 p.
3. Monomers: sb. stories: 2 part. Part 2/ed. V.V. Korshak. – M.: Publisher foreign lit-ry. 1953. 270 p.
4. Payne, G. F. Technology Organic Coatings: 2 part. Part 1: oils, resins, varnishes and polymers/G. F. Payne; translated from English. M. D. Gordonova [et al.] ed. E. F. Belenky. – L.: Goskhimizdat, 1959. 758 p.
5. Prokopchuk, N. R. Chemistry and technology of film-forming materials: Textbook. allowance for stud. Universities. N. R. Prokopchuk, E. T. Krutko. – Minsk: Belarusian State Technological University, 2004. 423 p.
6. Yukhnovskii, G. L. Synthesis and study benzine alkyd-styrene resins. G. L. Yukhnovskii, L. A. Sumtsova, N. P. Ternovaya. Journal of Applied Chemistry. 1970. № 11. pp. 2494–2499.
7. Coating materials: raw materials and intermediate products: handbook. ed. I. N. Sapgir. – M.: Goskhimizdat, 1961. 506 p.
8. Yukhnovskii, G. L. On the reaction of styrene with vegetable oils. G. L. Yukhnovskii, R. R. Popenker. Journal of Applied Chemistry. 1957. № 4. pp. 603–612.
9. Shneyderova, V. V. Investigation of the copolymerization of styrene with linoleic acid/V. V. Shneyderova. Coating materials and their application. 1962. № 4. pp. 33–34.
10. United States Patent № 2639270. Styrenated alkyd resins. Griess G. A., Strandskov C. V.; applicant Dow Chemical Company, 1953.
11. Drinberg, A. Ya. Copolymers of vegetable oils with styrene. A. Y. Drinberg, B. M. Fundyler, L. A. Lifits. Journal of Applied Chemistry. 1954. № 6. pp. 618–624.
12. Patton, T. K. Technology of alkyd resins. Formulation and calculations/T. K. Paton; translated from English. I. E. Samolubova; ed. K. P. Belyaev. – M.: Chemistry, 1970. 128 p.
13. Konovalov, P. G. Laboratory workshop on the chemistry and technology of film-forming varnishes and paints. P. G. Konovalov, V. V. Zhebrovsky, V. V. Shneyderova. – M.: Rosvuzizdat, 1963. 203 p.
14. Sorokin, M. F. Chemistry and technology of film-forming. M. F. Sorokin, L. G. Shode, Z. A. Kochnova. – M.: Chemistry, 1981–448 p.

CHAPTER 7

PECULIARITIES OF MINERAL NUTRITION OF CEREALS IN ALUMINUM-ACID SOIL CONDITIONS

LYUDMILA N. SHIKHOVA,[1] EUGENE M. LISITSYN,[1, 2] and GALINA A. BATALOVA[1, 2]

[1]Vyatka State Agricultural Academy, 133 Oktyabrsky Ave., Kirov, 610010, Russia

[2]North-East Agricultural Research Institute, 166-a Lenin St., Kirov, 610007, Russia; E-mail: edaphic@mail.ru

CONTENTS

ABSTRACT

Influence of aluminum ions on requirements of plants of cereal crops (wheat *Triticum aestivum* L. and oats *Avena sativa* L.) in macronutrients is studied. In the first series of experiments it is revealed that doubling of a dose of phosphorus in the acid media has caused doubling of its inclusion in metabolic processes in aluminum-resistant wheat Irgina on early stages of development already, but in aluminum-sensitive wheat Priokskaya – only on a flowering stage. Development of root systems in wheat variet- ies differ in their Al-resistance level in the conditions of full supply with nitrogen and potassium does not depends on presence of ions of hydro- gen, aluminum or phosphorus in soil. Aluminum-sensitive wheat variety is characterized by considerable dependence of photosynthesis on conditions of a mineral nutrition whereas intensity of photosynthetic processes of aluminum-resistant variety is influenced basically by a development stage.

In the second series of experiments rather different reaction of the investigated wheat varieties on input of calcium into acid growth media is pointed out. These differences are related with transformation of absorbed phosphorus into mineral and organic acid-soluble fractions. Decreasing of ratio organic: mineral phosphorus indicates less intensiveness of metabolic processes in aluminum-resistant wheat Irgina at early stage of growth and following increasing of these processes under liming. Opposite, sensitive wheat Priokskaya displays sharply decreasing of metabolic activity by the termination of experiment under calcium input.

In the third plants of aluminum-resistant oats variety Krechet were capable to support metabolic processes with participation of nitrogen, phosphorus and potassium in the conditions of action of the stressful fac- tor at the same level as in the neutral growth media. Changes of relative requirements in macronutrients specifies in considerable reorganizations of metabolic reactions in plants of Al-sensitive oats variety Argamak under stressor action. The biochemical processes related with action of the photosynthetic apparatus of leaves in stressful conditions have a priority in their supply with macronutrients, that is specified with smaller variability of the triple ratio N:P:K necessary for the maximum development of the leaves of oats in comparison with root systems at aluminum influence.

7.1   INTRODUCTION

The stresses limiting plant growth on acid soils consist of proton rhizotoxicity (low pH), nutrient deficiency (primarily phosphorus but also potassium, calcium, and other minerals), and metal toxicity (aluminum and manganese) [1]. Among these constraints, toxicity of exchangeable Al^{3+} ions is considered to be the major limiting factor at cultivation of plants on acid soils [2], a major factor reducing efficiency of plants on 67% of all acid soils [3].

Adequate entering of nutrients is necessary for efficient crop production. As the majority of acid sod-podzolic soils in natural state are deficient in primary nutrients particularly nitrogen and phosphorus [4], and the part of these elements become gradually depleted by crops removal, it is necessary to fill this shortage at the expense of external sources. However, this strategy has become economically less feasible with increase of production cost, especially in the soils demanding phosphoric fertilizers. Therefore efforts of scientists are directed toward breeding of plants, which are capable to receive the maximum nutrient elements from the soil and/or make this process more efficient.

Aluminum interferes with absorption, translocation and utilization of many essential nutrients necessary for a plant, such as, nitrogen, calcium, potassium, magnesium, and phosphorus [5, 6].

Nitrogen fertilizers are widely used for increase productivity of grain crops and the protein content of grain in cereal crops. However, it is necessary to optimize their use in order to decrease the risks of environmental contamination and production costs [7]. For that reason, efficiency of use of nitrogen fertilizers by plants becomes a very important trait in studying and breeding of plants, including cereal [8, 9]. Increase of accumulation of nitrogen in plants not at the expense of increase of entering of nitrogen fertilizers, but at the expense of creating genotypes with the better ability of their root systems to absorb nitrogen from soil becomes the core task in the decision of the problem. On the other hand, in order to get higher values of grain yield, that process should be followed by an increased intensity of photosynthesis. Otherwise high concentration of nitrogen in grain and straw will be reached with lowering in efficiency of nitrogen utilization [10]. As percent of acid soils is high throughout the world, so there is plenty of references dealing with parameters of metabolism of nitrogen on such soils [11] and considerable efforts is directed to establishment of genotypic specificity of parameters of a nitrogen metabolism [7, 12].

Aluminum toxicity and phosphorus deficiency are two common constraints limiting crop production in acid soils [13]. Understanding the mechanisms underlying aluminum and phosphorus interactions will help to develop management principles to sustain production of agricultural plants in acid soils. Effects of phosphorus-aluminum interaction on adaptation of plants to toxicity of acid soils were studied in many researches [14–17]. Fukuda et al. [18] have assumed that a common metabolic system is responsive to both deficiency of phosphorus and to toxicity of aluminum in rice.

Increasing phosphorus supply substantially decreased extractable Al in bulk soil [19]. This decrease in aluminum ions extractability in soil is likely to have resulted from the chemical sedimentation of aluminum with the added phosphorus that leads to lowering the activity of trivalent aluminum ions in the soil solution [20, 21].

The increasing entering of phosphorus can ameliorate the toxic effect of aluminum ions on root growth, but degree of the amelioration is dependent on the severity of the stress. High doses of phosphorus were more effective on influence on development of shoots than on the root systems. Very strong positive effect of entering of raising doses of phosphorus on shoot growth was found out at high level of stressful influence (200 µM aluminum) [19]. The key moment was that moving of aluminum ions from the roots to the shoot was markedly reduced at entering of 80 mg of phosphorus per kg of soil. Irrespectively of the level of added aluminum wheat seedlings were able to absorb more phosphorus from the soil, translocated more phosphorus to the shoots and use phosphorus more effectively for shoot growth and development with increase of level of entering phosphorus although total absorption of phosphorus decreased with the aluminum influence. Similar findings have been received previously in the conditions of nutrient solution cultures [20, 22].

It has been suggested that immobilization of aluminum by phosphorus in cell wall of roots is a potential mechanism for Al-tolerance in buckwheat [17], barley [23], and maize [22]. Under Al stress in hydroponics phosphorus application was shown to stimulate malate exudation (an indicator of aluminum resistance) from the tap root tip of the P-efficient soybean (*Glycine max*) genotypes compared with the P-inefficient genotypes [11]. In field experiments with soybean [22] it is shown that phosphorus addition to acid soils stimulates aluminum resistance, especially for the genotypes capable to absorb this macroelement effectively. Subsequent studies in hydroponic culture conditions have shown that solution pH, levels of aluminum and phosphorus coordinately changed growth of roots of soybean and exudation of malate anions (as a basic mechanism of plant Al-resistance).

Iqbal et al. [17] assume at least four different ways in which phosphorus can lower toxicity of aluminum. First, phosphorus directly reacts with aluminum in soil forming Al-P precipitates, and thus reduces activity of Al^{3+} ions in a soil solution. Second, phosphorus reduces the amount of apoplastic aluminum, that was bound to the root cell walls and this binding was around 37% of the total Al uptake by the root. Third, high doses of entering of phosphorus reduce the total absorption of aluminum by plant (to 50%), thus simultaneously its concentration in roots decreases to a lesser degree (on 12%), than in shoots (on 88%) with high degree of stress. And, at last, fourth, phosphorus reduces moving of aluminum from roots to shoots by up to 90% in high doses of phosphorus and aluminum.

The aluminum can reduce the absorption of potassium by competitive inhibition [23]. It has been found that the potassium deficiency, induced by aluminum, affects nitrogen metabolism by stimulating the accumulation of putrescine [24]. At *Stylosanthes* aluminum also increases the adsorption of nitrogen in tolerant species [25], leading to increases of nitrate, free amino acids and proteins contents in tissues [26]. This increase in nitrogen metabolism in tolerant species subjected to aluminum is related to the synthesis of specific proteins [27, 28] that provides differential tolerance of plants to aluminum. Some researchers [26] have found that the toxic effects of aluminum are shown only when nitrogen is presented in nitric form. However, results of study [29] indicate that aluminum does not statistically influenced the growth rates of two *Stylosanthes* species when supplied with nitrate form of nitrogen fertilizer although under these conditions the tendency of decrease in relative growth rate in specie *S. guianensis* is observed. In the presence of nitric nitrogen, aluminum increased potassium concentrations in *S. macrocephala* plants, but not in *S. guianensis*. When the nitrogen source was supplied by ammoniacal form, aluminum did not influence the adsorption of potassium in both species. In the presence of the nitric source, the aluminum increases the potassium concentrations only in the tolerant specie, the *S. macrocephala*.

Thus, although mechanisms of interaction of aluminum and elements of a mineral nutrition of plants has been taken into consideration in a few studies on plant adaptation to acid soils, this subject is still quite poorly understood.

Agricultural use of fertilizers must be as optimal as possible for ensuring increase of plant productivity with the same or higher quality of products, and minimizing environment contamination by their surplus. Therefore, it is necessary to optimize the ratio of all three basic macronutrients (N, P, and K) in fertilizer simultaneously. However, this problem is connected with a considerable amount of variants under investigation. In a case of acid soils presence of ions of exchangeable aluminum increases twice amount of variants of the experience necessary for finding-out of the put question.

Plant physiologists offered principally new approach in decision of the question – estimation of total N+P+K doze in fertilizer and an optimum of N:P:K ratio within this doze. This approach gives an exact approximation for optimum parity of nutrients in fertilizer and demands a little number of variants (10–15). Complexity of such research can be lowered even more, if a method of systematic variants [30] is used. The main advantage of a method is that only 3 variants are necessary for estimation the optimum N:P:K ratio. At studying influence of the stressful factor 3 more variants are added and, thus, total number of variants for each investigated variety of plants makes six.

7.1.1 PROCESS OF A MINERAL NUTRITION OF PLANTS IS CLOSELY RELATED WITH PHOTOSYNTHESIS

A mineral nutrition provides the growing photosynthetic apparatus with building elements. Three main components of plant productivity of cereal crops – number of productive stems, number of grains per ear (panicle) and 1000 grain mass – positively influence increase in photosynthesis intensity at stages of formation of these components. The positive interrelation between intensity of CO_2 digestion and the specific leaf area (SLA) is an important regularity of varietal variability of photosynthesis. The heightened content of chlorophyll in plant leaves is also related with the heightened plant productivity or quality of product [31]. These characters are stable enough for a genotype, but under influence of edaphic factors they change in different degree. Change of the ratio of mineral elements in fertilizer can change considerably reaction of the photosynthetic apparatus to stressful influence of aluminum. Moreover, differences in aluminum tolerance level between plants are often related with their ability to use phosphorus fertilizer under acid growth conditions.

Thus, the main tasks of a given study were: (a) to establish aluminum influence in acid soil on modification of requirement of plants of grain crops in the basic macronutrients, and (b) to estimate influence of soil acidity on uptake of phosphorus by plants and its transformation into organic and mineral compounds within whole plants.

7.2   MATERIALS AND METHODOLOGY

7.2.1   INFLUENCE OF SOIL ACIDITY ON THE CONTENT OF PHOSPHORIC COMPLEXES IN WHEAT PLANTS

In the conditions of acid reaction of growth medium pot experiment with two spring wheat varieties is put: Irgina – aluminum-resistant variety, and Priokskaya – aluminum-sensitive variety. Reaction of the varieties on aluminum influence was estimated in laboratory with a method described earlier [32]. The natural sod-podzolic soil (pH$_{KCl}$ 3.97, hydrolytic acidity = 2.48 mg-equivalent/100 g of soils) containing 16 mg of exchangeable aluminum per 100 g of soils and 1.7 mg of phosphorus per 100 g of soils (at extraction with 0.2 H HCl) was used. Plants grew up in pots with 4 kg of soil in triple replications on four nutrition backgrounds.

Optimum doses of nitrogen and potassium were calculated according to [33]: nitrogen – 120 mg/L, potassium – 150 mg/L of a nutrient solution. Phosphorus was brought in quantity equal to the aluminum content in soil (variant NP$_1$K), at the double by aluminum content in soil (variant NP$_2$K). For calculation of quantity of the phosphorus precipitated with aluminum, tables of recalculation were used [34]. Considering that 4 kg of soil correspond to 10 l of a solution, following quantities of nutritious salts (g/pot) were taken (Table 7.1).

Fertilizers were input in the form of chemically pure salts one week prior to sowing. Sowing was carried out with 15 dry seeds per pot and 10 most vigorous seedlings have been left after germination. Each variant had 12 pots in total. Samples for estimation of a chlorophyll content and fractional structure of phosphates consisted of the mixed plant sample (3 pots with 10 plants each per each variant of study). Estimation

TABLE 7.1   Entering of Nutritious Salts For Creation of Nutrient Backgrounds, g/pot

Nutrient background	NH_4NO_3	KCl	KH_2PO_4
The control (natural soil)	-	-	-
The control + NK	3.43	1.43	-
The control + NP_1K	3.43	-	2.59
The control + NP_2K	3.43	2.87	5.17

of the basic fractions of phosphorus was spent according to Ref. [35] after wet combustion. Selection of plants for the analysis was carried out at three growth stages: tillering, leaf-tube formation, and flowering. The content of a chlorophyll defined in acetone extract with "SHIMADZU UVmini-1240" spectrophotometer by a [36] technique.

7.2.2   INFLUENCE OF LIMING ON UPTAKE AND TRANSFORMATION OF PHOSPHORUS COMPOUNDS WITHIN WHEAT PLANTS

Liming is one of the most effective means of decreasing of negative influence of soil acidity on agricultural crops, so as a extension of a previously described experiments additional variant with use of liming according to hydrolytic acidity of soil was used. Estimation of basic phosphorus fraction was spent three times per growth season – at tillering, leaf-tube formation, and flowering stages on the same spring soft wheat varieties: Irgina – aluminum-resistant variety, and Priokskaya – aluminum-sensitive variety.

7.2.3   INFLUENCE OF ALUMINUM-ACID GROWTH MEDIA ON MODIFICATION OF REQUIREMENT OF OATS PLANTS IN THE BASIC MACRONUTRIENTS

Plants were grown up in the conditions of sand culture. 15 dry seeds were sown in pots with 4 kg of sand each and 8 most vigorous seedlings have been left after germination. In this study 2 oat varieties were investigated – Krechet (Al-resistant) and Argamak (Al-sensitive) [37]. Duration of the study has made 30 days according to the literary data [38, 39].

The content of micronutrients was estimated in mg-atoms, instead of weight quantities of "acting matter" (N, P_2O_5, K_2O) as it is accepted in agronomical practice. Content of real atoms (ions NO^{3-} NH^{4+}, PO_4^{3-} K^+, or atoms N, P, K) absorbed by roots is more important for plants than the contents of conditionally "acting matters". One kg of each of "acting matter" will contain the different quantity of atoms. Application of step schemes of change of quantity of input fertilizers (for example, 30:30:30, 60:60:60 and so on) leads to disturbed of the requirement of "only distinction" between variants as it changes not only quantity of input substances, but also ratio in number of input atoms of each element. Besides, "acting matters" include different amount of oxygen atoms, which is not taken into account. Therefore D.B. Vakhmistrov, V.A. Vorontsov [40] suggest to count a ratio of elements in mg-atoms, and in all variants of experiment the total content of elements should be identical and correspond to their sum in standard Hogland-Arnon1 medium (22 mg-atom per 1 L of a solution).

Thus research consisted of three variants at neutral reaction of growth medium (pH = 6.5), and the same three variants at acid reaction (pH = 4.3 + 1 mM aluminum in the form of sulfate): N:P:K = 70:15:15 atomic %, N:P:K = 15:70:15 atomic %, and N:P:K = 15:15:70 atomic %. Total N+P+K content are equal to 22 mg-atom per 1 kg of substrate. Each variant of experiment is put in four replications. Upon termination of experiments dry weight of roots and shoots, the total and specific leaf area, and the chlorophyll content were estimated.

Each experiment is repeated twice within two years. Thus, the data resulted in tables, represents average value from 12–16 replications (3–4 biological replications x 2 replications in each year x 2 years). Statistical calculations and an estimation of an optimum ratio of macronutrients in growth media are spent according to [41]. Accuracy of experiment made 2.6–3.2% depending on the year.

7.3   RESULTS AND DISCUSSION

7.3.1   INFLUENCE OF PHOSPHORUS ON PLANT RESISTANCES TO HIGH SOIL ACIDITY

Results of estimation of content of various fractions of phosphoric complexes in leaves of spring wheat are presented in Table 7.2.

TABLE 7.2 Influence of Nutrient Backgrounds on the Content of Phosphorus Forms in Leaves of Spring Wheat (g/g of Fresh Mass)

Nutrient background	Tillering stage		Leaf-tube formation stage		Flowering stage	
	1*	2	1	2	1	2
Mineral phosphorus						
Control	0.04	0.04	0.07	0.03	0.02	0.04
Control + NK	0.04	0.02	0.03	0.05	0.03	0.04
Control + NP$_1$K	0.05	0.04	0.06	0.04	0.04	0.03
Control + NP$_{2K}$	0.02	0.05	0.03	0.05	0.02	0.01
Organic acid-soluble phosphorus						
Control	0.55	0.35	0.20	0.24	0.24	0.28
Control + NK	0.34	0.22	0.50	0.41	0.27	0.28
Control + NP$_1$K	0.32	0.43	0.39	0.33	0.52	0.43
Control + NP$_{2K}$	0.20	0.40	0.34	0.44	0.34	0.26
Organic acid-insoluble phosphorus						
Control	0.14	0.11	0.11	0.05	0.08	0.07
Control + NK	0.11	0.08	0.11	0.11	0.09	0.06
Control + NP$_1$K	0.12	0.09	0.10	0.09	0.03	0.03
Control + NP$_{2K}$	0.12	0.09	0.10	0.07	0.02	0.02
Total phosphorus						
Control	0.73	0.50	0.37	0.33	0.34	0.39
Control + NK	0.49	0.32	0.64	0.57	0.36	0.38
Control + NP$_1$K	0.49	0.56	0.55	0.45	0.59	0.49
Control + NP$_{2K}$	0.34	0.53	0.47	0.56	0.37	0.29

Note: * 1 – variety Irgina, 2 – variety Priokskaya.

The data about a ratio of organic and inorganic forms of phosphoric complexes which serves as an indicator of intensity of inclusion of phosphorus in exchange processes of an organism is most interesting to researchers and breeders. It is possible to indicate that input of nitrogen, phosphorus and potassium into acid soil at tillering stage of growth has lowered intensity of phosphorus metabolization though for Al-sensitive variety it has occurred only at the doubled dose of phosphorus. At the following stages of development, in contrary, improvement of a mineral nutrition promotes the strengthened inclusion of phosphorus in a metabolism, and it is manifested much more strongly it Al-resistant variety.

Doubling of a dose of phosphorus has caused doubling of its inclusion in exchange processes in a resistant variety at early growth stages, but in sensitive variety – only on a flowering stage. This finding indicates higher level of a metabolism in Al-resistant variety at early stages of growth. At first two stages of development input of nitrogen and potassium into soil was more effective for phosphorus metabolization than input of phosphorus at nitrogen and potassium background.

Synthesis of organic acid-soluble phosphorus (in % of the total phosphorus content) in the course of growth raises constantly, and at tillering stage it is more strongly shown in aluminum-sensitive variety, but at the following stages of growth the resistant variety catches up sensitive variety and even overtakes it a little on the given parameter.

If to compare the content of total phosphorus in shoots of contrast varieties in absolute (g) instead of relative (g/g of dry matter) values, that is to consider higher shoot mass of Al-resistant variety, is possible to suggest that resistant variety takes significantly more amount of phosphorus from soil, than sensitive variety.

At a tillering stage there are no distinctions between varieties on biomass accumulation on all nutrient backgrounds, thus improvement of nutrition leads to double increase of a biomass of plants (Table 7.3).

At a leaf-tube formation stage differences between varieties were not revealed also, but differences on nutrition backgrounds were considerably

TABLE 7.3 Influence of Nutrition Backgrounds on Accumulation of the Total Biomass by Spring Wheat (grams of Dry Mass Per 1 Plant)

Nutrition background	Tillering stage		Leaf-tube formation stage		Flowering stage	
	1*	2	1	2	1	2
Control	0.095	0.112	0.327	0.271	0.527	0.438
Control + NK	0.185	0.206	0.391	0.399	0.691	0.479
Control + NP_1K	0.172	0.202	0.439	0.468	0.710	0.562
Control + NP_{2K}	0.202	0.199	0.608	0.624	0.906	0.655

Note: * 1 – variety Irgina, 2 – variety Priokskaya.

showed. The doubled dose of phosphorus has led 1.5–2 fold increase in plant productivity. However, plants of a control background have strongly grown up by this stage also that was especially showed in aluminum-resistant variety. Varietal differences were especially showed at flowering stage on all nutrition backgrounds; resistant variety has appeared much more productive than sensitive one.

The particular interest represents the fact that entering of phosphoric fertilizers both into one- and in double dose has led to almost identical strengthening of accumulation of dry matter by both varieties (both NPK backgrounds in comparison with NK background). In other words, differences in resistance to aluminum have not affected action of phosphoric fertilizers. The same conclusion arises at the analysis of accumulation of a biomass of an underground part of plants that is development of root systems of plants contrast on Al-resistance level in the conditions of supply with nitrogen-and-potassium nutrition depends a little on presence of ions of hydrogen, aluminum or phosphorus in soil. The analysis of growth of plants on natural acid soil (control background) has allowed to show differences in schemes of development of the investigated varieties.

Though both varieties as a whole have shown great advance of a gain of roots from tillering stage to leaf-tube formation stage, and in a gain of shoot – from tillering stage to flowering phase, absolute value of a gain both roots and shoots is much more for aluminum-resistant variety. In other words, in acid soil resistant variety increases root mass more intensively at the beginning of growth that, possibly, allows it to increase shoot mass more intensively too.

Summarizing the above-stated it is possible to point out that selection of plants on acid soils can be conducted on development of root system at leaf-tube formation stage already, but on development of shoot mass – not earlier than a flowering stage.

It is known that autotrophic growth of organisms is provided, on the one hand, at the expense of root nutrition and, on the other hand, at the expense of assimilation of carbon of air in the course of photosynthesis. These two processes in an organism are interrelated and interdependent; synthesis of elements of the photosynthetic apparatus depends on absorption of necessary mineral substances from soil.

At the analysis of wheat varieties contrast on Al-resistance level it is revealed that in the control background (acid soil) resistant variety synthesized higher quantity of chlorophylls (chlorophyll *a, b* and their sums), than sensitive variety at all investigated stages of development (Table 7.4).

Obviously given parameter (content of a chlorophyll per gram of dry mass) as index of resistance to lowered pH of soil solution requires more detailed research on more number of varieties in field conditions.

Other interesting fact which is necessary to noting in respect of influence of nutrition backgrounds: at input of phosphoric fertilizers into soil Al-sensitive variety synthesizes much more quantity of all forms of chlorophyll than resistant variety at all stages of development. In other words, at improvement of conditions of mineral nutrition compensation mechanisms of plants limit excessive power consumption and plastic substances on construction of the photosynthetic apparatus as receipt of substances and energy raises at the expense of root system.

Thus if the resistant variety reduces the content of chlorophyll already at tillering stage then in sensitive variety decrease begins only at leaf-tube formation stage, besides decrease in synthesis of chlorophyll in resistant variety occurs much more sharply. It can testify higher plasticity of a metabolism of aluminum-resistant variety – it reacts on change of environment conditions much faster.

The parameter of specific leaf area (SLA) indirectly characterizes a thickness of leaf and a share a dry matter in it. Influence of input of phosphorus on the given parameter in our experiments was differing for the

TABLE 7.4   Influence of Nutrition Backgrounds on the Total Content of a Chlorophyll in Leaves of Spring Wheat (mg/g of Dry Mass)

Nutrition background	Tillering stage		Leaf-tube formation stage		Flowering stage	
	1*	2	1	2	1	2
Control	7.51	6.80	6.30	5.59	3.63	3.24
Control + NK	6.99	8.30	3.59	3.85	3.31	2.87
Control + NP$_1$K	6.85	9.40	3.61	4.78	2.01	2.89
Control + NP$_{2K}$	6.98	7.37	3.03	4.11	2.09	2.50

Note: * 1 – variety Irgina, 2 – variety Priokskaya.

investigated varieties. Strengthening of a phosphoric nutrition has led to increase of an average SLA of all leaves of plants of both varieties at tillering stage (a background with a double dose of phosphorus), to its increase in resistant variety at leaf-tube formation stage and to drop at flowering phase whereas in sensitive variety, on the contrary, strengthening of a phosphoric nutrition has lowered this parameter at leaf-tube formation stage and has strengthened at flowering stage (in comparison with natural acid background and background with one dose of phosphorus) (Table 7.5).

This data indicates considerable differences in a metabolism of the studied varieties: at flowering stage resistant variety has strengthening of outflow of plastic substances from leaves into generative organs whereas a sensitive variety has not complete formation of the leaf apparatus by this time and continued to increase it to the detriment of generative organs.

Results of the two-way ANOVA have allowed to calculate shares of influence of growth stage and nutrition backgrounds on change of SLA. Nutrition backgrounds have rendered twice a greater influence on SLA variability in sensitive variety (influence of factors "nutrition background" + "nutrition background x growth stage" = 61/9% against 27.5% in resistant variety). Thus, sensitive variety is characterized by considerable dependence of photosynthesis on conditions of mineral nutrition whereas intensity of photosynthetic processes in resistant variety is influenced basically by growth stage.

TABLE 7.5   Influence of Nutrition Backgrounds on Specific Leaf Area of Spring Wheat Varieties, Contrast on Al-Resistance (mg/sm^2)

Nutrition background	Tillering stage		Leaf-tube formation stage		Flowering stage	
	1*	2	1	2	1	2
Control	2.51	2.67	3.47	3.49	4.18	2.41
Control + NK	2.33	2.37	4.21	4.83	4.19	4.54
Control + NP_1K	2.43	2.47	3.76	3.20	4.27	2.91
Control + NP_{2K}	3.28	3.06	3.94	3.17	3.06	4.45

Note: * 1 – variety Irgina (Al-resistant), 2 – variety Priokskaya (Al-sensitive).

7.3.2 INFLUENCE OF LIMING ON UPTAKE AND TRANSFORMATION OF PHOSPHORUS COMPOUNDS WITHIN WHEAT PLANTS

The content of the total phosphorus during growth season at the backgrounds without fertilizers and liming decreased gradually at aluminum resistant wheat Irgina (except mineral phosphorus) whereas at sensitive wheat Priokskaya it raised in time at the expense of almost all forms of complexes, except for acid-insoluble fraction (Table 7.6).

At plants of aluminum resistant wheat the share of mineral complexes in total phosphorus content increased from 17 (tillering stage) up to 30% (leaf-tube formation stage), at sensitive wheat – remained a constant at level of 24%. Liming increased the absolute content of mineral

TABLE 7.6   Input and Transformation of Phosphorus Compounds Into Wheat Plants Under Action of Lime, % of Total Dry Mass

Variety	Addition of Ca	P_2O_5				
		Total phosphorus	Mineral phosphorus	Organic		Organic/ mineral
				Acid-soluble phosphorus	Acid-insoluble phosphorus	
Tillering stage						
Irgina	0	0.364	0.064	0.193	0.107	4.7
Priokskaya	Ca	0.348	0.088	0.134	0.126	3.0
	0	0.329	0.078	0.141	0.110	3.2
	Ca	0.256	0.078	0.085	0.093	2.3
Leaf-tube formation stage						
Irgina	0	0.348	0.067	0.192	0.089	4.2
Priokskaya	Ca	0.392	0.092	0.170	0.126	3.2
	0	0.370	0.089	0.166	0.111	3.1
	Ca	0.392	0.096	0.189	0.107	3.1
Flowering stage						
Irgina	0	0.326	0.097	0.158	0.071	2.4
Priokskaya	Ca	0.396	0.114	0.193	0.089	2.5
	0	0.462	0.108	0.250	0.104	3.3
	Ca	0.416	0.136	0.176	0.104	2.1

phosphates, thus a little if their relative share in phosphatic fund of a resistant variety increased only in a tillering stage (up to 25%) then at a sensitive variety growth of a share of mineral phosphorus up to 30–33% is marked both in the beginning, and in the end of experiment.

Organic acid-soluble phosphorus complexes are most closely links with synthetic processes at plants; they are the most mobile complexes of this element and are presented with nucleotides and carbohydrate–phosphates. On an early stage of development of aluminum resistant wheat the content of this fraction reaches 53% from the general pool of phosphorus, whereas at sensitive variety – 42% only. Liming reduces a share of these compounds to 38 and 33% accordingly. In later stages the quantity of acid-soluble organic fractions at resistant variety decreases gradually in a control variant (to 48%), and its share raises to the same level at liming. At sensitive variety by the end of experiment the content of acid-soluble phosphorus though increases, but in a variant with liming it is remain on level 10% below the control.

Dynamics of influence of calcium on the content of acid-insoluble fractions of phosphorus at both varieties is similar as a whole: it decreases gradually; at the first stage of growth calcium considerably strengthens phosphorus transformation into acid-insoluble compounds, and by the end of experiment there is an alignment of indicators of a share of this fraction in the total phosphorus in both variants of experiment.

Thus, distinctions in reaction of wheat varieties with various level of aluminum resistance on calcium entering are reflected mainly in the content of mineral and organic acid-soluble phosphorus fractions.

Decrease in sizes of ratio of organic and mineral forms testifies to smaller intensity of exchange processes under liming conditions at aluminum resistant wheat variety in a tillering stage and gradual increase of these processes under the influence of calcium at later stages. Unlike it, the sensitive variety by the end of experiment had a sharp decrease in intensity of a metabolism at calcium entering.

7.3.3   INFLUENCE OF ALUMINUM ON MODIFICATION OF OAT PLANTS' REQUIREMENT IN THE BASIC MACRONUTRIENTS

The study of mechanisms and genetic basis of aluminum resistance of agricultural plants and creating of varieties with high level of this type of resistance becomes more actual in relation with a drop in volumes of soil liming and, accordingly, increase in the areas of acid soils all over the world. However, presence of exchangeable Al^{3+} ions in growth medium determined low pH of these soils is not the sole stressful edaphic factor. Not the smaller role is played by shortage of some elements of a mineral nutrition and absence of a proper ratio between them. The sand culture allows to simulate soil key parameters, abstracting from complexity of this natural medium of plant habitat linked with features of mineral, salt structure, presence of organic structures and microbiological activity. The estimation of relative level of aluminum resistance in laboratory and greenhouse experiments was carried out by different researchers, which used various nutrient solutions. The content and ratio of elements of a mineral nutrition in used media is so differ sometimes that can mask exact action of the aluminum. So, the particular ratio of nitrogen, phosphorus and potassium (in mg-atoms per liter of solution) makes the following values: [42] – 54:0:46; [43] – 77:5:18; [44] – 64.5:0.5:35; [45] (Steinberg's solution) – 87:1:12. Standard Hogland-Arnon1 medium [46] contains the specified elements in the ratio 68:5:27. Probably, such considerable distinctions in the ratio of the basic macronutrients can affect a comparative estimation of Al-resistance level of plant varieties.

Using of Homes' method [30] for an estimation of optimum N:P:K ratio in a condition of sand culture for oat varieties differing in their reaction to aluminum, has shown that resistance can be explained by a higher degree of resistance of metabolic processes (maintenance of balance between synthetic and catabolic reactions) which was testified by smaller changes of relative need in macronutrients under stress (Table 7.7).

Al stress had little influence on the ratio of metabolic/catabolic processes in the roots of the resistant oats variety Krechet. Of course, the total level of reactions of synthesis and destruction can change (increase or decrease), but in equal degrees. Metabolic activity of roots of a sensitive variety Argamak has changed considerably: the requirement for phosphorus has increased by 2.5 times, and requirement in potassium has decreased by 1.5 times. These changes can mean that some biochemical processes have been switched from the basic metabolic pathway to some other.

Development of aboveground mass under stressful influence has also shown distinction between the varieties. If the resistant variety demanded increase in a relative share of phosphorus (at the expense of

TABLE 7.7   Optimum Ratio of Nitrogen, Phosphorus and Potassium For Some Indicators of Development of Plants of Two Oats Varieties, Contrast on Aluminum-Resistance (%)

pH of growth media	Root dry mass	Shoot dry mass	Total dry mass	Specific leaf area	Total leaf area	Total content of chlorophyll
Variety Krechet (Al-resistant)						
pH 6.5	46:25:29	45:18:37	45:19:36	30:43:28	46:15:39	49:20:31
pH 4.3	43:25:32	29:34:37	33:31:36	24:41:35	45:32:23	47:17:36
Variety Argamak (Al-sensitive)						
pH 6.5	46:9:45	36:10:54	39:9:52	33:35:32	23:30:47	51:22:27
pH 4.3	45:25:30	46:22:32	48:21:31	29:44:27	47:18:35	54:23:23

nitrogen) at constant requirement in potassium the sensitive variety has changed relative shares of all three elements (requirement for nitrogen and phosphorus have increased by 1.3 and 2.2 times, accordingly, at the expense of requirement reduction in potassium by 1.7 times). The forming of the maximum leaf area of a resistant variety has demanded relative increase in a share of phosphorus and reduction in share of potassium in the growth medium at the constant content of nitrogen, but for a sensitive variety the requirement for nitrogen has increased, and the requirement for other elements has decreased. Modification of requirements of the leaf apparatus in macronutrients has not affected processes of synthesis of a chlorophyll neither at resistant, nor at a sensitive variety. Thus the interrelation between N, P and K at which the process of synthesis of photosynthetic pigments was maximum at a resistant variety actually coincides with the ratio of the macronutrients necessary for the maximum development of root system both in neutral, and in the acid growth condition. It can indicate the greater level of coordination of processes of a mineral nutrition and photosynthesis at aluminum-resistant oat variety than at a sensitive one.

However, from our point of view, more interesting is the fact that mathematically it is possible to calculate that ratio between macronutrients at which plants will not suffer depression of growth under the influence of the stressful factor. If we take into account the relation of dry weights of different parts of a plant (i.e., root-to-shoot ratio) or all plant in the control and in the test treatment thus their maximum ratio (100%) will correspond to optimum N:P:K ratio.

In our case the following optimum N:P:K ratio are received for oat varieties Krechet and Argamak: for absence of depression of root growth – 30:34:36 and 36:10:54, accordingly; for absence of depression of shoot growth – 18:52:30 and 31:11:58, accordingly; for a total mass of plants – 21:49:30 and 32:11:57, accordingly.

7.4 CONCLUSION

Doubling of a phosphorus dose in the growth medium has caused doubling of its inclusion in metabolic processes in Al-resistant wheat variety at early growth stages already, but in sensitive variety phosphorus inclusion amplified only at flowering stage. At first two stages of development addition of nitrogen and potassium into soil is more effective for phosphorus metabolization than input of phosphorus salt at nitrogen and potassium background.

Development of root systems in plants of wheat varieties contrast on aluminum resistance level in the conditions of supply with nitrogen and potassium nutrition depends a little on presence of ions of hydrogen, aluminum or phosphorus in soil.

Aluminum-sensitive variety of wheat is characterized by considerable dependence of photosynthesis on conditions of mineral nutrition whereas intensity of photosynthetic processes in aluminum-resistant variety is influenced basically by a growth stage.

Different reaction of the investigated wheat varieties on input of calcium into acid growth media is related with transformation of absorbed phosphorus into mineral and organic acid-soluble fractions. Decreasing of ratio organic: mineral phosphorus indicates less intensiveness of metabolic processes in aluminum-resistant wheat Irgina at early stage of growth and following increasing of these processes under liming. Opposite, sensitive wheat Priokskaya displays sharply decreasing of metabolic activity by the termination of experiment under calcium input.

Al-resistant oat variety maintained relative levels of N, K and P metabolism (increased or decreased them by equal extent) in roots under stress condition. Possibly, it occurs at the expense of modification of nitrogen and phosphorus metabolism in shoots. The Al-sensitive variety of oat under conditions of aluminum influence keeps N-metabolism level in roots, but levels of phosphorus and potassium metabolism are exposed to significant modifications. In shoots there is a reorganization of the metabolic processes occurring to participation of all three elements.

Choosing a certain ratio of elements of a mineral nutrition it is possible to reach a situation when aforementioned parameters will not expose negative action of aluminum on their development.

KEYWORDS

- aluminum
- nitrogen
- oats
- phosphorus
- potassium
- resistance
- spring wheat

REFERENCES

1. Kochian, L. V., Hoekenga, O. A., Pineros, M. A. How do crop plants tolerate acid soils? Mechanisms of aluminum tolerance and phosphorous efficiency. Annu. Rev. Plant Biol. 2004. Vol. 55. p. 459–493.
2. Jayasundara, H. P. S., Thomson, B. D., Tang, C. Responses of cool season grain legumes to soil abiotic stresses. Adv. Agron. 1998. Vol. 63. p. 77–151.
3. Eswaran, H., Reich, P., Beinroth, F. Global distribution of soils with acidity. Braz. Soil Sci. Soc. 1997. p. 159–164.
4. Lisitsyn, E. M., Shchennikova, I. N., Shupletsova, O. N. Cultivation of barley on acid sod-podzolic soils of north-east of Europe. Barley: Production, Cultivation and Uses. (Ed. S. B. Elfson) New York: Nova Publ. 2011. p. 49–92.
5. Guo, T. R., Zhang, G. P., Zhou, M. X., Wu, F. B., Chen, J. X. Influence of aluminum and cadmium stresses on mineral nutrition and root exudates in two barley cultivars. Pedosphere. 2007. Vol. 17. p. 505–512.
6. Olivares, E., Pena, E., Marcano, E., Mostacero, J., Aguiar, G., Benitez, M., Rengifo, E. Aluminum accumulation and its relationship with mineral plant nutrients in 12 pteridophytes from Venezuela. Env. Exp. Bot. 2009. Vol. 65. p. 132–141.

7. Le Gouis, J., Fontaine J-X., Laperche, A., Heumez, E., Devienne-Barret, F., Brancort-Hulmel, M., Dubois, F., Hirel, B. Genetic analysis of wheat nitrogen use efficiency: coincidence between QTL for agronomical and physiological traits. Proceedings of the 11th International Wheat Genetics Symposium. 2008 (URL: http://hdl.handle.net/2123/3217).

8. Hirel, B., Le Gouis, J., Ney, B., Gallais, A. The challenge of improving nitrogen use efficiency in crop plants: towards a more central role for genetic variability and quantitative genetics within integrated approaches. J. Exp. Bot. 2007. Vol. 58. p. 2369–2387.

9. Deletić, N., Stojković, S., Djurić, V., Biberdžić, M., Gudžić, S. Genotypic specificity of winter wheat nitrogen accumulation on an acid soil. Research Journal of Agricultural Science. 2010. Vol. 42, N. 1. p. 71–75.

10. Stojković, S. Genotipska variranja nekih pokazatelja akumulacije i iskorišćavanja suve materije i azota kod ozime pšenice. Magistarska teza, Poljoprivredni fakultet, Univerzitet u Prištini. 2001. [in Serbian].

11. Bednarek, W., Reszka, R. The influence of liming and mineral fertilization on the utilization of nitrogen by spring barley. Annales Universitatis Mariae Curie–Skłodowska, Lublin – Polonia. 2009. Vol. 64. N. 3. p. 11–20.

12. Habash, D. Z., Bernard, S., Schondelmaier, J., Weyen, J., Quarrie, S. A. The genetics of nitrogen use in hexaploid wheat: N utilization, development and yield. Theor. Appl. Genet. 2007. Vol. 114. p. 403–419.

13. Liao, H., Wan, H., Shaff, J., Wang, X., Yan, X., Kochian, L. V. Phosphorus and aluminum interactions in soybean in relation to aluminum tolerance: exudation of specific organic acids from different regions of the intact root system. Plant Physiol. 2006. Vol. 141. p. 674–684.

14. Dong, D., Peng, X., Yan, X. Organic acid exudation induced by phosphorus deficiency and/or aluminum toxicity in two contrasting soybean genotypes. Physiol. Plant. 2004. Vol. 122. p. 190–199.

15. Jemo, M., Abaidoo, R. C., Nolte, C., Horst, W. J. Aluminum resistance of cowpea as affected by phosphorus-deficiency stress. J. Plant Physiol. 2007. Vol. 164. p. 442–451.

16. Sun, Q. B., Shen, R. F., Zhao, X. Q., Chen, R. F., Dong, X. Y. Phosphorus enhances Al resistance in Al-resistant *Lespedeza bicolor* but not in Al-sensitive, *L. cuneata* under relatively high Al stress. Ann. Bot. (Lond). 2008. Vol. 102. p. 795–804.

17. Zheng, S. J., Yang, J. L., He, Y. F., Yu, X. H., Zhang, L., You, J. F., Shen, R. F., Matsumoto, H. Immobilization of aluminum with phosphorus in roots is associated with high aluminum resistance in buckwheat. Plant Physiol. 2005. Vol. 138. p. 297–303.

18. Fukuda, T., Saito, A., Wasaki, J., Shinano, T., Osaki, M. Metabolic alterations proposed by proteome in rice roots grown under low P and high Al concentration under low pH. Plant Sci. 2007. Vol. 172. p. 1157–1165.

19. Iqbal, T., Sale, P., Tang, C. Phosphorus ameliorates aluminum toxicity of Al-sensitive wheat seedlings. 19th World Congress of Soil Science, Soil Solutions for a Changing World 1–6 August 2010, Brisbane, Australia. 2010. P. 92–95.

20. Nakagawa, T., Mori, S., Yoshimura, E. Amelioration of aluminum toxicity by pretreatment with phosphate in aluminum-tolerant rice cultivar. J. Plant Nutr. 2003. Vol. 26. p. 619–628.

21. Silva, I. R., Smyth, T. J., Israel, D. W., Rufty, T. W. Altered aluminum inhibition of soybean root elongation in the presence of magnesium. Plant Soil. 2001. Vol. 230. p. 223–230.
22. Gaume, A., Machler, F., Frossard, E. Aluminum resistance in two cultivars of *Zea mays* L: root exudation of organic acid and influence of phosphorus nutrition. Plant Soil. 2001. Vol. 234. p. 73–81.
23. McCormick, L. H., Borden, F. Y. Phosphate fixation by aluminum in plant roots. Soil Sci. Soc. Am. J. 1972. Vol. 36. p. 779–802.
24. Liang, C., Piñeros, M. A., Tian, J., Yao, Z., Sun, L., Liu, J., Shaff, J., Coluccio, A., Kochian, L. V., Liao, H. Low pH, aluminum, and phosphorus coordinately regulate malate exudation through *GmALMT1* to improve soybean adaptation to acid soils. Plant Physiol. 2013. Vol. 161. p. 1347–1361.
25. Malavolta, E., Vitti, G. C., Oliveira, S. A. Avaliação do estado nutricional das plantas: princípios e aplicações. 2.ed. Potafos, Piracicaba. 1997. 319 p. [in Portuguese].
26. Basso, L. H. M., Lima, G. P. P., Gonçalves, A. N., Vilhena, S. M. C., Padilha, C. do, C. F. Efeito do alumínio no conteúdo de poliaminas livres e atividade da fosfatase ácida durante o crescimento de brotações de *Eucalyptus grandis x, E. urophylla* cultivadas in vitro. Sciencia Forestalis. 2007. Vol. 75. p. 9–18. [in Portuguese].
27. Cordeiro, A. T. Efeito de níveis de nitrato, amônio e alumínio sobre o crescimento e sobre a absorção de fósforo e de nitrogênio em *Stylosanthes guianensis* e *Stylosanthes macrocephala*. Viçosa: 1981. 53 f. Dissertação (Mestrado em Fisiologia Vegetal) – Universidade Federal de Viçosa. [in Portuguese].
28. Amaral, J. A. T. do, Cordeiro, A. T., Rena, A. B. Efeitos do alumínio, nitrato e amônio sobre a composição de metabólitos nitrogenados e de carboidratos em *Stylosanthes guianensis* e, *S. macrocephala*. Pesquisa Agropecuária Brasileira, 2000. Vol.35. N. 2. p. 313–320. [in Portuguese].
29. Amaral, J. A. T. do, Rena, A. B., Cordeiro, A. T., Schmildt, E. R. Effects of aluminum, nitrate and ammonium on the growth, potassium content and composition of amino acids in *Stylosanthes*. IDESIA (Chile). 2013. Vol. 31, N. 2. p. 61–68.
30. Homes, M. V. L. Alimentation minerale equilibree des vegetaux. Wetteren: Universa, 1961. Vol. 1. 55 p. [in French].
31. Zelensky, M. I. Photosynthetic characteristics of major agricultural crops and prospects of their breeding application. Physiological principles of plant breeding. St. Petersburg: Publishing office of VIR, 1995. p. 466–554.
32. Lisitsyn, E. M. Intravarietal Level of Aluminum Resistance in Cereal Crops. J. Plant Nutrition. 2000. Vol. 23. N. 6. p. 793–804.
33. Rinkis, G.Ya., Nollendorf, V. F. Balanced nutrition of plants with macro and microelements. Riga: Apgāds "Zinātne", 1982. 301 p. [in Russian].
34. Arinushkina, E. V. Guide on chemical analysis of soil. Moscow: Moscow University Press, 1970. 491 p.
35. Pronina, N. B., Ladonin, V. F. (Eds.) Physiological-and-biochemical methods of studying of action of chemical means complex on plants (Methodical recommendations). Moscow: All Russian Research Institute of Agrochemistry, 1988. 68 p. [in Russian].

36. Lichtenthaler, H. K., Buschmann, C. Chlorophylls and carotenoids – Measurement and characterization by UV-VIS. Current Protocols in Food Analytical Chemistry. John Wiley & Sons, Madison, 2001. P. F4.3.1-F4.3.8. [Nr. 107].

37. Lisitsyn, E. M., Batalova, G. A. Aluminum resistance of direct and reciprocal oats F_2 hybrids. Herald of Russian Academy of Agricultural Sciences. 2007. N. 1. p. 47–49. [in Russian].

38. Foy, C. D. Tolerance of barley cultivars to an acid, aluminum-toxic subsoils related to mineral element concentrations in their shoots. J. Plant Nutrit. 1996. Vol. 19. p. 1361–1380.

39. Blamey, F. P. C., Edmeades, D. C., Wheeler, D. M. Empirical models to approximate calcium and magnesium ameliorative effects and genetic differences in aluminum tolerance in wheat. Plant Soil. 1992. Vol. 144. p. 281–287.

40. Vakhmistrov, D. B., Vorontsov, V. A. Selective ability of plants is not directed on providing of their maximum growth. Russian Plant Physiol. 1997. Vol. 44. N. 3. p. 404–412. [in Russian].

41. Vakhmistrov, D. B. Separate estimation of optimal doze N+P+K and N:P:K ratio in fertilizer. 1. Foundation of problem. Agrochemistry. 1982. N. 4. p. 3–12. [in Russian].

42. Somers, D. J., Gustafson, J. P. The expression of aluminum stress induced polypeptides in a population segregating for aluminum tolerance in wheat (*Triticum aestivum*, L.). Genome. 1995. Vol. 38. p. 1213–1220.

43. Wagatsuma, T., Kawashima, T., Tamaraya, K. Comparative stainability of plant root cells with basic dye (methylene blue) in association with aluminum tolerance. Commun. Soil Sci. Plant Anal. 1988. Vol. 19. p. 1207–1215.

44. Grauer, U. E., Horst, W. J. Effect of pH and N source on aluminum tolerance of rye (*Secale cereale*, L.) and yellow lupin (*Lupinus luteus*, L.). Plant Soil. 1990. Vol. 127. p. 13–21.

45. Fleming, A. L., Foy, C. D. Root structure reflects differential aluminum tolerance in wheat varieties. Agron. J. 1968. Vol. 60. p. 172–176.

46. Hoagland, D. R., Arnon, D. I. The water-culture method for growing plants without soil. Circ. 347, Calif. Agric. Exp. Stn., Berkeley, CA. 1950.

CHAPTER 8

PHASE EQUILIBRIUM AND DIFFUSION IN THE SYSTEMS OF ETHYLENE-AMINOPROPYLTRIETHOXYSILANE COPOLYMERS

N. E. TEMNIKOVA,[1] O. V. STOYANOV,[1] and A. E. CHALYKH,[2] V. K. GERASIMOV,[2] S. N. RUSANOVA,[1] and S. YU. SOFINA[1]

[1]Kazan National Research Technological University, K. Marx str., 68, Kazan, 420015, Tatarstan, Russia, E-mail: ov_stoyanov@mail.ru

[2]Frumkin Institute of Physical Chemistry and Electrochemistry, Russian Academy of Sciences, Leninskii pr. 31, Moscow, 119991, Russia, E-mail: vladger@mail.ru

CONTENTS

ABSTRACT

The solubility of components has been studied in a wide range of temperatures and compositions in the systems copolymers of ethylene – aminoalkoxysilane. Phase diagrams have been constructed. Temperature and concentration areas of changes in solubility have been identified and structure of the modified copolymers has been studied.

8.1   INTRODUCTION

Copolymers of ethylene are widely used for obtaining materials and products for various purposes including coatings and adhesives. In this connection there is a need for continuous improvement in the properties of existing materials, because the synthesis of new polymers is difficult. Thus, to extend the scope of industrial-produced copolymers of ethylene is possible by their modification. One of the effective ways of modification is the introduction of organosilicon compounds. Introduction of such additives allows to achieve various changes in polymer properties [1–3], including adhesive characteristics [4, 5].

So γ-aminopropyltriethoxysilane (AGM-9) is used in fiberglass and paint industries to improve the adhesion of different polymers and coatings (acrylates, alkyds, polyesters, polyurethanes) to inorganic substrates (glass, aluminum, steel, and others) and to increase water resistance and corrosion stability of paint materials. AGM-9 is also used as pigmenting additives (enhance the interaction of the pigment with the polymeric matrix of composite material or paint material).

In order to optimize the composition of the polymer compounds and conditions of the structure formation of their mixtures the information about the phase organization of these systems is of considerable interest [5].

Despite the fact that there are papers devoted to improving adhesion to various substrates when modifying copolymers of ethylene by aminosilanes [6, 7], the information about the influence of monoaminofunctional silane on phase balance and phase structure of the polyolefin compositions in the scientific literature is not available.

The aim of this work was to study the formation of the phase structure of silanol-modified ethylene copolymers with vinyl acetate and vinyl acetate and maleic anhydride in a wide range of temperatures and compositions.

8.2 SUBJECTS AND METHODS

As the objects of study were used copolymers of ethylene with vinyl acetate Evatane 2020 (EVA20) and Evatane 2805 (EVA27) with vinyl acetate content of 20 and 27 wt%, respectively; copolymers of ethylene with vinyl acetate and maleic anhydride brand Orevac 9307 (EVAMA13) and Orevac 9305 (EVAMA26) with vinyl acetate content of 13 and 26 wt%. Main characteristics of the copolymers are given in Table 8.1.

As the modifier was used silane, containing an amino group – γ-aminopropyltriethoxysilane (AGM-9). Transparent, colorless liquid with a molecular weight 221. Density is 962 kg/m^3. Refractive index n^d_{20} = 1.4178, the content of amine groups is 7–7.5%. The melting temperature is −70°C.

Determination of the composition of coexisting phases and the inter-diffusion coefficients was carried out by a processing of series of interferograms obtained by microinterference method. Interferometer ODA-3 was used for the measurements. Measurements were performed at a range of temperatures from 50 to 150°C. To construct profiles of concentrations by interference patterns the temperature dependencies of the refractive index of the components are required [6–8]. Refractive index measurements were carried out by an Abbe refractometer IRF-454 BM at a range of temperatures from 20 to 150°C.

The structure of the modified copolymers was investigated by transmission electron microscopy. Identification of the phase structure of the samples was carried out by etching of the surface in high-oxygen plasma discharge with the subsequent preparation of single-stage carbon-platinum replicas. View of samples was performed on PEM EM-301 ("Philips ", Holland).

TABLE 8.1   Characteristics of the Copolymers of Ethylene

Polymer	Symbol	VA content, %	MA content, %	Melting temperature, °C	M_v	MFR, g/10 min 125°C	Density, g/cm³
Evatane 2020	EVA20	20	-	80	44,000	2.23	0.936
Evatane 2805	EVA27	27	-	72	57,000	0.74	0.945
Orevac 9305	EVAMA26	26	1,5	47	20,000	11.13	0.951
Orevac 9307	EVAMA13	13	1,5	92	73,000	1.1	0.939

8.3   RESULTS AND DISCUSSION

8.3.1   KINETICS OF MIXING THE COMPONENTS

Typical interferograms of interdiffusion zones of the systems EVA
(EVAMA) – modifier are shown in Figure 8.1.

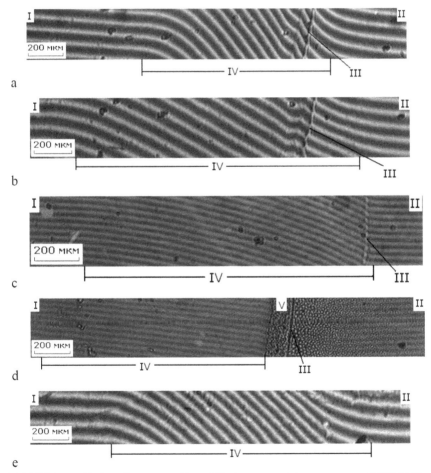

FIGURE 8.1   The interferograms of interdiffusion zones of the systems: a – EVAMA26–
AGM-9 (100°C); b – EVAMA26–AGM-9 (60°C – cooling); c – EVA20–AGM-9 (120°C);
d – EVA20–AGM-9 (90°C – cooling), e – EVAMA26–AGM-9 (140°C).

Preliminary studies have shown that at high temperatures, the inter-diffusion process is completely reversible, that is, the phase structures, occurred when the temperature is lowered, dissolved again when the temperature rises. This means that the net of the diffusion experiment is not formed or it is formed but broken during the diffusion of the modifier, which is unlikely.

It is known that during the interaction of polymers with a modifier transitional zones appear, within which the structure, composition and properties vary continuously at the transition from one phase to another.

For all systems the general picture is characteristic for partially compatible systems with a primary dissolution of alkoxysilanes in the melt of copolymers.

Phase boundary (III), a region of diffusion dissolution of modifiers in the melt of copolymers (IV) and phases of a pure copolymer (I) and the modifier (II) are clearly expressed on interferograms. Situation in the systems changes with temperature decreasing: near the interface there is a region of opacity with separation of the dispersed phase in the melt of the copolymer (V) and in the modifier. However, when the temperature rises again, the region of opacity disappears, indicating the reversibility of phase transformations occurring in the systems.

However, there were registered also the compatible systems and the temperatures corresponding to this dissolution. AGM-9 is compatible in the systems EVA27 – AGM-9, EVAMA26 – AGM-9 (see Figure 8.1e) at a temperature above 100°C, in the system EVA20 – AGM-9 (at a temperature above 120°C) and EVAMA13 – AGM-9 (at a temperature above 150°C).

Typical profiles of the concentration distribution in these systems are shown in Figure 8.2.

The size of the diffusion zone is influenced by several factors: the temperature and the time of observation.

It can be seen that the sizes of the diffusion zones on both sides of the phase boundary increases in time, whereas the values of concentrations near the interphase boundary in isothermal process conditions do not change their values.

The influence of temperature has a number of characteristic features. The higher the temperature, the greater the distance the molecules of the

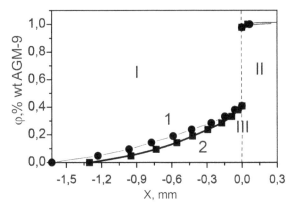

FIGURE 8.2 Profiles of the concentration distribution in the system EVAMA13 – AGM-9 at 135°C. Diffusion time: 1–11 min, 2–7 min. I – diffusion zone of AGM-9 in EVAMA13, II – diffusion zone of EVAMA13 in AGM-9, III – phase boundary.

modifier diffuse for equal periods of time and the larger the diffusion zone. This distribution of the concentration profiles when the temperature changes indicates that the system belongs to a class of systems with upper critical point of mixing [8].

Figure 8.3 shows the kinetic curves of isoconcentration planes moving at different temperatures for the systems studied.

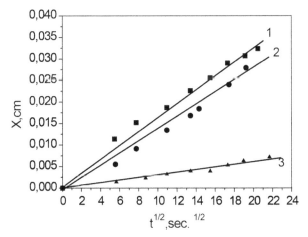

FIGURE 8.3 The time dependence of the size of the diffusion zone EVAMA26 – AGM-9 at various temperatures: 1–140°C 2–120°C, 3–100°C.

Despite the fact that the kinetics of movement of the modifier front in the matrix of the copolymer has a linear dependence, IR-spectroscopy showed that all of the systems chemically react [9]. We can assume that the method of interferometry was unable to fix a chemical reaction under the given observation time. The chemical reaction rate is comparable or slightly higher than the rate of diffusion, and the movement of the modifier already occurs into a chemically modified matrix, which is the reason for the lack of bending motion of the modifier front into the copolymer. The resulting matrix is soluble in the copolymer, so we do not observe the phase decomposition by reheating.

As the temperature increases, the nature of the concentration distribution in the zone of diffusion mixing of the components is maintained. Only the velocity of the isoconcentration planes movement changes. The angle of inclination of these relationships varies with temperature: the higher the temperature, the greater the angle of inclination of the line in the coordinates $X - t^{1/2}$. The slope of the kinetic lines is proportional to the coefficient of the modifier diffusion into the matrix. Therefore, the greater the angle of inclination, the higher the numerical value of the diffusion coefficient.

8.3.2 PHASE EQUILIBRIA

Consideration of diffusion zones of interacting copolymers and modifiers allows us to obtain not only the concentration profiles, but also the phase diagrams of the studied systems by quantitative analysis of interferograms obtained at different temperatures.

There are two binodal curves on all phase diagrams: the right branch of the binodal corresponds to the solubility of the copolymer in the modifier and is located in the area of infinitely dilute solutions. The second binodal curve represents the solubility of the modifier in the copolymers and is located in a fairly wide concentration area. The solubility of the modifier in the copolymer is increased as the temperature rises.

In all phase diagrams (see Figures 8.4 and 8.5) there are the areas corresponding to temperature-concentration areas of the components mixing (area III), and the areas corresponding to the structure of the compounds and their physical properties (area IV). The diagrams show that the preparation

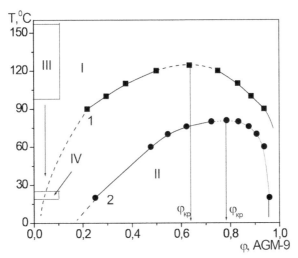

FIGURE 8.4   Phase diagrams of the systems a copolymer of ethylene with vinyl acetate –
AGM-9: 1 – EVA20; 2 – EVA27: I, II – the areas of true solutions, heterogeneous condition;
III – the area of preparation of the compositions; IV – the area of study of the structure and
physical properties.

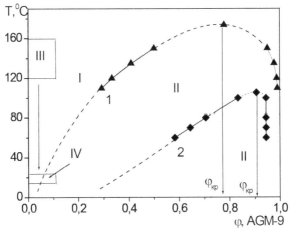

FIGURE 8.5   Phase diagrams of the systems a copolymer of ethylene with vinyl acetate
and maleic anhydride – AGM-9: 1 – EVAMA13; 2 – EVAMA26: I, II – the areas of true
solutions, heterogeneous condition; III – the area of preparation of the compositions; IV –
the area of study of the structure and physical properties.

of the mixtures takes place in the single-phase area (area I in the diagrams). As the temperature decreases the figurative point of the systems crosses the binodal curve and the system goes into a heterogeneous area (area II in the diagrams). The phase decomposition takes place and it is uniquely fixed in electron microscopic images (see Figures 8.6 and 8.7). Judging from the microphotographs, in the mixtures with EVAMA13, precipitated phases have a size of from 0.1 to 1 micron, whereas for EVAMA26 the dispersed particles have a size of from 50 to 100 micron.

Particles, protruding from the surface, etched in plasma of high-oxygen discharge, have smaller etching rate compared with the dispersion

FIGURE 8.6   Microphotograph of EVAMA13 – AGM-9 (10%).

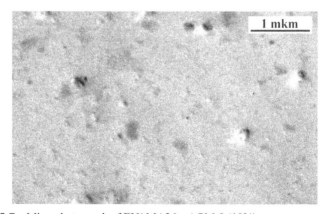

FIGURE 8.7   Microphotograph of EVAMA26 – AGM-9 (10%).

medium. It has been shown previously [10] that the lowest etching rate among carbo- and heterochain polymers have polysiloxanes. Thus, it can be concluded, that the dispersed phase is enriched by siloxanes.

With the increase of the content of vinyl acetate groups both in EVA and in EVAMA solubility of AGM-9 increases. In this case, the tendency in the change of solubility is the same for high and low temperature areas.

8.4 CONCLUSION

Thus, comprehensive studies of diffusion, phase and structural-morphological characteristics of the compositions allow us to identify the contribution of chemical reactions into the change of phase equilibrium and into the phase structure formation.

ACKNOWLEDGEMENT

This work was financially supported by the Ministry of Education and Science of Russia in the framework of the theme №693 "Structured composite materials based on polar polymer matrices and reactive nanostructured components"

KEYWORDS

- copolymers of ethylene
- aminoalkoxysilane
- phase diagrams

REFERENCES

1. Stoyanov, O. V. Modification of industrial ethylene vinyl acetate copolymers by limit alkoxysilanes. O. V. Stoyanov, S. N. Rusanova, R. M. Khyzakhanov, O. G. Petykhova, A .E. Chalykh, V. K. Gerasimov. Bulletin of Kazan Technological University. 2002. № 1–2. p. 143–147.

2. Phase structure of silanol-modified ethylene-vinyl acetate copolymers. Chalykh A.E., Gerasimov V. K., Petukhova O. G., Kulagina G. S., Pisarev S. A., Rusanova S. N. Polymer Science Series A. 2006. T.48 – №10. p. 1058–1066.

3. Stoyanov, O. V., Rusanova, S. N., Khuzakhanov, R. M., Petuhova, O. G., Deberdeev, T. R. Russian Polymer News, 7 (4), 7 (2002).

4. Temnikova, N. E. The Effect of Aminoalkoxy and Glycidoxyalkoxy Silanes on Adhesion Characteristics of Double and Triple Copolymers of Ethylene. N. E. Temnikova, S. N. Rusanova, S.Yu. Sof'ina, O. V. Stoyanov, R. M. Garipov, A. E. Chalykh, V. K. Gerasimov. Polymer Science Series D. 2014. Vol. 7 № 3. p. 84–187.

5. Temnikova, N. E. Influence of aminoalkoxy- and glycidoxyalkoxysilanes on adhesion characteristics of ethylene copolymers. N. E. Temnikova, S. N. Rusanova, S.Yu. Sofina, O. V. Stoyanov, R. M. Garipov, A. E. Chalykh, V. K. Gerasimov, G. E. Zaikov. Polymers Research Journal. 2014. Vol. 8 № 4. p. 305–310.

6. Osipchik, V. S. Development and investigation of the properties of silanol-linked polyethylene. V. S. Osipchik, E. D. Lebedeva, L.G Vasilets. Plastic masses. 2000. № 9. p. 27–31.

7. Kikel, V. A. Comparative analysis of the structure and properties of polyethylenes cross-linked by different methods. Kikel V. A., Osichik V. A., Lebedeva E. D. International Polymer Science and Technology. 2006. №4. P.15–20.

8. Polymer mixtures. In two volumes. Edited by D. R. Pola and K. B. Baknella. Translation from English edited by Kuleznev V. N. Petersburg: Fundamentals and Technologies, 2009. 1224 p.

9. Rusanova, S. N. IR spectroscopic study of the interaction of glycidoxy silane and copolymers of ethylene. S. N. Rusanova, N. E. Temnikova, O. V. Stoyanov, V. K. Gerasimov, A. E. Chalykh. Bulletin of Kazan Technological University. 2012. № 22. p. 95–96.

10. Temnikova, N. E. Effect of amino and glycidoxyalkoxysilanes on the formation of the phase structure and properties of ethylene copolymers: Thesis PhD. N. E. Temnikova. Kazan. 2013. 154 p.

CHAPTER 9

TECHNOLOGY OF RESTORATION OF OIL PRODUCTION ON ABANDONED OIL FIELDS

E. ALEKSANDROV, S. VARFOLOMEEV, M. CHERTENKOV,
G. E. ZAIKOV, V. ZAVOLZHSKY, V. LIDGI-GORYAEV, and
A. PETROV

N.M. Emanuel Institute of Biochemical Physics, Russian Academy of Sciences, 4, Kosygin st., Moscow 119334, Russia,
E-mail: 28en1937@mail.ru

CONTENTS

ABSTRACT

The modern Binary Mixtures (BM) technology developed by the authors since 1997 at present differs from all the rest of similar technologies in the heat emission optimization in the course of reagent injection. For this reason the technology ensures a much more thorough elimination of the skin layer around the producing oil wells. Accumulated during tens of years the skin layer restricts the profitable extraction to less than a half of the deposited oil.

9.1   INTRODUCTION

The cornerstone problem of modern oil production is water intrusion in the productive layers resulted from water flood displacement (water introduction from water injecting to oil producing wells aimed at oil displacement and subsequent oil recovery). Water is typically introduced not earlier than the end of active and most cost effective stage of development (10–20 years). The water introduction is aimed at the increase of layer pressure. In Russia after pumping millions of tons of water and extracting about 40% of oil deposits the process gradually turns from oil extraction into water extraction. This process is accompanied by the formation of skin layer consisting of viscous heavy oil, which more easily adheres to the surface of pores and cracks as compared to light oil. Because of oil hydrophobic properties, the water driven from water injection to oil recovery wells moves along the washouts thus displacing oil on a very limited scale. As a result of this about 60% of Russian known oil reserves stay underground labeled difficult or impossible to extract. These difficulties are further aggravated by the ever-increasing cost of geological survey of new oil fields, almost all deposits at small depth (up to 5 km) being already known. Still during the recent years there has evolved a way to solve the above-described problems by means of thermochemical technology of Binary Mixtures (BM) [1–6].

9.2   TECHNOLOGY DEVELOPMENT MILESTONES

Binary mixtures are water solutions of petersalts (ammoniac and organic ones) with the reaction initiator of petersalt decomposition (metal hydrides or sodium nitrite) [1, 2]. The two components are injected into the oil well via two separate channels and the reagents react upon contact opposite or inside the productive layer emitting heat and gas. The binary mixtures found their application in the oil extraction domain starting from 1982. It should be noted that for the reason of considerable explosion hazard the Russian Technical Supervision Service used to restrict the injection amount to 1 ton of saltpeter per oil well. Up to the year of 2010 the BM reaction was maintained in an uncontrollable mode with its efficiency approximately amounting to 0.4 [3–5].

The year 2010 saw the development of the system of optimization and control over the heat emission inside the well, the BM reaction efficiency having increased from 0.4 to 0.8. The BM technology architects obtained the permission of the Russian Technical Supervision Service (№ 25–ID–19542–2010) allowing them to inject the unlimited quantities of petersalt into oil wells. The system of reaction control has greatly contributed to spot and study the exothermic reaction of saltpeter decomposition in a heated layer with the emission of heat (Q1) and oxygen oxidizing a small part of layer oil accompanied by heat emission (Q_2). As $Q_2 \approx 2Q_1$ the progress of heat front along the layer containing petersalt water solution in its cracks and pores is as a rule a self-maintaining process.

The whole process is explosion-proof as once in the productive layer petersalt emits heat produced during the reaction which is further absorbed by the rock, the petersalt and rock mass ratio amounting to 1:20.

By increasing the BM reagent mass injected into the oil well and the productive layer by tens of times the authors mastered the heating process as well as stage-by-stage removal of the skin layer accumulated near the wells as a result of their long performance.

In 2011, the partial removal of skin layer blocking the Usinsk oil field No. 1242 and No. 3003 by means of BM [6] led to the increase in oil extraction by 4.95 and 8.44 t/day, respectively. The amount of additional oil extracted from the mentioned oil wells in 2012 amounted to nearly 3.4 thousand of tons, that is – approximately 1.7 thousand per oil well.

In 2012 in the course of treatment of the Usinsk oil fields No. 6010, No. 600, No. 1283, No. 7169 and No. 8198 (see Table 9.1) managed by E. Alexandrov and V. Zavolzhsky the technology was further improved. The reaction catalyst of petersalt decomposition (sodium nitrite) was abandoned for thermal catalyst. The water vapor heat preliminary injected into the layer at the temperature of 250°C promoted saltpeter decomposition and oxidation of small part of oil by the emitted oxygen. During the year following the treatment the extraction rate from the five described oil wells amounted to 12658 tons of additional oil. It is likely in this case that the treatment process promoted more thorough cleaning of the skin layer than the one described in oil wells No. 1242 and No. 3003. The increase of the average annual growth of oil extraction per one oil well from 1700 tons (2011) to 2532 tons (2012) is a safe indicator.

In July 2013 oil wells No. 8 and No. 10 of the Eastland oil field (Texas, USA) were treated by the BM reaction products. The mentioned oil field was abandoned in 1994 labeled unprofitable. Starting from July, 1 to July, 7

TABLE 9.1 Production Data For Wells of the Usinsk Oil Field After Stimulation

Pump model	Well no	Production restart	Baseline oil production t/d	Number of days	Average oil production t/d	Incremental oil t	Year
EVNT-25–1500	6010	27.05.2012	3.6	547	10.2	5566	2012–2013
	600	04.05.2012	Was taken as zero	81	5.0	405	2012
	1283	15.05.2012	4.5	225	9.6	2160	2012
	7169	12.12.2012	4.5	194.9	8.4	1637	2012–2013
	8198	14.01.2012	0.8	307.7	9.1	2800:	2012
	Total					12568	

just before treatment, the layer fluid was pumped from the well. The fluid consisted of water (99.999%) and oil film (< 0.001%). On July, 8 after injecting saltpeter solutions (about 35 tons) and sodium nitrite (about 12 tons) into the oil well both oil wells produced oil with the industrial ratio of 30% of oil and 70% of water. Although the production performance proved profitable solely from oil well No. 8 the customer company Viscos Energy put an advertisement mentioning the successful oil recovery at an abandoned field with the application of the Russian technology. The difference in the oil recovery rate at oil well No. 8 and oil well No. 10 is explained by the difference in their location. Oil well No. 8 is located in the center of the field while oil well No. 10 is situated on its border.

9.3   MAIN RESULTS AND PROSPECTS OF THE BM TECHNOLOGY

1. The modern BM technology developed by the authors since 1997 at present differs from all the rest of similar technologies in the heat emission optimization in the course of reagent injection. For this reason the technology ensures a much more thorough elimination of the skin layer around the producing oil wells. Accumulated during tens of years the skin layer restricts the profitable extraction to less than a half of the deposited oil.

2. Defined by the market competition if compared to other technologies in respect to their cost the modern BM technology holds the second place and as related to the rate of oil extraction increase trails only the hydro fracturing technology (USA and Canada). Starting from the first patent (RU 2126084 97111229/031997.06.30) and up to the latest ones of 2010–2012 (Patent WO 2010/043239, April 22, 2010 и Patent WO 2012/025150, March 1, 2012) the efficiency increase of the technology application during the last 16 years is clearly seen from the increase in treatment cost of one oil well:

   - $\approx$ \$5 000 (1997, Customer – the owner of the Vostochno-Poltavskoye oil field, Ukraine);
   - $\approx$ \$40 000 (2010–2011, Customer – LLC LUKOIL-KOMI);
   - $\approx$ \$60 000 (2013, Customer – Viscos Energy Holding ltd. USA).

3. We tend to interpret the facts described in points 1–2 as a turning point from over the century accumulation of nonrecoverable oil deposits to their profitable extraction. It should be noted that at present the amount of hydrocarbons still deposited underground is far greater than the deposits at the oil fields under current treatment. For this reason discovering the efficient technology for nonrecoverable oil extraction equals discovering a new large-scale oil field.

4. The oil recovery revival of the previously unprofitable Russian oil fields by means of the BM technology can be regarded a new direction of field thermochemistry development, which can promote a radical improvement of Russian economy as a great energy extracting power.

KEYWORDS

- **fields**
- **oil**
- **production**
- **restoration**
- **technology**

REFERENCES

1. Alexandrov, E., Kuznetsov, N. Science and Research journal (in Rus.), 2007, № 4, pp.113–127.

2. Aleksandrov, E. N., Aleksandrov, P. E., Kuznetsov, N. M., Petrov, A. L., Lidzhi-Goryaev, V. Yu. Advances in Sustainable Petroleum Engineering Science, Vol. 6, No 1, 2014, pp. 93–108.

3. Merzhanov, A., Lunin, V., Aleksandrov, E., Lemenovsky, D., Petrov, A., Lidgi-Goryaev, V. Science and technology applied to industry journal (in Rus.), 2010.vol. 2. pp. 1–6.

4. Aleksandrov, E., Varfolomeev, S., Lidgi-Goryaev, V., Petrov, A. Research and Development journal (in Rus.), 2012, № 158, pp. 14–15.

5. Aleksandrov, E., Aleksandrov, P., Kuznetsov, N., Lunin, V., Lemenovsky, D., Rafikov, R., Chertenkov, M., Shiryaev, P., Petrov, A., Lidgi-Goryaev, V. Petrochemistry journal (in Rus.), 2013, vol. 53, № 4, pp. 312–320.

6. Russian Patent. Gas evolving oil viscosity diminishing compositions for stimulating the productive layer of an oil reservoir, Aleksandrov, E., Lemenovsky, D., Koller, Z. Patent WO2010/025150 A1.ww

CHAPTER 10

ALKALI METAL METAPHOSPHATES AS INORGANIC POLYMERS

B. S. ALIKHADZHIEVA

Chechen State Pedagogical Institute, 33, Kievskaj str., 364037, Grozny, Russia, E-mail: belkaas52@list.ru

CONTENTS

ABSTRACT

Metaphosphates alkali metals of inorganic polymers, which in the molten state retain polymer structure as electrolytes with high ionic conductivity and have a substantially viscosity.

Due to introduction to polytungstate systems the phosphates alkali metals the alkaline tungsten bronzes were obtained, which are used as catalysts.

Experimentally state diagrams were investigated involving metaphosphates of sodium (potassium) with tungsten alkali metals (Na, K).

The systems metaphosphate-tungstate sodium and potassium can be used to create the crystallization resistant vitreous semiconductor materials.

10.1 INTRODUCTION

The study of physicochemical processes in the melts of multicomponent systems is the basis for developing new materials and inventing technological processes. Composite materials containing transition metal compounds are widely used in modern technology. The study of phase balance and chemical reactions in the oxide-saline systems can also solve the problem of synthesis of new stoichiometric – and nonstoichiometric compounds with a wide range of physical and chemical properties which are perspective for application in modern technology [1].

High requirements which are currently set for the quality of oxide and oxide-saline materials-powders, ceramics, films and fibers have led to the development of fundamentally new methods for their preparation. One of the major problems of modern inorganic chemistry is to obtain new polymeric and composite materials with desired properties.

Tungsten-containing complex oxide phases with unique physical and chemical properties are perspective inorganic materials for creating new techniques and technologies. Therefore, considerable attention is paid to the improvement and development of theoretical and practical bases for their preparation [2].

Currently, one of the main methods for their preparation is the electrolysis of molten oxide and tungsten salts with using electrolytes of lower fuse components as solvents. However, in this case coarse-crystalline deposits are formed, but for using, for example, oxide-tungsten bronzes (OTB) as catalysts high dispersion powders are required [3].

One of the possible solutions of the problem is the selection of high-viscous melt-electrolytes, which in particular, is achieved by loading alkali metal phosphate into politungsten systems [4].

Depending on the amount of the embedded metal their physicochemical properties change: color, texture, electrical conductivity, etc. The crystal lattice of oxide-tungsten bronzes is constructed of tungsten trioxide octahedrons interconnected in various ways.

Between the octahedrons there are voids where one ion of the size equal or less than the oxygen one can locate without any lattice distortion. Depending on how the tungsten trioxide octahedrons are connected with each other, and what kind of voids are formed the structures of particular crystallographic symmetry can be formed. In particular, for oxide-tungsten bronzes cubic, tetragonal, hexagonal, orthorhombic, monoclinic structures are known [5].

Wide range of oxide-tungsten bronzes compositions opens the possibility to vary their valuable physicochemical properties. The most explored of all the oxide-tungsten bronzes are alkali tungsten bronzes. The study of acid based properties of melts in the system Na_2WO_4-$NaPO_3$ also showed that potential of platinum-oxygen electrode, placed in the studied melts shifted to the positive area with an increase of sodium met phosphate concentration in them [6]. It is also associated with the anionic polymerization of tungsten groups and thus, with decreasing the activity of oxygen ions:

$$2WO_4^{2-}+PO_3^- \leftrightarrow PO_4^- +W_2O_7^2$$

$$PO_3^- + O_2 \leftrightarrow PO_4^{3-} \qquad (9)$$

Anion PO_3^-, being a strong acceptor of oxygen ions, when infused into the tungstate melts shifts their action [7] to the right and causes polymerization of tungstate ions, just as it happens in these melts when infusing the tungsten oxide into them (VI).

Thus, the infusing alkali metal metaphosphate into the tungstate melt leads to an increase of ditungstate-ions $W_2O_7^2$ concentration in the melt. These ions "deliver" particles of WO_3 into the melt, that is, the melt is a source of tungsten oxide (VI). Therefore, it is possible to synthesize by chemical means powders of oxide-tungsten bronzes (OTB) in tungstate phosphate system melts, but in the absence of tungsten oxide (VI).

Phosphorus has the ability to form different polycompounds with wide range of properties.

The most complete information about the interaction of condensed phosphates can be obtained by studying the phase diagrams of complex physicochemical methods of analysis, including DTA, RPA. These data give an idea of the state, the properties of the solid and liquid phases, areas of glass formation. This approach avoids unnecessary wastage of time and materials selecting practical and important compositions.

It seemed interesting to study state diagrams involving inorganic polymers – sodium metaphosphate (potassium) with alkali metals wolframites (Na, K), which have-not still been well explored properly [2].

Alkali metal metaphosphates are part of the class of inorganic poly-mers, which retain the polymer structure in the molten state, being elec-trolytes of high ionic conductivity. An important advantage of alkali metal metaphosphates of polymeric structure is the ability to dissolve many metal oxides.

Like many other inorganic polymers, alkali metal metaphosphates in the molten form have considerable viscosity, which is due to pecu-liarities of the polymeric structure of these compounds. According to research [3], the degree of alkali metal metaphosphates polymerization increases in the series: $LiPO_3$, KPO_3. The viscosity, according to research [4], decreases while the radius of the cation increases. When interacting with some oxides, for example, V_2O_5, metaphosphates form complexions. The formation of complexions itself explains the increase in density and viscosity of melts. Alkali metal metaphosphates form glass with oxides of some metals.

Alkali metal metaphosphates, according to researches [5, 6] are ther-mally stable up to temperatures of 100–150°C above their melting point.

10.2 EXPERIMENTAL PART

We obtained the initial condensed met phosphates by solid-phase reactions of saline (sodium carbonate, potassium) with orthophosphoric acid:

$$H_3PO_4 + Na_2CO_3 = 2NaH_2PO_4 + CO_2 + H_2O$$

$$NaH_2PO_4 = NaPO_3 + H_2O$$

$$H_3PO_4 + K_2CO_3 = 2KH_2PO_4 + CO_2 + H_2O$$

$$KH_2PO_4 = KPO_3 + H_2O$$

The phase identification has been performed by X-ray analysis. The results of studies of the interaction of sodium metaphosphate with sodium tung-state by complex of physicochemical analysis methods showed that in this system $NaPO_3 - Na_2WO_4$ form two distektiks (D_1) with the melting point 680°C, (D_2) with the melting point 570°C, four eutectics e_1, e_2, e_3, e_4 with the melting points 612C°, 580°C, 540°C, 500°C, respectively.

The system $KPO_3 - K_2WO_4$ has been studied too. Its components form congruent compounds $K_2WO_4 \cdot 2KPO_3$ (D_3) with the melting point 646°C and the eutectic points correspond to the compositions 55 (620°C) and 75-mole% (618°C) KPO_3.

10.3 RESULTS AND DISCUSSIONS

From the experimentally obtained data has been constructed the melting diagram of these systems and the fields of crystallizing phases as well as the nature of nonvariant points has been delineated. The metaphosphates of monovalent metals show the greatest ability to form doubles a line. The inclination of metaphosphates of one- and divalent metals to form limited solid solutions should be mentioned too.

10.4 CONCLUSION

The study of the crystallization ability of these systems in metaphosphate-tungstate sodium and potassium melts is of great interest for creating of crystallization resistant glassy semiconductor materials.

KEYWORDS

- electrochemical synthesis
- eutectic
- oxide tungsten bronzes
- phase transitions
- two-component systems

REFERENCES

1. Alikhadzhieva, B. S. Thesis of PhD, *Alkali Metal Metaphosphates* (in Rus.) As Inorganic Polymers, Dagestan State Pedagogical University. Makhachkala City. 2011. 118 pages.

2. Proceeding of IV All-Russian scientific Bergmanovsky conference (in Rus.), Dagestan State Pedagogivcal University Publishing House. Makhachkala City, 13–14 April 13–14. 2012. 250 pages.
3. Mandelkorn, L. Nonstoichiometric compounds. Chemistry Publishing House (in Rus.), Moscow, 1971. p. 607 pp.
4. Scheibler, C. Uber wolframoxyd verbindungen. J. Ract. Chem. (in German) 1861, B.183. p. 320–324.
5. Ozerov, R. P. Tungsten and vanadium bronzes. Herald of Academy of sciences of USSR (in Rus) 1954. V. 99, №1. p. 93–95.
6. Spitsin, V. I. Oxide bronzes. M. Nauka (Science Publishing House) in Rus., Moscow 1982. 350 pp.
7. Kollong, R. Nonstoichiometry. Mir Publishing House (in Rus), Moscow, 1974. p. 287.

CHAPTER 11

INORGANIC POLYMERS SEMICONDUCTOR MATERIALS

B. S. ALIKHADZHIEVA

Chechen State Pedagogical Institute, 33, Kievskaj str., 364037, Grozny, Russia; E-mail: belkaas52@list.ru

CONTENTS

ABSTRACT

Metaphosphates alkali metals of inorganic polymers, which in the molten state retain polymer structure as electrolytes with high ionic conductivity and have a substantially viscosity.

Due to introduction to polytungstate systems the phosphates alkali metals the alkaline tungsten bronzes were obtained, which are used as catalysts.

Experimentally state diagrams were investigated involving metaphosphates of sodium (potassium) with tungsten alkali metals (Na, K).

The systems metaphosphate-tungstate sodium and potassium can be used to create the crystallization resistant vitreous semiconductor materials.

11.1 INTRODUCTION

One of the most important problems of modern inorganic chemistry is the obtaining of new polymer and composite materials with predetermined properties. In nowadays, there are high requirements to the quality of oxide and oxide-salt materials such as powders, ceramics, pellicles and fibers have led to the development of principally new methods of obtaining them [1].

Complex oxide tungsten containing phases, with unique physical and chemical properties, are promising inorganic materials for creation of new engineering and technologies. Therefore, considerable attention is paid to the improvement and development of theoretical and practical bases of obtaining them [2].

Currently, one of the main methods of obtaining them is electrolysis of oxide melts and tungsten salts with implementation as thinners the electrolytes more low-melting component. However, there is the formation of macrocrystalline precipitation, but for use, for example, oxide tungsten bronzes (OGB) as catalysts are required powders of high dispersity [3].

One of the possible solution is to set a high-viscosity melts such as electrolytes, which in particular achieved by the introduction in polytungstate system phosphates alkaline metals [4].

Physicochemical properties vary depending on the amount of metal embedded: color, texture, conductivity, etc. Crystal lattice of the oxide tungsten bronzes is built from octahedra three-tungsten oxide, interconnected in a variety of ways. There are some voids between octahedra where an ion can fit with no distortion of the lattice; the size is equal to or less oxygen. Depending on how connected octahedra of tungsten trioxide with each other, and which kind of voids with the form, we can obtain the structures of one or another crystal or a crystallographic symmetry. In particular, for oxide tungsten bronzes currently known cubic, tetragonal, hexagonal, orthorhombic, monoclinic structure [5].

A wide range of compositions oxide tungsten bronzes opens the possibility to vary with valuable physical and chemical properties. The most studied of all the oxide tungsten bronzes are alkaline tungsten bronze. Research of acid-base properties of melts in the system Na_2WO_4 - $NaPO_3$ also showed that the potential platinum oxigen electrode, immersed in the study melts, is moving to the positive area while increasing them in concentration metaphosphate sodium [6], which obviously, is also associated with the anionic polymerization groups of tungsten, consequently, reduced activity of oxygen ions:

$$2WO_4^{2-} + PO_3^- \leftrightarrow PO_4^- + W_2O_7^2$$

$$PO_3^- + O_2 \leftrightarrow PO_4^{3-}$$

Anion PO_3^- as strong acceptor ions of oxygen, with the injection of tungstate melts shifts the reaction to the right, and induces the polymerization tungstate – ions as well as in these melts with the injection the oxide tungsten (VI) [7].

Thereby, injection to tungsten melts of metaphosphate of alkali metal leads to an increase in the melt concentration double tungstate ion $W_2O_7^{2-}$ that "supply" in the melt WO_3 particles, that is, the melt is source of tungsten (VI). Therefore, in melts tungsten phosphate systems in the chemical way it is possible to synthesize in oxide tungsten bronzes powders (OGB), but in the absence of tungsten (VI). Phosphorus has the ability to form the various polymer compounds with the range of properties.

The most complete information about the interaction of condensed phosphates is available by examining the phase diagram of the complex of methods of physical-chemical analysis, including the DTA, the ARF. These data give an idea about the state, the properties of the solid and liquid phases, the areas of glass formation. Such approach allows to avoid unnecessary losses of substances and time when selecting a practical important compositions.

It seemed interesting to study of phase diagrams with the participation of inorganic polymers-metaphosphate sodium (potassium) with tungstate of alkali metals (Na, K), which still not enough studied [2].

Metaphosphates of alkali metals belong to the class of inorganic polymers, which in the molten state retain the polymer structure as electrolytes with high ionic conductivity. One of the most important advantages of metaphosphates of alkali metals with polymeric structure is the ability to dissolve the oxides of many metals. Like many other inorganic polymers, metaphosphates of alkali metals in molten form have considerable viscosity, which is conditioned by the peculiarities of polymer structure of these compounds [2].

According to the data of [3] the degree of polymerization of metaphosphates of alkaline metal minerals increased in the range: $LiPO_3$, KPO_3. Viscosity according to the data of [4] with increasing radius cation decreases. At interaction with some oxides, for example, V_2O_5, the metaphosphates form complex ions. The increase in density and viscosity in melts is explained by formation of complex ions. With oxides of some metals, the metaphosphates of alkaline metals are forming the glass.

Metaphosphate of alkaline metals, according to data of Refs. [5, 6] thermally resistant to temperatures of 100–150°C above their melting point.

11.2 EXPERIMENTAL PART

Initial condensed metaphosphates were received by us with the method of solid-phase reactions of interaction of salt (soda, a potassium) with orthophosphoric acid:

$$H_3PO_4 + Na_2CO_3 = 2NaH_2PO_4 + CO_2 + H_2O$$

$$NaH_2PO_4 = NaPO_3 + H_2O$$

$$H_3PO_4 + K_2CO_3 = 2KH_2PO_4 + CO_2 + H_2O$$

$$KH_2PO_{4-} = KPO_3 + H_2O$$

Identification of phases was carried out by the X-ray phase analysis. The results of the research of interaction of metaphosphate of sodium with sodium tungstate and complex of methods of the physical and chemical analysis allowed to establish that in this $NaPO_3 - Na_2WO_4$ system form two dystectics (D1) with melting point 680°C, (D2) with melting point 570°C, forms four eutectics e_1, e_2, e_3, e_4 with melting points 612°C, 580°C, 540°C, 500°C, respectively.

The $KPO_3 - K_2WO_4$ system is also studied. Its components form congruent connections of $K_2WO_4 \cdot 2KPO_3$ with melting point 646°C, and to the eutectic points there correspond structures 55 (620°C) and 75 mol. % (618°C) KPO_3.

11.3 RESULTS AND DISCUSSIONS

The chart of fusibility of data of systems was constructed of experimentally obtained data and fields of crystallizing phases and character of non-variant points are outlined. The greatest ability to formation of double salts metaphosphates of monovalent metals is differ. It should be noted the tendency of metaphosphates mono and divalent metals to formation of restricted solid solutions.

11.4 CONCLUSION

Studying of crystallization ability in melts metaphosphate – sodium tungstate and a potassium of data of systems is of interest to creation of crystallization and steady vitreous semiconductor materials.

KEYWORDS

- **electrochemical synthesis**
- **eutectic**
- **oxide tungsten bronzes**
- **phase transitions**
- **two-component systems**

REFERENCES

1. Alikhadzhieva, B. S. Thesis of PhD, *Alkali Metal Metaphosphates* (in Rus.) As Inorganic Polymers, Dagestan State Pedagogical University. Makhachkala City. 2011. 118 pages.
2. Proceeding of IV All-Russian scientific Bergmanovsky conference (in Rus.), Dagestan State Pedagogivcal University Publishing House. Makhachkala City, 13–14 April 13–14. 2012. 250 pages.

3. Mandelkorn, L. Nonstoichiometric compounds. Chemistry Publishing House (in Rus.), Moscow, 1971. p. 607 pp.

4. Scheibler, C. Uber wolframoxyd verbindungen. J. Ract. Chem. (in German) 1861, B.183. p. 320–324.

5. Ozerov, R. P. Tungsten and vanadium bronzes. Herald of Academy of sciences of USSR (in Rus) 1954. V. 99, №1. p. 93–95.

6. Spitsin, V. I. Oxide bronzes. M. Nauka (Science Publishing House) in Rus., Moscow 1982. 350 pp.

7. Kollong, R. Nonstoichiometry. Mir Publishing House (in Rus), Moscow, 1974. p. 287.

CHAPTER 12

INCOMBUSTIBLE POLYETHERFORMALTEREPHTALOYL-DI(N-OXIBENZOAT)

Z. S. KHASBULATOVA

Chechen State Teacher Institute, 33, Kievskaj str., 364037, Grozny, Russia; E-mail: hasbulatova@list.ru

CONTENTS

ABSTRACT

The chapter describes synthesis of new polyether, received from diane and phenolphthalein oligoformals and diacylhloride terephtaloyl-di(n-oxiben-zoat) in conditions of acceptor-cathalitical polycondensation, and shows data of synthesized polyetherfromals incombustibility.

12.1 INTRODUCTION

Synthesis and properties of polyethers, based on chemically active bifunctional oligomers, capable to come into the polycondensation reaction are of special academic and practical interests.

12.2 EXPERIMENTAL PART

Oligoformal synthesis [1] is conducted by method of high-temperature polycondensation by aprotic dipolar thinner of dimethylsulfoxid (DMSO) in noble gas (nitrogen) atmosphere. Oligoformals with different polycondensation level synthesized by interaction diphenol (diane or phenolphthalein) dihalidmethylene overflow with molecular ratio – 2:1, 6:5, 11:10, 21:20 by following scheme:

$$nHO-Ar-OH + 2nNaOH \xrightarrow[-2nH_2O]{} nNaO-Ar-ONa$$

$$(n+1)NaO-Ar-ONa +_nCH_2Cl_2 \xrightarrow[-2_nNaCl]{} NaO-Ar-(O-CH_2-O-Ar)_nONa \xrightarrow[-NaOOC-COONa]{HOOC-COOH}$$

$$\xrightarrow{\hspace{2cm}} HO-Ar-(O-CH_2-O-Ar)_n OH$$

где Ar = [structure] или [structure]

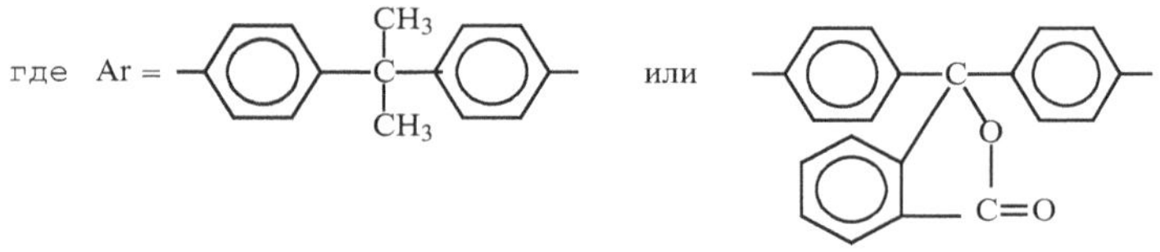

As of acidic component in the process of polyethers synthesis new comonomer had been used (i.e., diacylhloride terephtaloyl-di(n-oxibenzoat)). That diacylchloride [2] produced in two phases. The first phase is where terephtaloyl-di(n-oxibenzoat) interreacted with n-oxibenzoat and terephtaloyl chloride, and the second is chloration of produced acid by thionil chloride (SOCl$_2$).

The most exit end of diacylhloride terephtaloyl-di(N-oxibenzonat) comes with 10:1 ratio rating of SOCl$_2$ and terephtaloyl-di(n-oxibenzoat).

Polyethers' synthesis had been conducted in conditions of acceptor-cathalitical polycondensation [3]. Optimal conditions for polyether synthesis are: thinner – 1,2-dicholrethain; reaction temperature – 20°C; reaction duration – 1.5 h; thriethylamine quantity – double overflow to oligomers – (2:1); oligomers optimal concentration – 0.3 mol/L.

The actual physical, mechanical and chemical methods of analysis for synthesized polyethers are thermal stability of polymers, dielectric properties and stress-stain properties, reagent resistance, solvability in different thinners and incombustibility complexes had been examined [4]. Further the detailed view of synthesized polymers incombustibility research will be described.

The biggest interest in the field of synthesis and polymers research is paid to the technology of flameproof material production, since the polymer ware is used into the different spheres of life necessities.

Polymer incombustibility test run methods are various. Frequently the same polymer can be referred to noncombustible, self-extinguishing or even combustible polymer type, depends on used test-run method. For complete combustibility characteristics of polymeric material the combustion, smolder, self-ignition and self-heating temperature, smoke emission capability, glowing melt setting up and toxic level of degradation product should be determined [5].

Therefore, incombustibility test runs of aromatic polyetherformals, based on diane and phenolphthalein olygoformals, synthesized, as it was said earlier, by acceptor-cathalitical polycondensation method, main chain of which includes new comonomers, consisted of terephtaloyl-di(n-oxibenzoat) parts, had been taken. Received polyetherformals incombustibility evaluation based on time of sample self-extinguishing from the moment of ejection from blowpipe, and also by oxygen index (OI), which is characterizing the lowest content of oxygen leading to inflammation of polymer.

The lower OI the easier the samples of polymer inflame and the compounds with less than 21 oxygen index are burning in air atmosphere (if atmosphere contains 21% of oxygen) [6, 7].

The burning rate of polyether samples based on diane oligoformals in the process of ejection from blowpipe doesn't exceed 2 seconds, it shows self-extinguishing properties of polyethers. Table shows that synthesized polyetherfromals based on phenolphthalein oligoformals with parts of terephtaloyl-di(n-oxibenzoat) have high incombustibility, in contrast to diane oligoformals. It is possibly the result of card group inputted in poly-ether structure.

However, polyetherformal by incombustibility yields to halogenic polymers – polyarylate, polyarylsulphones and others.

12.3   RESULTS AND DISCUSSIONS

Polyethers incombustibility research results confirmed that OI value of polyethers based on diane oligoformals lies in interval 34.0–35.0% and based on phenol-phtalein formal in interval – 35.5–36.5% (Table 12.1).

12.4   CONCLUSION

The research has shown that synthesized polyetherformalterephtaloyl-di (n-oxibenzoat) based on diane and phenolphthalein oligoformal doesn't burn in air atmosphere and is self-extinguishing polymeric material. It means that the polymers can be used in incombustible constructional and filmy materials, for those spheres of technique, where incombustibility of polymers is high demanded.

TABLE 12.1   Polyetherformalterephtaloyl-di(n-oxibenzoat) Incombustibility*

#	Starter compounds**		Oxygen index %
1	OF-1D	DCHTOB	34.0
2	OF-5D	DCHTOB	34.0
3	OF-10D	DCHTOB	34.5
4	OF-20D	DCHTOB	35.0
5	OF-1P	DCHTOB	35.5
6	OF-5P	DCHTOB	35.0
7	OF-10P	DCHTOB	35.5
8	OF-20P	DCHTOB	36.5

* Dichloride terephtaloyl-di(n-oxibenzoat) is used as an acid component;

**Digits in oligomer are average value of *n* polycondensation level; D is diane; P is phenolphthalein.

KEYWORDS

- oligoformals
- polyetherformals
- diacylhloride terephtaloyl-di(n-oxibenzoat)
- oxygen index
- incombustibility

REFERENCES

1. Khasbulatova, Z. S., Asueva, L. A., Nasurova, M. A., Shustov, G. B., Kharaeva, R. A., Ashibokova, O. R. "Synthesis and properties of aromatic plyoethers. International conference dedicated to 145 years anniversary of the structure of organic compounds of A.M. Butlerov and 100 years anniversary of memory about F.F. Belstain (in Rus)" Nauka Publishing House S. Petersburg, 2006. 793–794.
2. Shepelevskiy, A. Y., Savinova, A. A., Skorokodov, T. E. S. S., Terephtaloyl-di (n-oxibenzoat) or its dichloride acid as monomers for synthesis of thermally stable polymers (in Rus): Soviet Certificate 792834 (USSR), Parts C07C63/06, C08K5/09, B. I., 1982. p. 2.
3. Khasbulatova, Z. C. Polyethers based on derivatives of n-oxibenzoat and phtalein acids (in Rus). Thesis Doctor of Chemistry – Kh.M. Berbekov Kabardino-Balkarian State University Nalchik City. 2010. 306 p.
4. Nasurova, M. A. Polyethers based on oligoformals and terephtaloyl-di(n-oxibenzoat) (In Rus). Kh.M. Berbekov, PhD. Kabardino-Balkarian State University Nalchik City. 2010. 154 p.
5. Tishenko, A. M., Popov, L. K., Gorbunov, B. N. Methods of bromophenol synthesis and its derivatives as antipyrene (flame retarder) for polymeric materials. (in Rus) Science Publishing House Moscow. 1982. 250 pp.
6. Aseeva, R. M., Zaikov, G. I. Polymeric materials combustion (in Rus) Science Publishing House Moscow. 1981. 281 p.
7. Khasbulatova, Z. S., Asueva, L. A., Nasurova, M. A., Shustov, G. B., Mikitaev, A. K. Aromatic polyformals. Plastic compounds journal (in Rus). 2008. № 8, 31–34.

CHAPTER 13

POLYESTERS AND THEIR APPLICATION

Z. S. KHASBULATOVA

Chechen State Teacher Institute, 33, Kievskaj str., 364037, Grozny, Russia; E-mail: hasbulatova@list.ru

CONTENTS

ABSTRACT

Aromatic polyesters are polycondensation organic compounds containing in their macromolecule ester groups, ether linkages, and aromatic fragments in various combinations.

13.1   INTRODUCTION

Aromatic polyesters are thermally stable polymers. They are thermoplastic products suitable for producing products and materials by the method so forming from solutions and melts.

Aromatic copolyesters are mainly used for producing plastic and films. They can also be used as varnishes, fibrous binders for synthetic paper, membranes, hollow fibers, as additives and intermediates for obtaining materials based on other polymers.

Many aromatic polyesters-based products are produced in industry. The world production of polyesters in 2008 as compared to 2004 increased from 38 to 50 million tons, that is, 32%. Among various polyesters there are polymers belonging to the class of constructional plastics.

Currently, technological progress in many industries, especially in mechanical and instrument engineering is determined precisely by using of engineering plastics. Such performance properties of polymers as strength, thermal stability, electrical insulation and antifriction properties, optical transparency, etc. stipulate their application instead of ferrous and nonferrous metals, alloys, wood, ceramics, glass [1]. A ton of polymer replaces 5–6 tons of ferrous and nonferrous metals and 3–3.5 tons of wood, and labor savings can reach up to 800 man-hours per a ton of polymers. Of the total amount of polymers used in engineering about 50% is consumed in electrotechnology and electronics. In electrotechnology polymers are used in 80% of production, and in instrument engineering – up to 95%.

The application of constructional plastics allows creating a quite new technology of manufacturing components, machine assemblies, devices, and that provides high economic efficiency. Constructional polymers are well processed by modern methods casting and extrusion into products working under alternating loads at temperatures of 100–200°C.

Modern chemical industry has developed constructional thermoplastic materials that reduce material consumption and weight of machines, appliances, machinery, energy consumption and labor-intensiveness in production and operation, and also ensure longer service life.

Currently, radio electronics, electrotechnical, aviation, shipbuilding, automotive and other industries cannot thrive without modern advanced polymers such as polyarylates, polyethersulphones, polyetherketones, polyethersulphon ketones, etc., which are promising structural materials. The domestic industry alone uses more than 50 types of plastics, which include more than 850 brands and different modifications [2]. As a result of this, the share of production of mechanical engineering and some other branches of industry, where plastics are used, increased from 32–35% in 1960 to 85–90% in 1990.

In 1975 the share of engineering (constructional) plastics in the total world plastics production was only 5% [3]. But in 1981–1985 these plastics dominated in general plastic production [4].

Introduction of polymers not only has a positive effect on the state of the existing traditional industries. It provided technical progress in the missile and nuclear industries, aircraft, television, plastic surgery and medicine in general, etc. Currently the world production is more than 10 million tons.

A special place among polymeric materials is held by thermal and heat-resistant engineering (constructional) plastics. The need to create such polymers is dictated by the fact that the use of traditional industrial-use polymeric materials is limited by low heat-resistance, which, as a rule, does not exceed 103–150°C.

13.2 CONCLUSION

To obtain large-capacity heat-resistant plastics two classes of polymers are used: polyamides, aromatic polyarylates [5]. On the basis of these polymers, complementing their various properties, materials with high performance and, especially, heat-resistant constructional plastics can be obtained.

KEYWORDS

- aromatic polyesters
- ester groups
- ether linkages
- polymers

REFERENCES

1. Didrusco, G., Valvaszori, A. Prospettivenel campo beitecnopolimeri. – Tecnopolimeresine, 1982. № 5. 27–30.
2. Abramov, V. V., Zharkova, N. G. Baranova, N.S. Effectiveness and scope of structural plastics application in engineering (in Rus). Proceedings of the All-Union

Conference "The operational properties of structural polymeric materials."
Kabardino-Balkarian State University Publishing House Nalchik-city, 1984. p. 5.

3. Tebbat Tom. Engineering plastics Wonder materials of expensive polymer plautih-ings. Eur. Chem. News. 1975. vol. 27. p. 707.

4. Stoenesou, F. A. Tehnopolimeri. Rev. Chem. 1981. v.32, № 8. – C. 735–759.

5. Nevsky, L. B., Gerasimov, V. D., Naumov, V. S. Heat-resistant engineering plastics. Proceedings of the All-Union Conference "The operational properties of structural polymeric materials (in Rus)." Kabardino-Balkarian State University Publishing House. Nalchik, 1984. p.3.

CHAPTER 14

ENTROPIC CRITERIA IN ECONOMICS AND PHYSIC CHEMISTRY

N. G. PETROVA[1], G. A. KORABLEV,[2] and G. E. ZAIKOV[3]

[1]Specialist-Expert, Department of Information Security and Communications, Ministry of Informatization and Communications, Udmurt Republic, Vadima Sivkova St., 186, 426057 Izhevsk, Russian Federation, E-mail: biakaa@mail.ru

[2]Doctor of Chemical Sciences, Professor, Head of Department of Physics,Izhevsk State Agricultural Academy, Studencheskaya St., 11, 426069 Izhevsk, Russian Federation, E-mail: korablevga@mail.ru

[3]N.M. Emanuel Institute of Biochemical Physics of Russian Academy of Sciences, Moscow – 119334, Kosygin st., 4, Russian Federation, E-mail: chembio@sky.chph.ras.ru

CONTENTS

14.1   INTRODUCTION

The idea of entropy appeared based on the second law of thermodynamics as the criterion of the process directedness and system irregularity degree.

In actual processes in the isolated system the entropy growth is inevitable – disorder and chaos increase in the system, the quality of internal energy goes down.

Since the system degradation degree is not connected with the physical features of the systems, the entropy statistic concept can also have other applications and demonstrations.

"It is clear that out of the two systems completely different by their physical content, the entropy can be the same if their number of possible microstates corresponding to one macroparameter (whatever parameter it is) coincides. Therefore the idea of entropy can be used in various fields. The increasing self-organization of human society leads to the increase in entropy and disorder in the environment that is demonstrated, in particular, by a large number of disposal sites all over the earth" [1].

In this research we are trying to multidimensionally apply the concept of entropy in economics, in particular, to evaluate the efficiency of business structures.

The literature overview contains some formation principles of entropy and spatial-energy parameter (P-parameter) concepts, substantiates the possibilities of these criteria transformation for other natural and economic processes, gives the examples of their multidimensional manifestations.

The method of obtaining optimal and critical values of entropy for specific practically important situations can be of special interest.

14.2 THERMOPHYSICAL AND STATISTIC ENTROPIES

The idea of adduced quantity of heat is applied with the mathematical arrangement of the entropy thermophysical concept.

The adduced quantity of heat in isothermal process is the relation between the quantity of heat Q obtained by the system and temperature T of heat-release body.

The entropy is the function S of the system state whose differential in the elementary reversible process equals the relation between the infinitely little quantity of heat transferred to the system and its absolute temperature:

$$\Delta S = (\Delta Q)/T$$

Using such thermophysical definition we can calculate only the difference between entropies. The entropy itself can only be found with the accuracy to the constant summand (integration constant).

In statistic thermodynamics the entropy of the isolated and equilibrious system equals the logarithm of the probability of its definite macrostate:

$$S = k \ln W \qquad (1)$$

where W – number of available states of the system or degree of the degradation of microstates; k – Boltzmann's constant.

Or: $$W = e^{S/k}$$

These correlations are general assertions of macroscopic character, they do not contain any references to the structure elements of the systems considered and they are completely independent from microscopic models [2].

Therefore the application and consideration of these laws can result in a large number of consequences, which are most fruitfully used in statistic thermodynamics.

In statistic thermodynamics the entropy is a function of the system state, which helps estimating the process directions and possible changes in them.

At any spontaneous changes in the isolated system the entropy always increases: $\Delta S > 0$.

The sense of the second law of thermodynamics comes down to the following:

The nature tends from the less probable states to more probable ones. Thus, the most probable is the uniform distribution of molecules through the entire volume. From the macrophysical point, these processes consist in equalizing the density, temperature, pressure and chemical potentials, and the main characteristic of the process is the thermodynamic probability – W.

At the same time, the thermodynamic probability equals the number of microstates corresponding to the given macrostate.

14.3   ON TWO PRINCIPLES OF ADDING ENERGY CHARACTERISTICS OF INTERACTIONS

The analysis of the kinetics of various physicochemical processes demonstrates that in many cases the reciprocals of velocities, kinetic or energy characteristics of the corresponding interactions are added up.

Here are some examples: ambipolar diffusion, total rate of topochemical reaction, change in the light velocity when transiting from vacuum into the given medium, effective permeability of biomembranes.

Also: It is known from the traditional mechanics that the relative motion of two particles with the interaction energy U(r) is the same as the motion of a material point with the reduced mass μ:

$$\frac{1}{\mu} = \frac{1}{m_1} + \frac{1}{m_2} \tag{2}$$

in the field of central force U(r), and the total translational motion – as the free motion of the material point with the mass:

$$m = m_1 + m_2 \tag{3}$$

Such situation can be also found in quantum mechanics" [3].

The problem of two-particle interactions flowing by the bond line was solved in the time of Newton and Lagrange:

$$E = \frac{m_1 v_1^2}{2} + \frac{m_2 v_2^2}{2} + U\left(\overline{r_2} - \overline{r_1}\right) \tag{4}$$

where E – system total energy, first and second components – kinetic energies of the particles, third component – potential energy between particles 1 and 2, vectors $\bar{r}_2$ and $\bar{r}_1$ characterize the distance between the particles in final and initial states.

For moving thermodynamic systems the first law of thermodynamics can be shown as follows [4]:

$$\delta E = d\left(U + \frac{mv^2}{2}\right) \pm \delta A \tag{5}$$

where: δE – amount of energy transferred to the system; component $d\left(U + \frac{mv^2}{2}\right)$ – characterizes changes in internal and kinetic energies of the system; $+\delta A$ – work performed by the system; $-\delta A$ – work performed on the system.

Since the work value numerically equals the change in the potential energy, then:

$$+\delta A = -\Delta U \tag{6}$$

and

$$-\delta A = +\Delta U \tag{7}$$

Probably not only the value of potential energy but its changes are important in thermodynamic and also in many other processes In the dynamics of interactions of moving particles. Therefore, by the analogy with Eq. (4) the following should be fulfilled for two-particle interactions:

$$\delta E = d\left(\frac{m_1 v_1^2}{2} + \frac{m_2 v_2^2}{2}\right) \pm \Delta U \tag{8}$$

Here

$$\Delta U = U_2 - U_1 \tag{9}$$

where U_2 and U_1 – potential energies of the system in final and initial states.

At the same time, the total energy (E) and kinetic energy $\left(\dfrac{mv^2}{2}\right)$ can be calculated from their zero value. In this case, only the last component is modified in the Eq. (4).

The character of the changes in the potential energy value (ΔU) has been analyzed by its index for different potential fields as given in Table 14.1.

From the table it is seen that the values of $-\Delta U$ and consequently $+\delta A$ (positive work) correspond to the interactions taking place by the potential gradient, and ΔU and $-\delta A$ (negative work) take place during the interactions against the potential gradient.

The solution of two-particle problem of the interaction of two material points with masses and obtained under the condition of no external forces available corresponds to the interactions taking place by the gradient, the positive work is performed by the system (similar to the attraction process in the gravitation field).

The solution for this equation through the reduced mass (μ) [5] is Lagrangian equation for the relative motion of the isolated system of two interacting material points with masses m_1 and m_2, in coordinate x it looks as follows:

$$\mu \cdot x'' = -\frac{\partial U}{\partial x}; \frac{1}{\mu} = \frac{1}{m_1} + \frac{1}{m_2}$$

Here: U – mutual potential energy of material points; μ – reduced mass. At the same time, $x'' = a$ (characteristic of system acceleration). For elementary regions of interactions Δx can be taken as follows:

$$\frac{\partial U}{\partial x} \approx \frac{\Delta U}{\Delta x}$$

That is:w

$$\mu a \Delta x = -\Delta U$$

Then:

$$\frac{1}{1/(a\Delta x)}\frac{1}{(1/m_1 + 1/m_2)} \approx -\Delta U; \quad \frac{1}{1/(m_1 a\Delta x) + 1/(m_2 a\Delta x)} \approx \Delta U$$

TABLE 14.1 Directedness of Interaction Processes

No	Systems	Potential field type	Process	U	r_2/r_1 $\left(x_2/x_1\right)$	U_2/U_1	Index ΔU	Index δA	Process directedness in the potential field
1	Opposite electric charges	Electrostatic	Attraction	$-k\dfrac{q_1 q_2}{r}$	$r_2 < r_1$	$U_2 > U_1$	$-$	$+$	By gradient
			Repulsion	$-k\dfrac{q_1 q_2}{r}$	$r_2 < r_1$	$U_2 < U_1$	$+$	$-$	Against gradient
2	same electric charges	Electrostatic	Attraction	$k\dfrac{q_1 q_2}{r}$	$r_2 < r_1$	$U_2 > U_1$	$+$	$-$	Against gradient
			Repulsion	$k\dfrac{q_1 q_2}{r}$	$r_2 < r_1$	$U_2 < U_1$	$-$	$+$	By gradient
3	Elementary masses and m_2	Gravitational	Attraction	$-\gamma\dfrac{m_1 m_2}{r}$	$r_2 < r_1$	$U_2 > U_1$	$-$	$+$	By gradient
			Repulsion	$-\gamma\dfrac{m_1 m_2}{r}$	$r_2 < r_1$	$U_2 < U_1$	$+$	$-$	Against gradient

TABLE 14.1　Continued

No	Systems	Potential field type	Process	U	r_2/r_1 $\left(x_2/x_1\right)$	U_2/U_1	Index ΔU	Index δA	Process directedness in the potential field
4	Spring deformation	Field of spring forces	Compression	$k\dfrac{\Delta x^2}{2}$	$x_2 < x_1$	$U_2 > U_1$	$+$	$-$	Against gradient
			Stretching	$k\dfrac{\Delta x^2}{2}$	$x_2 < x_1$	$U_2 > U_1$	$+$	$-$	Against gradient
5	Photoeffect	Electrostatic	Repulsion	$k\dfrac{q_1 q_2}{r}$	$r_2 < r_1$	$U_2 < U_1$	$-$	$+$	By gradient

Or:

$$\frac{1}{\Delta U} \approx \frac{1}{\Delta U_1} + \frac{1}{\Delta U_2} \qquad (10)$$

where ΔU_1 and ΔU_2 – potential energies of material points on the elementary region of interactions, ΔU – resulting (mutual) potential energy of these interactions.

Thus:

1. In systems in which the interaction takes place by the potential gradient (positive work), Lagrangian is performed and the resultant potential energy is found by the principle of adding the reciprocals of the corresponding energies of subsystems [6]. The reduced mass for the relative motion of the isolated system of two particles is calculated in the same way.

2. In systems in which the interaction takes place against the potential gradient (negative work), their masses and corresponding energies of subsystems are added algebraically (similar to Hamiltonian).

14.4 SPATIAL-ENERGY PARAMETER AND ITS ENTROPIC CHARACTERISTIC

"Electron with the mass m moving near the proton with the mass M is equivalent to the particle with the mass: $m_{\text{пр}} = \dfrac{mM}{m + M}$ " [7].

Therefore modifying the Eq. (10), we can assume that the energy of atom valence orbitals (responsible for interatomic interactions) can be calculated [6] by the principle of adding the reciprocals of some initial energy components based on the equations:

$$\frac{1}{q^2 / r_i} + \frac{1}{W_i n_i} = \frac{1}{P_M} \quad \text{or} \quad \frac{1}{P_0} = \frac{1}{q^2} + \frac{1}{(Wrn)_i}; \quad P_M = \frac{P_0}{r_i}$$

$$(11)-(12)-(13)$$

Here: W_i – orbital energy of electrons [8]; r_i – orbital radius of i orbital [9]; $q = Z^*/n^*$ – by [10, 11], n_i – number of electrons of the given orbital,

$Z*$ and $n*$ – nucleus effective charge and effective main quantum number, r – bond dimensional characteristics.

P_O is called a spatial-energy parameter (SEP), and P_E – effective P-parameter (effective SEP). Effective SEP has a physical sense of some averaged energy of valence orbitals in the atom and is the direct characteristic of electron density in the atom at the given distance from the nucleus (r_i).

During the formation of a solid solution and in other structural equilibrious-exchange interactions the same electron density is established in the contact spots of atoms-components. This process is accompanied by the electron density redistribution between valence zones of both particles and transition of some electrons from some external spheres into the neighboring ones.

Obviously, with the electron density proximity in free atoms-components the transfer processes between the boundary particle atoms are minimal that will favor a new structure formation. Thus, the task of evaluating the degree of such structural interactions often comes down to the comparative assessment of electron density of valence electrons in free atoms (on averaged orbitals) participating in the process by the following equations:

$$\alpha = \frac{P_0'/r_i' - P_0''/r_i''}{\left(P_0'/r_i' + P_0''/r_i''\right)/2} \cdot 100\% \quad \text{or} \quad \alpha = \frac{P_c' - P_c''}{P_c' + P_c''} \cdot 200\% \qquad (14), (15)$$

where P_c – structural parameter found by the following equation:

$$\frac{1}{P_c} = \frac{1}{N_1 P_9'} + \frac{1}{N_2 P_9''} + \dots \qquad (16)$$

here N_1 and N_2 – number of homogeneous atoms in subsystems.

Applying all the obtained data we construct the nomogram of structural interaction degree dependence (ρ) on coefficient α, the same for a wide range of structures (Figure 14.1).

This approach gives the possibility to evaluate the degree and direction of the structural interactions of phase formation, isomorphism and solubility processes in multiple systems, including molecular ones.

Such nomogram can be demonstrated as a linear logarithmic dependence:

$$\alpha = \beta \left(\ln \rho \right)^{-1} \qquad (17)$$

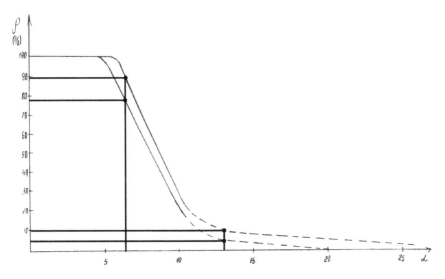

FIGURE 14.1   Nomogram of structural interaction degree dependence (ρ) on coefficient α.

where coefficient β – the constant value for the given class of structures. β can structurally change mainly within ± 5% from the average value. Thus, coefficient α is reversely proportional to the logarithm of the degree of structural interactions and, therefore, can be characterized as the entropy of spatial-energy interactions of atomic-molecular structures.

Actually the more is ρ, the more probable is the formation of stable ordered structures (e.g., the formation of solid solutions), that is, the less is the process entropy. But also the less is coefficient α. Thus, the relative difference of spatial-energy parameters of the interacting structures can be a quantitative characteristic of the process entropy:

$$\alpha = S = \beta \left(\ln \rho \right)^{-1} \tag{17a}$$

The Eq. (17) does not have the complete analogy with Boltzmann's equation as in this case not absolute but only relative values of the corresponding characteristics of the interacting structures are compared which can be expressed in percent. This refers not only to coefficient α but also to the comparative evaluation of structural interaction degree (ρ), for example – the percent of atom content of the given element in the solid solution relatively to the total number of atoms.

Therefore, in Eq. (17) coefficient $k = 1$.

14.5 ENTROPY OF BUSINESS STRUCTURE AGGREGATION

The main properties of business systems providing their economic advantages are: (i) effective competition, and (ii) maximal personal interest of each worker.

But on different economy concentration levels these *ab initio* features function and demonstrate themselves differently. Their greatest efficiency corresponds to small business – when the number of company staff is minimal, the personal interest is stronger and competitive struggle for survival is more active. With companies and productions growth, the number of staff goes up, the role of each person gradually decreases, the competition slackens as new opportunities for coordinated actions of various business structures arise. The quality of economic relations in business goes down, that is, the entropy increases. Such process is mostly vivid in monostructures at the largest enterprises of large business (syndicates and cartels).

The concept of thermodynamic probability as a number of microstates corresponding to the given macrostate can be modified as applicable to the processes of economic interactions that directly depend on the parameters of business structures.

A separate business structure can be taken as the system macrostate, and as the number of microstates – number of its workers (N) which is the number of the available most probable states of the given business structure. Thus, it is supposed that such number of workers of the business structure is the analog of thermodynamic probability as applicable to the processes of economic interactions in business.

Therefore it can be accepted that the total entropy of business quality consists of two entropies characterizing: (i) decrease in the competition efficiency (S1), and (ii) decrease in the personal interest of each worker (S2), i.e.: $S = S_1 + S_2$. S1 is proportional to the number of workers in the company: $S \sim$ N, and S2 has a complex dependence not only on the number of workers in the company but also on the efficiency of its management. It is inversely proportional to the personal interest of each worker. Therefore it can be accepted that $S_2 = \frac{1}{\gamma}$, where γ – coefficient of personal interest of each worker.

By analogy with Boltzmann's Eq. (1) we have:

$$S = \left(S_1 + S_2 \right) \sim \left[\ln N + \ln \left(\frac{1}{\gamma} \right) \right] \sim \ln \left(\frac{N}{\gamma} \right)$$

or

$$S = k \ln\left(\frac{N}{\gamma}\right)$$

where k – proportionality coefficient.

Here N shows how many times the given business structure is larger than the reference small business structure, at which N = 1, that is, this value does not have the name.

For nonthermodynamic systems we take k = 1. Therefore:

$$S = \ln\left(\frac{N}{\gamma}\right) \tag{18}$$

In Table 14.2 you can see the approximate calculations of business entropy by the Eq. (18) for three main levels of business: small, medium and large. At the same time, it is supposed that number N corresponds to some average value from the most probable magnitudes.

When calculating the coefficient of personal interest γ it is considered that it can change from 1 (one self-employed worker) to zero (0), if such worker is a deprived slave, and for larger companies it is accepted as $\gamma = 0.1 - 0.01$.

Despite of the rather approximate accuracy of such averaged calculations, we can make quite a reliable conclusion on the fact that business entropy, with the aggregation of its structures, sharply increases during

TABLE 14.2 Entropy Growth With the Business Increase

Structure parameters	Business		
	Small	Medium	Large
N_1–N_2	10–50	100–1000	10,000–100,000
γ	0.9–0.8	0.6–0.4	0.1–0.01
S	2.408–4.135	5.116–7.824	11.513–16.118
$\langle S \rangle$	3.271	6.470	13.816

the transition from the medium to large business as the quality of business processes decreases. The application of more accurate initial data allows obtaining specific values of business entropy, above which the process of economic relations can reach a critical level.

Comparing the nomogram (Figure 14.1) with the data from Table 14.2, we can see the additivity of business entropy values (S) with the values of the coefficient of spatial-energy interactions (α).

And the diversity of business systems is expressed in small and medium business. Therefore the optimal criteria of a more qualitative business are defined by the maximum value of their entropy: $S = 6.47$ (in relative units).

In live systems the entropy growth is compensated via the negative entropy (negoentropy), which is formed through the interaction with the environment. That is a live system is an open one. And business cannot be an isolated system for a long period without the exchange process and interactions with the environment. The role of the external system diminishing the increase in the business entropy must be fulfilled, for example, by the corresponding state and public structures functionally separated from business. Probably, the demonopolization of the largest economic structures carried out from the "top" in the evolutionary way can be the inevitable process here.

But the increase in the personal interest of each worker is defined not only by the parameters of business systems but it also depends on the overall arrangement of these processes by the employer. For instance, Ford managed to find such ways of work organization, which sharply increased the personal interest of all his employees.

In thermodynamics it is considered that the uncontrollable entropy growth results in the stop of any macrochanges in the systems, that is, to their death. Therefore, the search of methods of increasing the uncontrollable growth of the entropy in large business is topical. At the same time, the entropy critical figures mainly refer to large business. A simple cut-down of the number of its employees cannot give an actual result of entropy decrease. Thus, the decrease in the number of workers by 10% results in diminishing their entropy only by 0.6% and this is inevitably followed by the common negative unemployment phenomena.

Therefore, for such super-monostructures controlled neither by the state nor by the society the demonopolization without optimization (i.e., without decreasing the total number of employees) is more actual to diminish the business entropy.

14.6   ENTROPY OF AN ELEMENTARY BUSINESS STRUCTURE

In the process of a new business structure formation we go through seeking and recruiting the personnel, at the same time, the number of its personnel should correspond to the most probability of this process and for the given system is N_0. Here the key role is played by the probability of random values of this process. The similar picture is also characteristic for informative events: "It appears that for the information characteristics it is also possible to introduce the notion of entropy. In the information theory we introduce such a value called the random value entropy:

$$H = \sum_n P_n \log_2\left(P_n^{-1}\right) \tag{19}$$

Here: value H equals the number of binary digits required for difference (record) of the allowed values of the random value x [9]"; P_n – probability of the appearance of each given record of the random value.

Following the Eq. (19), the entropy of a random value is proportional to the total of probabilities and inversely proportional to the logarithm of their probabilities.

For the characteristic of continuous random value we use the function of probability distribution density:

$$y = f(x) = P(\Delta x_i)/\Delta x_i,$$

where $P(\Delta x_i)$ – probability of the random value ingress into the interval Δx_i of its values.

In its sense, the random value entropy is reversely proportional to this function: $S_0 \sim \dfrac{1}{y}$

or

$$S_0 \sim \Delta x_i \left[P(\Delta x)\right]^{-1} \tag{20}$$

Modifying the Eqs. (19) and (20) and transferring from the binary system to the decimal one, we can have:

$$S_0 = \Delta x_i \ln \left[P(\Delta x)_i \right]^{-1} \tag{21}$$

Let us consider the application of the Eq. (21) to the distribution of the probabilities of random processes during the staff formation of an elementary business structure regarding the most probable value of N_0.

Each interval of the random values relatively to N_0 can be more or less than its value and equals by the module:

$$\Delta N = |N_0 - N_i|$$

For the probability of $P(x)$ we take the value:

$$P(\Delta x_i) = \frac{N_0}{N_0 + \Delta N}$$

Then the Eq. (21) looks as follows:

$$S_0 = \Delta N_i \ln \left[\left(\frac{N_0}{N_0 + \Delta N_i} \right) \right]^{-1} \tag{22}$$

The calculations of some points of graphic dependence S_0 on ΔN by the Eq. (22) at $N_0 = 20$ result in Figure 14.2.

From the calculations and Figure 14.2 it is seen how the entropy goes up with the deviation of the amount of personnel N from the optimally acceptable value N_0. At the same time, not only those events at which $N < N_0$ are irrational, but also those, at which $N > N_0$. Thus, the nonoptimal increase in the bureaucratic apparatus, which seems to lighten the management work, actually results in such business entropy increase.

Thus, the given technique of entropy evaluation in a separate business structure allows establishing the norms of acceptable deviations from the most probable values of random processes in it.

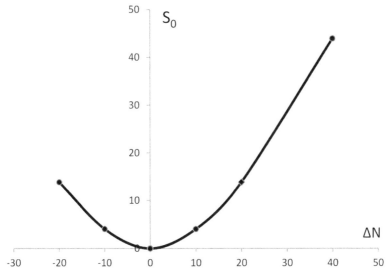

FIGURE 14.2 Entropy of random processes in the formation of the given business structure personnel.

14.7 ENTROPY OF RELATIVE MARGINAL PROFIT

Entrepreneurs try to find different ways to obtain the absolute profit: workday duration increase, labor utilization densification, labor productivity growth, etc. All these ways come to labor cost decrease. A profit obtained with the labor cost decrease is called a relative marginal profit. But there is another way to decrease the labor cost – direct reduction of a company product price – the way frequently applied under intensive competition.

A business structure profit can be considered as the main criterion defining the business quality, stability and concurrence of economic interactions in it. Therefore all the factors preventing such situation, especially if they are accumulated dynamically, define the profit entropy by their sense based on the equation:

$$S_M = \frac{1}{\Delta U} \tag{23}$$

where ΔU – relative marginal profit, S_M – its entropy.

Previously it was demonstrated that the comparison of process relative indexes allows taking coefficient k equaled to 1 in entropy equations that is assumed in Eq. (23).

In Refs. [12, 13] you can find specific calculations and comparison of the profit indexes of two segments (business structures) under the condition of distribution equal proportionality.

The example demonstrating the decrease in the labor cost between the segments under the condition of distribution equal proportionality is considered. First, the situation is designed in which both segments have similar indexes (option 1). Then we assume that the second segment, having diminished the prices, significantly increased its revenue, at the same time, raising the absolute marginal profit (option 2).

At the same time, two options of the situation are considered: without price decrease (option 1) and with price decrease in the second segment (option 2). From Table 14.3 it is seen that the absolute marginal profit of the first segment does not change, but it increases nearly by 1/3 in the second. The relative marginal profit does not change again in the first segment, but it decreases from 0.3800 to 0.2280 in the second, thus corresponding to entropy values 2.63 and 4.39, respectively, in Eq. (23).

According to nomogram (Figure 14.1), these figures are within 100% interaction, thus defining the business high quality. But nomogram 1 also

TABLE 14.3 Marginal Profit Calculations

	Option 1	Option 2
Revenue (W), including	800,000	1,400,000
– 1st segment revenue ($W1$)	400,000	400,000
– 2nd segment revenue ($W2$)	400,000	1,000,000
Variable costs (S), including	400,000	900,000
– 1st segment costs ($S1$)	200,000	200,000
– 2nd segment costs ($S2$)	200,000	700,000
Absolute marginal profit ($M = W - S$)	400,000	500,000
– 1st segment absolute marginal profit	200,000	200,000
– 2nd segment absolute marginal profit	200,000	300,000
Relative marginal profit ($Um^\wedge(1)$ and $Um^\wedge(2)$)	0.3800	0.2714
– 1st segment relative marginal profit	0.3800	0.3800
– 2nd segment relative marginal profit	0.3800	0.2280

defines the limiting values of ΔU above, which the business quality goes down. When $S_M = \alpha = 6$–7, the nomogram provides the values of the relative marginal profit decrease: $\Delta U_M = 0.167$–0.142.

Thus, the use of business entropy concept allows assessing the quality of business processes, in particular – to obtain the critical parameters of relative marginal profit.

14.8 ENTROPIC NOMOGRAM OF SURFACE-DIFFUSIVE PROCESSES

As an example, let us consider the process of carbonization and formation of nanostructures during the interaction of polyvinyl alcohol and metal phase in the form of copper oxides or chlorides in gels. At the first stage, small clusters of inorganic phase are formed surrounded by carbon containing phase. In this period, the main character of atomic-molecular interactions needs to be assessed via the relative difference of P-parameters calculated through the radii of copper ions and covalent radii of carbon atoms demonstrated in Table 14.4.

From Table 14.4 it is seen that, in this case, the coefficient $\alpha = 3.50$ corresponding to the complete structural interaction: $\rho = 100\%$. The process takes place only in the gel volume but not on the film surface, which is not formed yet.

In the next main carbonization period the metal phase is formed on the surface of the polymeric structures being formed by the reaction:

$$2CuCl + [-CH=CH-]_n \rightarrow 2Cu + 2HCl + [C_2]_n$$

From this point, the binary matrix of the nanosystem C→Cu is being formed. Let us consider the process of building up the film matrix of carbons in copper in the surface diffusion model. In this period of metal phase formation P-parameters calculated via the atom radii are valent-active.

In the liquid, the radius of molecular interaction sphere $R \approx 3r$, where r – molecule radius. Liquids are mainly formed by the elements of the system first and second periods. For the second period, the following can be written down: $R \approx 3r = (n+1)r$, where n – main quantum number. For both periods (first and second) we obtain $R = (\langle n \rangle + 1)r \approx 2.5r$.

TABLE 14.4 Structural Interactions During the Nanofilm Formation in the System C→Cu

Carbon atom			Copper atom				Interaction characteristics					
Orbital	$\dfrac{P_o}{R(n^*+1)}$ (eV)	$\dfrac{P_o}{r_u(n^*+1)}$ (eV)	K	Orbital	$\dfrac{P_o}{R(n^*+1)}$ (eV)	$\dfrac{P_o}{r_u(n^*+1)}$ (eV)	a (%)	$1/\alpha$ (%)	ρ (%)	t (hour)	ω (%)	Interaction type
$2P^2 2S^2$		3.1519	1	$4S^2$		3.0436	3.50	0.29	100	0	0	volumetric
$2P^2$	4.3554		1.6	$4S^1 3d^1$	2.2011		21.17	0.05	5–8	0	0	semisurface
$2P^2$	4.3554		1.7	$4S^1 3d^1$	2.2011		15.15	0.07	19–21	0.49	21.5	surface
$2P^2$	4.3554		1.8	$4S^1 3d^1$	2.2011		9.46	0.11	56–58	1.05	63.9	surface
$2P^2$	4.3554		1.9	$4S^1 3d^1$	2.2011		4.06	0.25	~98–100	1.6	95	surface
$2P^2$	4.3554		2.0	$4S^1 3d^1$	2.2011		1.07	0.93	100	2.0	98.3	surface

Let us assume that this principle with the definite approximation can be spread to different elements of the other periods, but taking the screening effects into account, introducing the value of the effective main quantum number (n*) instead of n. These values of n* and n*+1 taken by Slater [14] are given in Table 14.5.

Thus, let us assume that the sphere radius of atomic-molecular interaction during the particle diffusion is defined as:

$$R = (n^*+1)r \qquad (24)$$

where r – dimensional characteristic of the atomic structure. The total change of R is from 3r to 5.2r (from the second to sixth period).

The averaged value of the structural P_c-parameter falling on the radius unit of atomic-molecular interaction is defined by the following equation:

$$P_c = \frac{P_0}{KR} = \frac{P_0}{r(n^*+1)K} \qquad (25)$$

where K – coefficient taking into accounts the relative number of interacting particles and equal as follows (based on the calculations):

$$K = N/N \qquad (26)$$

Here N_0 – number of particles in the sphere volume of the radius R, N – number of particles or realized interactions depending on the process type (internal or surface diffusion).

Inside the liquid, the resultant of molecular interaction forces equals zero below the upper layer 2R thick (Figure 14.3).

Applying the initial analogy to the internal diffusion, we can accept that such equilibrious state corresponds to the equality $N_0=N$, then K=1.

On the upper part of the liquid surface layer, the sphere volume of atomic-molecular interaction and number of particles in it is practically

TABLE 14.5 Effective Quantum Number

n	1	2	3	4	5	6
n*	1	2	3	3.7	4	4.2
n*+1	2	3	4	4.7	5	5.2

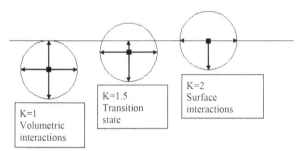

FIGURE 14.3   Relative number of the interacting particles.

twice less in comparison with the internal layers below 2R, that is, and K=2 – for surface diffusion (Figure 14.3).

Actually, the surface diffusion proceeds with the change of the coefficient K in the range from 1.5 to 2.0, which is taken into account in the calculations. Based on such initial ideas, the values of P-parameter and coefficient $\frac{1}{\alpha_2}$ are calculated by the Eqs. (25) and (26) for carbon and copper atoms (Table 14.4).

The values of the degree of structural interactions from coefficient are calculated, that is, – curve 2 given in Figure 14.4. Here, the graphical dependence of the degree of nanofilm formation (ω) on the process time is presented by the data from Ref. [5] – curve 1 and previously obtained nomogram in the form $\rho_1 = f(1/\alpha_1)$ – curve 3.

The analysis of all graphical dependencies obtained demonstrates the practically complete graphical coincidence of all three graphs: $\omega = f(t)$, $\rho_1 = f(1/\alpha_1)$, $\rho_2 = f(1/\alpha_2)$ with slight deviations in the beginning and end of the process. Thus, the carbonization rate, as well as the functions of many other physical-chemical structural interactions, can be assessed via the values of the calculated coefficient α and entropic nomogram.

14.9   NOMOGRAMS OF BIOPHYSICAL PROCESSES

1)   On the kinetics of fermentative processes
     "The formation of ferment-substrate complex is the necessary stage of fermentative catalysis… At the same time, n substrate molecules can join the ferment molecule" [16, p. 58].
2)   For ferments with stoichiometric coefficient n not equal one, the type of graphical dependence of the reaction product performance rate (μ)

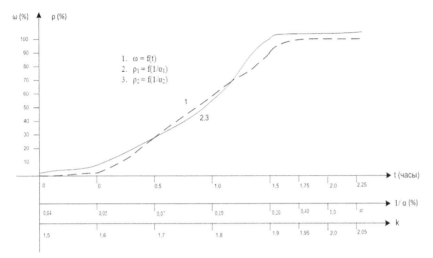

FIGURE 14.4 Dependence of the carbonization rate on the coefficient α.

depending on the substrate concentration (c) has [16] a sigmoid character with the specific bending point (Figure 14.5).

In Figure 14.5 it is seen that this curve generally repeats the character of the entropic nomogram in Figure 14.4.

The graph of the dependence of electron transport rate in biostructures on the diffusion time period of ions is similar [16, p. 278].

In the procedure of assessing fermentative interactions (similarly to the previously applied in par. 8 for surface-diffusive processes) the effective number of interacting molecules over 1 is applied.

In the methodology of P-parameter, a ferment has a limited isomorphic similarity with substrate molecules and does not form a stable compound

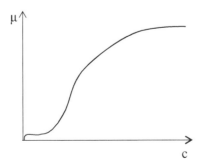

FIGURE 14.5 Dependence of the fermentative reaction rate (μ) on the substrate concentration (c).

with them, but, at the same time, such limited reconstruction of chemical bonds which "is tuned" to obtain the final product is possible.

3) Dependence of biophysical criteria on their frequency characteristics:

(a) The passing of alternating current through live tissues is characterized by the dispersive curve of electrical conductivity – this is the graphical dependence of the tissue total resistance (z-impedance) on the alternating current frequency logarithm (log ω). Normally, such curve, on which the impedance is plotted on the coordinate axis, and log ω – on the abscissa axis, formally, completely corresponds to the entropic nomogram (Figure 14.1).

(b) The fluctuations of biomembrane conductivity (conditioned by random processes) "have the form of Lorentz curve" [17, p. 99]. In this graph, the fluctuation spectral density (ρ) is plotted on the coordinate axis, and the frequency logarithm function (log ω) – on the abscissa axis.

The type of such curve also corresponds to the entropic nomogram in Figure 14.1.

14.10   LORENTZ CURVE OF SPATIAL-TIME DEPENDENCE

The intervals between the events in different coordinate systems are determined by Lorentz geometry of space-time. In this geometry, the velocity (β) is not additive by itself, therefore, the concept of the velocity parameter is introduced (θ). The connection between the velocity β and velocity parameter is simple: $\beta = th\Theta$, where $th\Theta$ means "hyperbolic tangent" and the law of adding two velocities is as follows:

$$th\Theta = th\left(\Theta_1 + \Theta_2\right) = \frac{th\Theta_1 + th\Theta_2}{th\Theta_1 th\Theta_2 + 1} \tag{27}$$

The dependence between the velocity parameter and velocity itself is demonstrated [18] with Lorentz curve in Figure 14.6. Both values are used in relative units in respect to the light velocity. The curve type is formally completely corresponds to the entropic nomogram in Figure 14.4.

Example: "Let the bullet be shot with the velocity $\beta' = 0.75$ from the rocket flying with the velocity $\beta_r = 0.75$. It is necessary to find the bullet speed β relatively to the laboratory system. We know that the velocity parameters are additive, but not the velocities. By the graph, for the point A we find $\theta' = \Theta_r = 0.973$. The addition produces $\theta = \theta' + \theta r = 1.946$. For

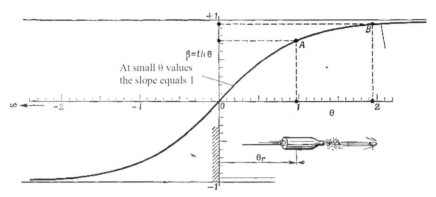

FIGURE 14.6   Connection between the velocity parameter θ and velocity itself $\beta = th\Theta$ obtained directly from the addition law.

this value of the velocity parameter we find the point B by the graph and velocity value $\beta = 0.96$" [18].

14.11   CONCLUSION

Comparing the nomogram (Figure 14.1) with the data from the Table 14.3, we can see the additivity of business entropy values (S) with the values of the coefficient of spatial-energy interactions (α), that is, $S = \alpha$.

For instance, the values of coefficient α for a high degree of structural interactions (up to 7%) coincide with the entropy values for small and medium business. Such regularity in the value changes is also preserved further: for the limited degree of structural interactions coefficient α rises sharply by the nomogram and similarly does the value S during the transition from the medium to large business.

Therefore as applicable to business processes the idea of business quality is similar to the concept of structural interaction degree (ρ).

All this allows approximately defining the critical values of these parameters using the initial nomogram in Figure 14.1. Thus, at $\rho \approx 10\%$ the value $S = \alpha \approx 12 -18\%$, that corresponds to the number of business structures in the range between 10000 and 100000 workers (in the average about 55000).

The optimal criteria of a more qualitative business are defined by the maximal values of their entropies: $S = 6 - 7$ (in relative units).

The same values of S have been obtained earlier and for more complete degree of structural interactions at the isomorphism, as continuous solid solutions correspond to the value $\alpha = 6 - 7$.

The entropic nomogram in Figure 14.1 illustrates the fact that when the values of $\alpha = S$ increase in two times, the efficiency of structural interactions and business quality decreases in 9–10 times. That is the nomogram provides the quantitative characteristic of the main period of business quality decrease during its aggregation.

It is known that the number of atoms in polymeric chain maximally acceptable for a stable system is about 100 units, which is 10^6 in the cubic volume. Then we again have $\lg 10^6 = 6 = S$, that is, this is the entropic characteristic of the optimal number of interacting subsystems.

Thus, the entropic nomograms given in Figures 14.1 and 14.4 are the common quantitative characteristic of many processes in physical chemistry, business and nature. The principles of similar graphical dependence of initial parameters in entirely different situations are obviously defined through the notion of entropy being modified.

KEYWORDS

- entropic criteria
- energy characteristics
- biophysical processes
- Lorentz curve
- thermophysical
- statistic

REFERENCES

1. Reif, F. Statistic physics. M.: Nauka, 1972, 352 p.
2. Gribov, L. A., Prokopyeva, N. I. Basics of physics. M.: Vysshaya shkola, 1992, 430 p.
3. Blokhintsev, D. I. Basics of quantum mechanics. M.: Vysshaya shkola, 1961, 512 p.
4. Yavorsky, B. M., Detlaf A.A. Reference-book in physics. M.: Nauka, 1968, 939 p.
5. Christy, R. W., Pytte, A. The structure of matter: an introduction to modern physics. Translated from English. M.: Nauka, 1969, 596 p.

6. Korablev, G. A. Spatial-Energy Principles of Complex Structures Formation, Netherlands, Brill Academic Publishers and VSP, 2005,426p. (Monograph).
7. Eyring, H., Walter, J., Kimball, J. Quantum chemistry. M., I. L., 1948, 528 p.
8. Fischer, C. F. Atomic Data, 1972, № 4, 301–399.
9. Waber, J. T., Cromer, D. T. J. Chem. Phys, 1965, V 42, -№ 12, 4116–4123
10. Clementi, E., Raimondi, D. L. Atomic Screening constants from, S. C. F. Functions, 1. J. Chem. Phys., 1963, v.38, №11, 2686–2689.
11. Clementi, E., Raimondi, D. L. J. Chem. Phys., 1967, V.47, № 4, 1300–1307.
12. Lumpov, N. A. Profit formula and its application at the distribution of constant costs. http//:referent.mubint.ru/security/8/4763/1, 14 p.
13. Lumpov, N. A. Profit formula: approach to the analysis and construction. Financial management. No 3, 2005.
14. Batsanov, S. S., Zvyagina, R. A. Overlap integrals and problem of effective charges. Novosibirsk: Nauka, 1966, 386p.
15. Kodolov, V. I., Khokhriakov, N. V., Trineeva, V. V., Blagodatskikh, I. I. Activity of nanostructures and its display in nanoreactors of polymeric matrixes and active media. Chemical physics and mesoscopy, 2008. V. 10. №4. P. 448–460.
16. Rubin, A. B. Biophysics. Book 1. Theoretical biophysics. M.: Vysshaya shkola, 1987, 319 p.
17. Rubin, A. B. Biophysics. Book 2. Theoretical biophysics. M.: Vysshaya shkola, 1987, 303 p.
18. Taylor, E., Wheeler, J. Space-time physics. Mir, M.: 1971, 320 p.

AUTHORS' PUBLICATIONS ON THE TOPIC

1. Korablev, G. A., Korableva, N. G., Zaikov, G. E. Dependence of characteristics upon spatial-energy parameter of free atoms. International Journal of Chemical Modeling. 2009, v. 2, pp. 295–312.
2. Korablev, G. A., Korableva, N. G., Zaikov, G. E., Mathematical modeling dependence of thermodynamic characteristics upon spatial-energy parameter of free atoms. Chemical and Biochemical Kinetics: New Perspectives. ed. by Gennady Zaikov; Russian Academy of Sciences. USA, 2011. 15–34.
3. Korablev, G. A., Petrova, N. G., Mathematical modeling of the dependencies of thermodynamic characteristics on spatial-energy parameters of free atoms. Scientific provision of AIC innovative development and agrarian education: Proceedings of Russian scientific-practical conference (Izhevsk, February 14–17, 2012). Izhevsk State Agricultural Academy. Izhevsk, 2012. v. 3. 301–303.
4. Korablev, G. A., Petrova, N. G., Korablev, R. G., Osipov, A. K., Lekomtsev, P. L., Business entropy. Bulletin of Izhevsk State Agricultural Academy. 2013. № 1 (34). P. 76–79.
5. Korablev, G. A., Petrova, N. G., Korablev, R. G., Osipov, A. K., On diversified demonstration of entropy. Encyclopedia of engineer-chemist. 2013. № 8. P. 36–41.
6. Korablev, G. A., Petrova, N. G., Korablev, R. G., Diversified factors of entropy demonstration. From nanostructures, nanomaterials to nanoindustry: IV International Conference: Proceedings, April 3–5, 2013. Kalashnikov Izhevsk State Technical University. 2013. P. 51–53.

7. Korablev, R. G., Petrova, N. G., Korablev, G. A., Osipov, A. K., Zaikov, G. E., Entropic factors in economics and physics. Wschodnie partnerstvo 2013: materialy IX Miedzynarodowej konferencji, 07–15 wrzesnia 2013 roku. Przemysl, 2013. Volume 4. P. 10–20.

8. Korablev, G. A., Korablev, R. G., Lekomtsev, P. L., Osipov, A. K., Petrova, N. G., Entropy of spatial-energy interactions. Agrarian science to AIC innovative development in modern conditions: Proceedings of Russian scientific-practical conference, February 12–15, 2013. Izhevsk State Agricultural Academy. Izhevsk: Izhevsk State Agricultural Academy, 2013. M. II. C.47–50.

9. Korablev, R. G., Korablev, G. A., Petrova, N. G., Entropy of elementary business structure formation. Agrarian science to AIC innovative development in modern conditions: Proceedings of Russian scientific-practical conference, February 12–15, 2013. Izhevsk State Agricultural Academy. Izhevsk: Izhevsk State Agricultural Academy, 2013. V. II. P. 64–66.

10. Petrova, N. G., Korablev, R. G., Osipov, A. K., Lekomtsev, P. L., Korablev, G. A., Entropy of business structure aggregation. Agrarian science to AIC innovative development in modern conditions: Proceedings of Russian scientific-practical conference, February 12–15, 2013. Izhevsk State Agricultural Academy. Izhevsk, 2013. V. II. P. 369–372.

11. Korablev, G. A., Petrova, N. G., Korablev, R. G., Osipov, A. K., Zaikov, G. E., Diversified Fenomene of Entropy. Quantitative Chemistry, Biochemistry and Biology. Complimentary Contributor Copy. Nova Science publishers. New York, 2013. P. 81–90.

12. Korablev, G. A., Petrova, N. G., Korablev, R. G., Zaikov, G. E., Quantitative Calculation of Spatial-Energy Interactions Entropy. Quantitative Chemistry, Biochemistry and Biology. Complimentary Contributor Copy. Nova Science publishers. New York, 2013. P. 91–98.

13. Petrova, N. G., Korablev, R. G., Korablev, G. A., Osipov, A. K., Entropic criteria of business. IX Mezinarodni věedesko-prakticka Konference. Věedesky prumysl evropskeho kontinentu. Praha, 2013. P. 15–21.

14. Korablev, G. A., Petrova, N. G., Korablev, R. G., Osipov, A. K., Zaikov, G. E., Diversified Demonstration of Entropy. Nev Book Announcement. Nanotechnologies to Nanoindustry. USA, Winter 2013/14. Chapter 8.

15. Korablev, G. A., Petrova, N. G., Korablev, R. G., Zaikov, G. E., Entropy of spatial-energy interactions. Encyclopedia of engineer-chemist, 2014, p. 2–6.

16. Korablev, G. A., Petrova, N. G., Korablev, R. G., Osipov, A. K., Zaikov, G. E., On Diversified Demonstration of Entropy. Polymers Research Journal, v.8, №3, Nova Science Publishers, Inc., 2014, pp.145–153.

17. Korablev, G. A., Petrova, N. G., Korablev, R. G., Zaikov, G. E., Entropy of spatial and power interactions. Science Journal of Volgograd State University. Technology and Innovations, 2013, №2 (9), pp. 21–26.

CHAPTER 15

THE SIMULATIONS OF CARBON NANOTUBES (NANOFILAMENTS) AS MACROMOLECULAR COILS: NANOCOMPOSITES REINFORCEMENT DEGREE

ABDULAKH K. MIKITAEV,[1] GEORGIY V. KOZLOV,[1] and GENNADY E. ZAIKOV[2]

[1] *Kh.M. Berbekov Kabardino-Balkarian State University, Chernyshevsky st., 173, Nal'chik-360004, Russian Federation*

[2] *N.M. Emanuel Institute of Biochemical Physics of Russian Academy of Sciences, Moscow – 119334, Kosygin st., 4, Russian Federation, E-mail: chembio@sky.chph.ras.ru*

CONTENTS

ABSTRACT

The carbon nanotubes (nanofilaments) in polymer nanocomposites simulation as macromolecular coils was performed. Such approach allows the estimation of the indicated nanofillers real anisotropy degree and prediction of the corresponding nanocomposites properties. The important role of division surfaces nanofiller-polymer matrix has been shown.

15.1   INTRODUCTION

As it is well-known [1], carbon nanotubes (nanofilaments) possess very high longitudinal elasticity modulus (1000–2000 GPa) and low transverse stiffness. These factors together with a large ratio length/diameter (high anisotropy degree), typical for the indicated nanofillers, result in formation by them ring-like structures, resembling outwardly macromolecular coils [2, 3]. This circumstance has already been noted in literature. So, the authors [2] supposed, that carbon nanotubes ring-like structures could be considered as macromolecular coils in semidilute solutions. The authors [4] used Flory formula for rod-like macromolecules in case of determination of carbon nanotubes percolation threshold in a polymer nanocomposite. Nevertheless, such examples are rare enough and do not have systematic character. This chapter purpose is Kuhn segment value (nanofiller real anisotropy degree) determination for carbon nanotubes (nanofilaments) and estimation on this basis of corresponding polymer nanocomposites reinforcement degree.

15.2   EXPERIMENTAL PART

Polypropylene (PP) "Kaplen" of mark 01030 of industrial fabrication, having average-weight molecular weight of $\sim (2\text{–}3) \times 10^5$ and polydispersity index of 4.5, was used as a matrix polymer. Two types of carbon nanotubes were used as a nanofiller. Nanotubes of mark "Taunite" have an external diameter of 20–70 nm, an internal diameter of 5–10 nm and length of 2 mcm and longer. Besides, multiwalled nanofilaments (CNF), having layers number of 20–30, diameter of 20–30 nm and length of 2 mcm and longer, have been used. The mass contents of carbon nanotubes of both types was changed within the limits of 0.15–3.0 mass %.

Nanocomposites PP/CNT and PP/CNF were prepared by the components mixing in melt on twin screw extruder Thermo Haake, model Reomex RTW 25/42 production of German Federal Republic. Mixing was performed at temperature 463–503 K and screw speed of 50 rpm during 5 min. Testing samples were obtained by casting under pressure method on a casting machine Test Samples Molding Apparate RR/TS MP of firm Ray-Ran (Taiwan) at temperature 483 K and pressure of 43 MPa.

Uniaxial tension mechanical tests have been performed on the samples in the shape of two-sided spade with sizes according to GOST 112 62–80. The tests have been conducted on universal testing apparatus Gotech Testing Machine CT-TCS 2000, production of German Federal Republic, at temperature 293 K and strain rate of $\sim 2 \times 10^{-3}$ s^{-1}.

15.3   RESULTS AND DISCUSSION

The authors [2] proposed the following equation for estimation of reinforcement degree E_n/E_m of polymer nanocomposites with anisotropic nanofiller:

$$\frac{E_n}{E_m} = 1 + 2\alpha \left(\frac{E_{nf} / E_m}{E_{nf} / E_m - 2\alpha} \right) \varphi_n \qquad (1)$$

where E_n, E_m and E_{nf} are elasticity moduli of nanocomposite, matrix polymer and nanofiller, respectively, α is anisotropic nanofiller aspect ration or its length and diameter ratio, φ_n is nanofiller volume contents.

In case of short fibers limit ($2\alpha << E_{nf}/E_m$) the equation becomes independent on nanofiller elastic properties [2]:

$$\frac{E_n}{E_m} = 1 + 2\alpha\varphi_n \qquad (2)$$

Let us consider methods included in the Eq. (2) parameter determination. At ring-like structures CNT (CNF) simulation as macromolecular coils

Kuhn segment with length A one should consider as a reinforcing filament, where the value A is determined according to the equation [5]:

$$R_{CNT}^2 = \frac{L_{CNT} A}{6}$$ (3)

where R_{CNT} is ring-like structures CNT (CNF) radius, L_{CNT} is carbon nanotube (nanofilament) length.

In its turn, the value R_{CNT} is determined with the aid of the following percolation equation [6]:

$$\varphi_n = \frac{\pi L_{CNT} r_{CNT}^2}{\left(2 R_{CNT}\right)^3}$$ (4)

where r_{CNT} is carbon nanotube (nanofilament) radius.

The value φ_n can be determined according to the well-known formula [7]:

$$\varphi_n = \frac{W_n}{\rho_n}$$ (5)

where W_n is nanofiller mass contents, ρ_n is its density, estimated for nanoparticles as follows [7]:

$$\rho_n = 188 \left(D_{CNT}\right)^{1/3}, \text{kg/m}^3$$ (6)

where D_{CNT} is carbon nanotube (nanofilament) diameter, which is given in nm.

Further the value α is determined as ratio:

$$\alpha = \frac{A}{D_{CNT}}$$ (7)

Since the value D_{CNT} for the studied CNT (CNF) is varied within wide enough limits and precise distribution of this parameter is unknown, then the value D_{CNT} for the determination α according to the Eq. (7) was chosen by method of theory and experiment the best correspondence. According to this method $D_{CNT} = 28$ nm for CNT and $D_{CNT} = 20$ nm for CNF, that is close enough to the indicated diameter lower limit.

In Figure 15.1 the dependence $\alpha(\varphi_n)$ for the considered nanocompos-ites is adduced, from which the fast decay α follows, that is, fast reduction of nanofiller anisotropy real degree, at φ_n growth. This effect is due to R_{CNT} corresponding reduction at φ_n growth: R_{CNT} decreases from 472 up to 206 nm for nanocomposites PP/CNT and from 326 up to 114 nm for 206 nm for nanocomposites PP/CNF, that is, defined by the formation of more "coagulated" CNT (CNF) structures.

The Eqs. (2) and (7) can be used for theoretical estimation of reinforce-ment degree E_n/E_m for the considered nanocomposites. In Figure 15.2 the comparison of the obtained experimentally and calculated according to the indicated mode dependences $E_n/E_m(\varphi_n)$ for nanocomposites PP/CNT and PP/CNF is adduced. As one can see, the proposed method gives a good cor-respondence to experiment – their mean discrepancy makes up less than 2%.

Let us consider the physical grounds of reduction of nanofiller anisotropy real degree, characterized by the parameter α, for the studied nanocompos-ites. As it is known [8], the conception of nanomaterials in general is based on decisive role of numerous division surfaces in the indicated nanomaterials as the basis for solids properties essential change, including polymer nanocom-posites. In conformity with these principles it is assumed, that such material can be considered as a nanomaterial, in which division surfaces fraction φ_d makes up in common volume of the material from 0.5 and larger [8]. The value φ_d can be estimated according to the following relationship [8]:

$$\phi_d = \frac{3s}{L} \tag{8}$$

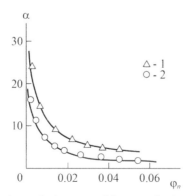

FIGURE 15.1   The dependence of anisotropy real degree α of carbon nanotubes (nanofilaments) on nanofiller volume contents φ_n for nanocomposites PP/CNT (1) and PP/CNF (2).

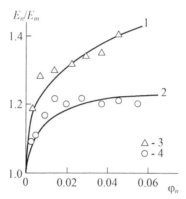

FIGURE 15.2   The comparison of calculated according to the Eqs. (2) and (7) (1, 2) and obtained experimentally (3, 4) dependences of reinforcement degree E_n/E_m on nanofiller volume contents φ_n for nanocomposites PP/CNT (1, 3) and PP/CNF (2, 4).

where s is boundary region width, L is grain size in nanomaterials.

The division surfaces fraction φ_d for nanocomposites polymer/carbon nanotubes (nanofilaments) can be estimated according to the Eq. (8) at the condition $s=l_{if}$ and $L=r_{CNT}$, where l_{if} is thickness of the interfacial layer on division boundary polymer matrix-nanofiller, r_{CNT} is a carbon nanotube (nanofilament) radius. The value l_{if} can be estimated with the aid of the equation [7]:

$$\varphi_{if} = \varphi_n \left[\left(\frac{r_{CNT} + l_{if}}{r_{CNT}} \right)^3 - 1 \right] \tag{9}$$

where φ_{if} is interfacial regions relative fraction, which is determined with the aid of the following percolation relationship [7]:

$$\frac{E_n}{E_m} = 1 + 11 \left(\varphi_n + \varphi_{if} \right)^{1.7} \tag{10}$$

In Figure 15.3 the dependence $\alpha(\varphi_d)$ for the considered nanocomposites is adduced, which is approximated well by the linear correlation and described analytically by the following empirical equation:

$$\alpha = 3.2\varphi_d \tag{11}$$

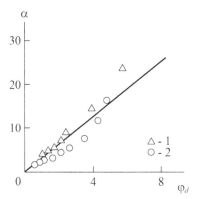

FIGURE 15.3   The dependence of anisotropy real degree α of carbon nanotubes (nanofilaments) on division surfaces fraction φ_d for nanocomposites PP/CNT (1) and PP/CNF (2).

At the boundary value $\varphi_d=0.5$, when nanocomposite has stopped to be true nanomaterial, the value $a \to 1$, that is, carbon nanotubes (nanofilaments) become practically isotropic nanofiller.

The Eqs. (2) and (11) combination allows to obtain the direct dependence of reinforcement degree on division surfaces fraction for the considered nanocomposites:

$$\frac{E_n}{E_m} = 1 + 6.4\varphi_d\varphi_n$$
$$(12)$$

The Eq. (12) allows to predict the value E_n/E_m with high enough precision (in this case mean discrepancy between theory and experiment makes up 3%). Let us pay attention to the fact, that the Eq. (12), as a matter of fact, gives the dependence of E_n/E_m on φ_{if} (the Eq. (9)), that confirms the treatment of interfacial regions in polymer nanocomposites as reinforcing element together with nanofiller.

15.4   CONCLUSIONS

Hence, the present work results have shown that carbon nanotubes (nanofilaments) ring-like structures simulation as macromolecular coils gives the correct estimation of the indicated nanofillers anisotropy real degree. Such approach allows nanocomposites polymer/carbon nanotubes (nanofilaments) reinforcement degree prediction with high enough precision. The considered nanofillers anisotropy real degree is defined by the division surfaces nanofiller-polymer matrix fraction.

KEYWORDS

- **anisotropy**
- **carbon nanotubes**
- **macromolecular coil**
- **nanocomposite**
- **reinforcement degree**
- **ring-like structures**

REFERENCES

1. Eletskiy, A. V. Uspechi Fizich. Nauk, 2007, 177, 223.
2. Schaefer, D. W., Justice, R. S. Macromolecules, 2007, 40, 8501.
3. Kozlov, G. V., Yanovskiy Yu. G., Zhirikova, Z. M., Aloev, V. Z., Karnet Yu. N. Mekhanika Kompozitsionnykh Materialov i Konstruktsii, 2012, 18, 131.
4. Komarov, B. A., Dzhavadyan, E. A., Irzhak, V. I., Ryabenko, A. G., Lesnichaya, V. A., Zvereva, G. I., Krestinin, A. V. Vysokomolek. Soed. A, 2011, 53, 897.
5. Budtov, V. P. Physical Chemistry of Polymer Solutions. Khimiya, Sankt-Petersburg, 1992.
6. Bridge, B. J. Mater. Sci. Lett., 1989, 8, 102.
7. Mikitaev, A. K., Kozlov, G. V., Zaikov, G. E. Polymer Nanocomposites: Variety of Structural Forms and Applications. Nova Science Publishers, Inc., New York, 2008.
8. Andrievskiy, V. A. Rossiiskiy Khimicheskiy Zhurnal, 2002, 46, 50.

CHAPTER 16

STRUCTURE, INTERACTIONS AND KINETICS OF RING-LIKE FORMATIONS OF CARBON NANOTUBES (NANOFIBERS) IN POLYMER NANOCOMPOSITES

A. K. MIKITAEV,[1] G. V. KOZLOV,[1] and G. E. ZAIKOV[2]

[1]Kh.M. Berbekov Kabardino-Balkarian State University, Chernyshevsky st., 173, Nal'chik-360004, Russian Federation

[2]N.M. Emanuel Institute of Biochemical Physics of Russian Academy of Sciences, Moscow – 119334, Kosygin st., 4, Russian Federation, E-mail: chembio@sky.chph.ras.ru

CONTENTS

ABSTRACT

It has been shown that carbon nanotubes (nanofibers) ring-like structures formation parameters can be described within the framework of fractal kinetics. In this case the main characteristic is interactions level of nanofiller in polymer matrix. The indicated model explained relatively large reinforcement degree of nanocomposites polymer/carbon nanotubes at ultra-small nanofiller contents.

16.1 AIMS AND BACKGROUND

As it is well-known [1], carbon nanotubes (nanofibers) possess very high longitudinal elasticity modulus (1000–2000 GPa) and low transverse stiffness. These factors together with a large ration length/diameter (high anisotropy degree), typical for the indicated nanofillers, result to formation by them ring-like structures, outwardly resembling macromolecular coils [2, 3]. This circumstance has already been noted in literature. So, the authors [2] assumed, that carbon nanotubes ring-like structures can be considered as macromolecular coils in semidiluted solutions. The authors [4] used Flory formula for rod-like macromolecules in case of carbon nanotubes percolation threshold in polymer nanocomposites determination. Nevertheless, such examples are rare enough and do not have systematic character.

It is obvious, that carbon nanotubes (nanofibers) ring-like structures formation process can not be instantaneous for example in virtue of the polymer melt high enough viscosity, in which these structures are formed. Hence, the indicated process has definite duration that makes necessary its kinetics study. Therefore the present work purpose is the study of intercommunication of interactions, formation kinetics and structure of ring-like formations of carbon nanotubes (nanofibers) in polymer nanocomposites.

16.2 EXPERIMENTAL PART

Polypropylene (PP) "Kaplen" of mark 01030 was used as matrix polymer for the studied nanocomposites. This PP mark has a melt flow index of 2.3–3.6 g/10 min, molecular weight of $\sim (2-3) \times 10^5$ and polydispersity index of 4.5.

Carbon nanotubes (CNT) of mark "Taunite", having an external diameter of 20–70 nm, an internal diameter of 5–10 nm and length of 2 mcm and more, were used as a nanofiller. In the studied nanocomposites PP/CNT taunite contents was varied within the limits of 0.25–3.0 mass %. Besides, the multiwalled nanofibers (CNF) were used, having a layers number of 20–30, diameter of 20–30 nm and length of the order of 2 mcm. In the nanocomposites PP/CNF CNF contents W_n was varied within the limits of 0.15–3.0 mass %.

Nanocomposites PP/CNT and PP/CNF were prepared by the components mixing in melt on twin-screw extruder Thermo Haake, model Reomex RTW 25/42, production of German Federal Republic. Mixing was performed at temperature 463–503 K and screw speed of 50 rpm during 5 min. Testing samples were prepared by casting under pressure method on a casting machine Test Samples Molding Apparate RR/TS MP of firm Ray-Ran (Taiwan) at temperature 503 K and pressure 43 MPa.

Uniaxial tension mechanical tests have been performed on the samples in the shape of two-sided spade with the sizes according to GOST 112 62–80. The tests have been conducted on the universal testing machine Gotech Testing Machine CT-TCS 2000, production of German Federal Republic, at temperature 293 K and strain rate of $\sim 2 \times 10^{-3}$ s^{-1}.

16.3 RESULTS AND DISCUSSION

CNT (CNF) ring-like structures radius R_{CNT} can be determined with the aid of the following percolation relationship [6]:

$$\varphi_n = \frac{\pi L_{\text{CNT}} r_{\text{CNT}}^2}{\left(2R_{\text{CNT}}\right)^3} \tag{1}$$

where φ_n is nanofiller volume contents, L_{CNT} and r_{CNT} are length and radius of carbon nanotube (nanofiber), respectively.

The value φ_n was calculated according to the well-known formula [7]:

$$\varphi_n = \frac{W_n}{\rho_n} \tag{2}$$

where W_n is nanofiller mass contents, ρ_n is its density, estimating for nanoparticles as follows [7]:

$$\rho_n = 188\left(D_{CNT}\right)^{1/3} \tag{3}$$

where D_{CNT} is a carbon nanotube (nanofiller) diameter, which is given in nm.

A CNT (CNF) ring-like formations structure can be characterized most exactly with the aid of its fractal dimension D_f, which is true structural characteristic, since it describes the distribution of CNT (CNF) ring-like formations elements in space [7]. The value R_{CNT} calculation according to the Eq. (1) has shown its reduction at φ_n growth. At the largest from the used φ_n values, corresponding to $W_n=3.0$ mass %, the indicated dependences have the tendency of asymptotic branch achievement, that supposes achievement by CNT or CNF ring-like structures of their R_{CNT} minimum values. By the analogy with macromolecular coils this means the achievement of maximally dense ring-like structure with the greatest limiting value of its fractal dimension D_f ($D_f^{\lim}$), which is determined according to the equation [8]:

$$D_f^{\lim} = \frac{4(d+1)}{7} \tag{4}$$

where d is the dimension of Euclidean space, in which a fractal is considered (it is obvious, in our case $d=3$). For $d=3$ the value $D_f^{\lim} =2.286$.

Further for the value D_f estimation the irreversible aggregation model can be used, which describes polymerization processes (macromolecular coil formation) and gives the following relationship for particles aggregates radius R_{agr} determination [9]:

$$R_{agr} \sim c_0^{-1/\left(d-D_f\right)} \tag{5}$$

where c_0 is aggregating particles initial concentration.

Coefficient in the relationship (5) can be determined at the following conditions: $R_{agr}=R_{CNT}$, $c_0=\varphi_n$ and $D_f = D_f^{\lim}$. The values R_{CNT} and φ_n were accepted for $W_n=3.0$ mass %. As the estimations according to the indicated relationship have shown, the value D_f grows at φ_n increasing (R_{CNT} reduction) from 1.91 up to 2.29 for nanocomposites PP/CNT and from 1.76 up to 2.21 for nanocomposites PP/CNF.

As it is known [5], the process rate in fractal-like medium is described by the following equation:

$$\vartheta \sim t^{-h} \tag{6}$$

where t is process duration, h is medium heterogeneity exponent ($0<h<1$), which is transformed into zero for homogeneous samples only [5].

The value h was calculated according to the equation [10]:

$$h = \frac{D_f - 1}{2} \tag{7}$$

In Figure 16.1 the dependences of R_{CNT} on CNT (CNF) ring-like structures formation process rate ϑ are adduced for the considered nanocomposites. As it was to be expected, the process rate ϑ increasing results in R_{CNT} growth, that is, the value ϑ characterizes not CNT (CNF) rolling up in ring-like structures, but their unrolling. The dependence $R_{CNT}(\vartheta)$ can be expressed analytically by the following empirical equations:

$$R_{CNT} = 90 + 5.0 \times 10^3 \, \vartheta \tag{8}$$

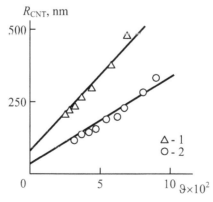

FIGURE 16.1 The dependences of CNT (CNF) ring-like formations radius R_{CNT} on their unrolling rate ϑ for nanocomposites PP/CNT (1) and PP/CNF (2).

for carbon nanotubes and

$$R_{CNT} = 45 + 2.70 \times 10^3 \, \vartheta \qquad (9)$$

for carbon nanofibers. The parameters t in the Eq. (6) and R_{CNT} in the Eqs. (8) and (9) are expressed in s and nm, respectively.

Let us note, that the smallest R_{CNT} value according to the Eqs. (8) and (9) is equal approximately to $2D_{CNT}$ at $\vartheta = 0$, since for the obvious reasons the condition $R_{CNT} = 0$ is impossible. The greatest value R_{CNT} in case of homogeneous mediums ($h=0$) is equal to 5090 nm for CNT and 2745 nm for CNF.

Let us consider the intercommunication of CNT (CNF) ring-like structures formation rate ϑ and interaction between them. At the indicated structure modeling as macromolecular coils the interaction between their elements and polymer matrix can be characterized by the parameter ε, determined as follows [10]:

$$\varepsilon = \frac{2 - D_f}{D_f} \qquad (10)$$

For macromolecular coils the value ε is varied within the limits of $-1/3 \div 1.0$. In the point $D_f = 2.0$ parameter ε changes its sign, that corresponds to the interactions type change from repulsion forces (positive ε) up to attraction forces (negative ε). In Figure 16.2 the dependence of CNT (CNF) ring-like structures formation rate ϑ on interaction parameter ε is adduced for the considered nanocomposites. As one can see, the linear dependence of CNT (CNF) ring-like structures unrolling rate ϑ at ε increasing is obtained, that is, the repulsion interaction intensification, which is described analytically by the following empirical equation:

$$\vartheta = 0.275(\varepsilon + 0.215) \qquad (11)$$

From the Eq. (11) it follows, that the value $\vartheta = 0$ is achieved at $\varepsilon = -0.215$, that is, at $D_f = 2.548$. The greatest value $\vartheta = 0.334$ is realized at $\varepsilon = 1.0$, that corresponds to h = 0.192 or $D_f = 1.384$.

As it is noted above, nanofiller contents ϑ_n increasing results in dimension D_f enhancement and, according to the Eq. (10), in the exponent h increasing,

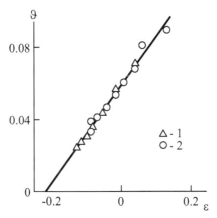

FIGURE 16.2 The dependence of CNT (CNF) ring-like structures unrolling rate ϑ on interaction parameter ε for nanocomposites PP/CNT (1) and PP/CNF (2).

that is, in medium heterogeneity degree enhancement. In Figure 16.3 the dependences of parameter ε on the value (such form of the indicated dependences was chosen for their linearization) are adduced for the considered nanocomposites. As one can see, ε linear reduction is observed, that is, the attraction interactions intensification, at nanofiller contents growth. This dependence can be expressed analytically by the following empirical equations:

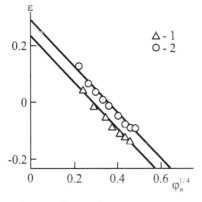

FIGURE 16.3 The dependences of interaction parameter ε on nanofiller volume contents ϑ_n for nanocomposites PP/CNT (1) and PP/CNF (2).

$$\varepsilon = 0.25 - 0.85\varphi_n^{1/4} \qquad (12)$$

for carbon nanotubes and

$$\varepsilon = 0.30 - 0.85\varphi_n^{1/4} \qquad (13)$$

for carbon nanofibers.

The Eqs. (11)–(13) combination demonstrates, that for the considered nanocomposites ε variation is realized within the range, which is smaller than theoretical one for macromolecular coils ($\varepsilon=-1/3\div1.0$), namely, within the limits of $\varepsilon=-0.215\div0.30$. The condition of full balance of attraction and repulsion forces for CNT (CNF) ring-like structures $\varepsilon=0$ is realized at $\varphi_n=0.0075$ for nanocomposites PP/CNT and $\varphi_n=0.0155$ for nanocomposites PP/CNF. In Figure 16.4 the dependences $R_{CNT}(\varphi_n)$ are adduced for the indicated nanocomposites, each one from which can be approximated by two straight lines with enough precision degree. The transition between these two parts of the dependences $R_{CNT}(\varphi_n)$ corresponds to $\varphi_n\approx0.0088$ for nanocomposites PP/CNT and $\varphi_n\approx0.0120$ for nanocomposites PP/CNF, that agrees well enough with the indicated above φ_n values, at which the condition $\varepsilon=0$ is achieved or, in other words, with values φ_n, at which the transition from attraction interaction up to repulsion interactions of CNT (CNF)

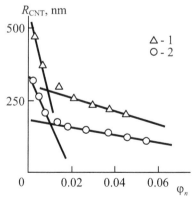

FIGURE 16.4 The dependences of CNT (CNF) ring-like formations radius R_{CNT} on nanofiller volume contents φ_n for nanocomposites PP/CNT (1) and PP/CNF (2).

ring-like structures is realized. The linear dependences $R_{CNT}(\varphi_n)$ slope is in about 15 times larger in case of the repulsion interactions in comparison with the attraction interactions. In other words, in case of interactions first type much more fast R_{CNT} growth at φ_n reduction is observed, that influences positively on the characteristics of nanocomposites polymer/carbon nanotubes [3]. This effect explains the indicated nanocomposites relatively high reinforcement degree at ultra-small concentrations ($\varphi_n \leq 0.0015$) of carbon nanotubes [4, 11].

16.4 CONCLUSIONS

Thus, in the present work the results have shown that the formation (unrolling) rate of carbon nanotubes (nanofibers) ring-like structures in polymer nanocomposites can be described within the framework of fractal kinetics, where the indicated rate is defined by interactions level in these structures. A medium heterogeneity level or interactions degree and sign is controlled by nanofiller contents. The transition from attraction interactions up to repulsion interactions defines carbon nanotubes (nanofibers) ring-like structures radius fast growth that results in nanocomposites relatively large reinforcement degree at ultra-small contents of the indicated nanofillers.

KEYWORDS

- carbon nanotubes (nanofibers)
- fractal kinetics
- interactions
- nanocomposite
- ring-like structure

REFERENCES

1. Moniruzzaman, M., Winey, K. I., Polymer Nanocomposites Containing Carbon Nanotubes. Macromolecules, 39(16), 5194 (2006).

2. Schaefer, D. W., Justice, R. S., How Nano are Nanocomposites? Macromolecules, 40(24), 8501 (2007).

3. Yanovsky, Yu. G., Kozlov, G. V., Zhirikova, Z. M., Aloev, V. Z., Yu. Karnet, N., Special Features of the Structure of Carbon Nanotubes in Polymer Composite Media. Intern. J. Nanomechanics Science and Technology, 3(2), 99 (2012).

4. Komarov, B. A., Dzhavadyan, E. A., Irzhak, V. I., Ryabenko, A. G., Lesnichaya, V. A., Zvereva, G. I., Krestinin, A. V., Epoxy Amine Composites with Ultrasmall Concentrations of One walled Carbon Nanotubes. Vysokomolek. Soed. A, 53(6), 897 (2011).

5. Kopelman, R., Excitons Dynamics Resembling Fractal one: Geometrical and Energetical Disorder. In book: Fractals in Physics. Ed. Pietronero L., Tosatti Amsterdam, E., Oxford, New York, Tokyo, North-Holland, 1986, 524.

6. Bridge, B., Theoretical Modeling of the Critical Volume Fraction for Percolation Conductivity of Fiber-Loaded Conductive Polymer Composites. Mater, J., Sci. Lett., 8(2), 102 (1989).

7. Mikitaev, A. K., Kozlov, G. V., Zaikov, G. E., Polymer Nanocomposites: Variety of Structural Forms and Applications. Nova Science Publishers, Inc., New York, 2008, 319 p.

8. Family, F., Fractal Dimension and Grand Universality of Critical Phenomena. Stat, J., Phys., 36(5/6), 881 (1984).

9. Shogenov, V. N., Kozlov, G. V., Fractal Clusters in Physics-Chemistry of Polymers. Polygraph service and T, Nalchik, 2002, 268 p.

10. Kozlov, G. V., Dolbin, I. V., Zaikov, G. E., The Fractal Physical Chemistry of Polymer Solutions and Melts. Apple Academic Press, Toronto, New Jersey, 2014, 316 p.

11. Blond, D., Barron, V., Ruether, M., Ryan, K. P., Nicolosi, V., Blau, W. J., Coleman, J. N., Enhancement of Modulus, Strength and Toughness in Poly(methyl methacrylate)-Based Composites by the Incorporation of Poly(methyl methacrylate)-Functionalized Nanotubes. Advanced Functional Mater., 16(6), 1608 (2006).

CHAPTER 17

STRUCTURAL, MORPHOLOGICAL AND OPTICAL PROPERTIES OF NANOPRODUCTS OF ZIRCONIUM TARGET LASER ABLATION IN WATER AND AQUEOUS SDS SOLUTIONS

V. T. KARPUKHIN, M. M. MALIKOV, T. I. BORODINA, G. E. VALYANO, O. A. GOLOLOBOVA, and D. A. STRIKANOV

Joint Institute for High Temperatures, Russian Academy of Science, 13(2) st. Izhorskaya, Moscow, 125412, Russian Federation, E-mail: vtkarp@gmail.com

CONTENTS

ABSTRACT

Structural, morphological and optical properties of nanoproducts of laser ablation of zirconium target in water and aqueous SDS solutions were investigated. Depending on experiment conditions the indicated products can appear as different crystalline phases of zirconia and organic-inorganic composites, which include SDS alkyl chains intercalated between layers of zirconium oxides or hydroxides. The formation of zirconium dioxide-based hollow nano and microstructures is demonstrated. It is suggested that ablation formed gas-vapor bubbles can serve as templates for generation of hollow structures.

17.1   INTRODUCTION

Numerous publications devoted to both synthesis and investigation of properties of nano dispersed zirconia have appeared by now [1–8]. The interest is determined by unique mechanical, chemical, optical and other properties that open way to a broad practical application of the material in science and technology [9–13]. Synthesis of oxides and other metal nanocompositions by laser ablation in a liquid environment is one of perspective methods [14–16]. However, the data on zirconia generation by the mentioned way is far not sufficient [17–21]. The experiments were carried out mainly at low laser pulses repetition rate (~ 10 Hz), that limits the perspective of establishing productive technologies, potential investigations of a number of physical and chemical processes accompanying the synthesis and affecting properties of the final product.

 Thus, the given work is devoted to the attempt of generating nanocrystalline zirconia by laser ablation of zirconium target in aqueous solutions of a surfactant. A copper vapor laser (CVL) with the power output 10–15 W, radiation pulse duration 20 ns, and pulse repetition rate 10 kHz was used as radiation source.

17.2   EXPERIMENTAL SECTION

Physical and technical descriptions of laser ablation in liquid are given in detail in numerous original articles and reviews [22–25]. CVL generation was performed at two wavelengths: 510.6 and 578.2 nm, line power ratio was correspondingly 2:1. UV light (255, 271 and 289 nm) was obtained by nonlinear conversion of the Cu laser radiation (510.6 and 578.2 nm) in a BBO crystal [26]. The laser beam was focused onto the target surface by an achromatic lens with focal distance of $f = 280$ mm, that provided the spot diameter of less than 100 µm. The target was placed into a glass cell with deionized water or aqueous solutions of a surfactant. The volume of the liquid in the cell was ~ 10 cm³. The cell was placed in a vessel with cooling water, its temperature was kept at ~ 300 K. The vessel was installed on a movable table that permitted to constantly move the focal spot over the target surface. The surfactant SDS ($C_{12}H_{25}SO_4Na$) of anionic group was used for the experiment.

The optical characteristics of colloidal solutions with nanostructures of zirconia were analyzed by the absorption spectral method in the range of 200–700 nm by SF-46 spectrophotometer with automatic data processing. RAMAN Spectra were recorded by double monochronomator KSVU-23. The structure and composition of the solid phase extracted from the colloid solution after centrifugation at 15000 rpm and evaporation at 320–330 K were analyzed at X-ray diffractometer DRON-2 (K_α copper line). Shapes and sizes of nanostructures were determined by Hitachi S405A scanning electron microscope at accelerating voltage of 15 kV, according to the standard technique.

17.3   RESULTS AND DISCUSSION

17.3.1   XRD DATA

The experiments were performed both in deionized water (SDS solution molarity M = 0) and at M – 0.0001, 0.01, 0.05 and 0.1. Exposition time of radiation onto the target τ_{exp} was varied within intervals from 5 to 180 min. In a number of cases properties of colloid and its solid phase were analyzed of tens hours later (the aging time). Figure 17.1 demonstrated typical XRD patterns of solid phase of colloids at M SDS = 0, 0.01, 0.05 and 0.1. After ablation of zirconium in deionized water the synthesized zirconia was mainly in X-ray amorphous state, that is, the size of the crystallites did not exceed 1–2 nm (curve 1). The crystalline part of the solid contains 5–9 vol. % of monoclinic phase, 4 vol. % of tetragonal phase and about 3 vol. % of metal zirconium. Sizes of metal zirconium crystallites make not less than 100 nm, of monoclinic phase ~ 40–75 nm and of tetragonal

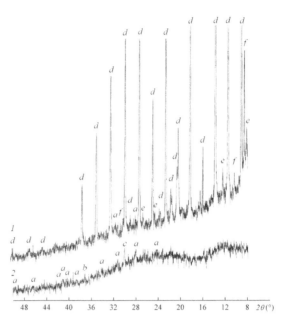

FIGURE 17.1 X-ray diffraction patterns of precipitates extracted from the colloidal solution, depending on the concentration (M) SDS: 0 (*1*), 0.1 (*2*). Phases: a – monoclinic, b – SDS, c – SDS·H$_2$O, d – system based on Zr and SDS (Zr – SDS nanocomposite).

phase ~ 60–100 nm. Sizes of nanoparticles become less with the increase of exposition time. Crystalline lattice parameters of dioxide tetragonal modification stay within the limits of $a = 0.05083$–0.5093 nm, $c = 0.5127$–0.5185 nm, with ratio c/a varies from 1.007 (close to cubic structure) to 1.019. It is supposed that ions of OH$^-$ appearing in the process of ablation can serve as a stabilization factor for tetragonal phase at M = 0. It is supported in other similar experiments [27, 28]. We can observe a tendency of extending monoclinic phase along with exposition time growth and reduction of metal zirconium amount.

At M = 0.01 a greater part of colloid solids is in the amorphous state, as well. The crystalline part consists of SDS, monoclinic, tetragonal phases and a small part of metal zirconium. The ratio of monoclinic and tetragonal phases makes ~ 1.6:1. Lattice parameters of tetragonal phase are: $a = 0.5115$ nm, $c = 0.5133$ nm, that is, the lattice is close to cubic. Sizes of monoclinic phase particles are 60–100 nm, 45–100 nm for tetragonal one.

The increase of ablation time up to 180 min the crystalline part enhances. The monoclinic phase makes about 8 vol. %, the tetragonal phase is about 7 vol. %. Tetragonal phase average size equals to about 60 nm that of monoclinic phase is about 80 nm. The increase of SDS concentration to M = 0.05 does not practically change XRD patterns structure. Similar to M = 0.01, about 2.5 vol.% of zirconium is presented. Average size of particles for both parts of dioxide and metal zirconium reduces to 40–45 nm. The reduction of crystallite average sizes at M = 0.01 and 0.05 with regards to crystallite sizes at M = 0 is caused by the influence of surfactant – SDS, which restricts nanoparticles growth [29, 30]. The presence of SDS in the initial solution creates prerequisites for the appearance of the third high temperature phase of zirconia, that is, a cubic phase. Thus, in experiments at M = 0.01, 0.05, 0.1 phases with lattice spacing of a = 0.5127 nm, 0.5115 nm and of c = 0.513 nm were presented, that is, phases close to cubic ones. Tetragonal and cubic phases can be stabilized by anions OH$^-$, anions SO$^-_2$, SO$^-_3$, SO$^-_4$ of SDS which appear during ablation time, interact with crystalline surface and prevent oxygen penetration, as well as cations Na$^+$ [27, 28]. At the concentration of SDS in solution up to M = 0.1 XRD patterns is observed as a spectrum of well crystallized SDS, a considerable part of which is oriented along the surface of glass plate (plane 00L), a smaller part is distributed chaotically (Figure 17.1, curve 2). Along with lines belonging to SDS and ZrO$_2$ the XRD spectrum has a number of other maxima, their intensity considerably surpasses intensity of zirconia lines, though it is weaker than SDS maxima. Some of those lines can be attributed to hydrate SDS·H$_2$O and Na$_2$S$_2$O$_3$·H$_2$O. The spectrum also clearly demonstrated the phase with interlayer distance d = 5.165 nm. Distances close to the indicated ones are typical of organic-inorganic composites with alkyl chains intercalated between oxide or hydroxide layers of transition metal Zn, Co, Fe, Cu, etc. [31]. The chains may be placed perpendicular to oxide or hydroxide layers and tilting, as well, besides, the chains are connected between themselves by hydrocarbon "tails". They may are partially interdigitated. Thus, formation of one-layer and two-layer structures with broad variation of parameter $d \sim 0.25$–6 nm is possible. The formulas of such compositions are as follows: (M)$_2$(OH)$_x$X·z·H$_2$O and (M)(OH)$_2$X·z·(H$_2$O) for hydroxides and X·M^{2+}·z·(H$_2$O); X·M$_2$O^{2+}·z·(H$_2$O) for oxides, where M is divalent metals (Zn, Cu, Zr), X is intercalated anion alkyl sulfate(C$_n$H$_{2n+1}$SO$_4$, where n = 12),

z is molecules number of (H_2O) [31, 32]. The authors have already obtained similar Zn and Cu composites during ablation of the indicated metal targets in aqueous SDS solutions [32, 33].

Long exposition of radiation in colloid may cause breaking of SDS molecules, that formed new phases – β-Zr$(SO_4)_2$ and β-Zr$(SO_4)_2 \cdot H_2O$ in the spectrum [34]. It should be emphasized that a specific feature of the experiment is predominance of synthesized zirconia in amorphous state mainly.

17.3.2 RAMAN SPECTRA

The spectrum obtained at M = 0 shows the presence of monoclinic, tetragonal and cubic phases of ZrO_2 in solid part of colloid (Figure 17.2). While comparing the spectra taken at M = 0.1 and deionized water (Figure 17.3), at M = 0.1 we see peaks in the region of 2000–2400 cm^{-1}

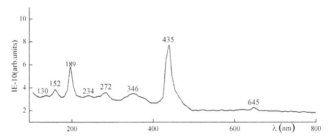

FIGURE 17.2   The Raman spectra of the precipitate colloid λ_{exc} = 510 nm. Phase ZrO_2: 130, 189, 346 – monoclinic; 152, 234, 272, 435 – tetragonal; 520–670 cubic.

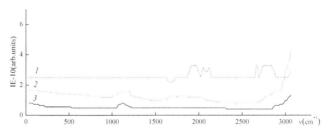

FIGURE 17.3   The Raman spectra of the precipitate colloid λ_{exc} = 271 nm, depending on the concentration (M) SDS: 0.1 (*1*), 0 (*2*), H_2O (*3*). 2000–2400 cm^{-1} – Zr(OH)$_4$, 2780–3000 cm^{-1} – Zr(OH)$_4$ and SDS.

and 2800–3000 cm^{-1}, that can be attributed to a vibrational structure of Zr(OH)$_4$ molecule. Fixation of zirconium oxide and hydroxide in RAMAN spectra at M ≥ 0.01 together with XRD data most obviously speaks in favor of synthesis of organic-inorganic composite Zr – SDS, consisting of zirconium oxide or hydroxide layers with intercalated SDS alkyl chains.

17.3.3 MORPHOLOGY OF SYNTHESIZED PRODUCTS

Figures 17.4 and 17.5 present images of solid phase structures of colloid at different τ_{exp} and SDS solution molarity M. The treatment of samples synthesized in deionized water (M = 0) at τ_{exp} = 5, 20, 180 min showed that they consist of separate large particles (up to 10 µm), aggregates of rounded particles (~ 100–500 nm) and dense mixture of grains with poorly defined boundaries (Figure 17.4a). The latter ingredient of the microstructure is more typical for τ_{exp} = 180 min, where it is main, X-ray amorphous component. At τ_{exp} = 5 and 20 min it can be observed in the structure among large and small particles spherical hollow units with sizes from tens nm to several µm, some of them concave and partially broken (Figures 17.4b and 17.4c). The cover of large hollow particles consist of separate layers, the thickness of the cover can be within 50–200 nm and more (Figures 17.4a and 17.4d). That correlates with the XRD data for an average size of nanoparticles (~ 40–100 nm) that build up microstructures.

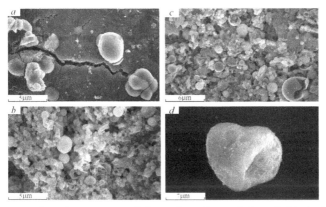

FIGURE 17.4 The microstructure of the surface phase of the colloid obtained in the concentration M SDS = 0 M, (a) τ_{exp} = 180 min (b) – (d) τ_{exp} = 20 min.

The crystalline part of the solid phase extracted from colloid is significantly changed when SDS is added to the initial solution. It is known that micellae (Hartley type) start to appear in the solution, if SDS concentration exceeds the critical level (M ~ 0.008 mol/L). With the increase of M crystals begin to grow. In accordance with Wolf principle chain-like structure of SDS molecule predetermines that crystals are formed either as thin elongated plates for SDS anhydrite or monohydrite, or as octahedral-like plates for $C_{12}H_{25}OSO_3Na \cdot 1/8H_2O$ [35]. After aging of colloid for tens hours the images show crystals in the shape of long sticks (Figure 17.5b). The above-mentioned plate units may serve as templets to form laminar structures of organic-inorganic composites based on zirconium oxides or hydroxides. When exposition exceeds 5 min pictures show zones of crystallized SDS structures, areas of amorphous SDS and zirconia and droplets of aggregates or separate large solid and hollow zirconia nanoparticles.

Hollow nano and microspheres are observed at all regimes of zirconium target ablation both in deionized water and in SDS solution (Figure 17.4c, Figure 17.5d). However, at M = 0 average sizes of hollow particles are noticeably larger than at M ≥ 0.01. Some publications consider adsorption of nanoparticles on the surface of vapor-gas bubbles to be a basic mechanism for formation of hollow structures, that is, the gas bubble serves as a

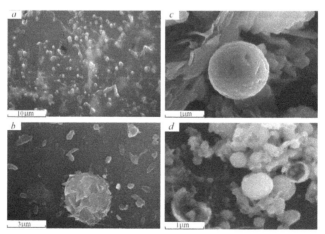

FIGURE 17.5   The microstructure of the surface phase of the colloid obtained in the concentration M > 0: (a) units of ZrO_2, (b) forming a shell of crystallites SDS, (c) melted shell (d) the structure of the "rattle".

"soft" template [36–39]. There might be several reasons for bubble formation. There are certain zones in the focal spot on the target surface due to
space and time power distribution over beam area: ablation ($\geq 10^8$ W/cm^2),
melting ($< 10^8$ W/cm^2), surface heating above liquid boiling point [40].
The ablation produces a plasma-vapor-gas bubble, its collapse may create
secondary gas (H_2, O_2) and vapor bubbles [40–44]. The process generates
both heated solid particles and metal droplets with a vapor cloud around
them [45]. Besides, in passing through the laser beam particles may get
heated up to temperatures above the melting point, break to small fragments, and it is accompanied by vapor bubbles formation [46]. The main
reason for appearance of a solid shell on gas bubble surface is the adsorption of solid particles on the gas – liquid boundary (capillary effect). It
leads to decreasing the surface energy of the bubble-particle system:

$$\Delta E = - \pi R^2 \gamma (1 - \cos \theta)^2,$$

where R is a particle radius, γ is surface energy of a gas bubble (water –
gas boundary), θ is a contact angle of the particle with the bubble surface.
Angle θ is < 90, as particles of metal oxides are hydrophilic. Calculation
shows that for R = 2 nm, $\gamma \approx 74 \times 10^{-3}$ H/m and $\theta \approx 40°$, $|\Delta E| \approx 12.3$kT, that
is, considerably higher of the particles heat energy at T $\sim$ 300 K [47]. In
other words, the separation of particles from the surface is hindered.

The initial layer of adsorbed particles grows due to Oswald effect and
finally forms a layered shell [48, 49]. Its growth rate depends on of particles diffusion rate in liquid. SDS presence in the solution will increase
its viscosity, retard particle diffusion and, therefore, will prevent the
grows of shell sizes. Shell formation does not occur only because of small
nanoparticles of zirconia. In the presence of SDS we can observe separate
attached SDS crystallites and, possibly, crystallites of Zr – SDS composite (Figure 17.5b). When drifting in the colloid nanostructures get to the
area of laser beam, it leads to fusion of their surface layer (Figure 17.5c)
[39, 50, 51]. It is noteworthy, that high pulses repetition rate makes the
generation of vapor-gas bubbles go very intensively, thus, the possibility of hollow nano and microstructure formation is increased. Kircandle
effect might be observed for particles with "nucleus-shell" structure, where
the nucleus may be metal and the shell made of another material (metal
oxide, etc.) [52–55], that leads to cavities formation inside particles due

to differences of component diffusion rate. The formation time may make up to several minutes [52] and it goes down noticeably when particles are heated. In this regard, the appearance of such particles in the experiments is highly probable. Both effects can be a cause for appearing "rattle" structures (Figure 17.5d) (smaller spheres are inside a lager one) [38]. It should be noted that the indicated hollow nano and microspheres at laser ablation of zirconium in water and aqueous solutions of SDS seem to be obtained for the first time.

17.3.4 OPTICAL CHARACTERISTICS: ABSORPTION SPECTRA

Zirconia is a wide band gap insulator with two direct transmissions between zones of valence and conductivity of 5.2 and 5.79 eV [56]. Upper levels of the valence zone are mainly of O(2p) type and lower levels of the conductivity zone are determined by 4d orbitals zirconium ion Zr^{4+} [57, 58]. Changes of phase composition of ZrO_2 and the synthesis conditions, presence of surfactant, in particular, influence the absorption spectra that carry information about the structure of the formed crystalline lattice and its intrinsic and surface defects, vacancies, other inclusions [59, 60]. The investigation of absorption curves (Figures 17.6 and 17.7) of the colloid solution obtained at 5 and 20 min target ablation in deionized water and aqueous SDS solution reveals some peculiar features of the spectrum:

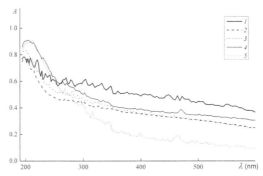

FIGURE 17.6   The absorption spectra of $A(\lambda)$ τ_{exp} = 5 min depending on the concentration (M) SDS: 0 (*1*), 0.01 (*2*), 0.01 (20 h aging) (*3*), 0.1 (*4*), 0.1 (48 h aging) (*5*).

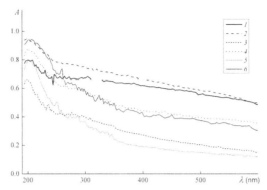

FIGURE 17.7   The absorption spectra of $A(\lambda)$ $\tau_{exp} = 20$ min depending on the concentration (M) SDS: 0 (*1*), 0.01 (*2*), 0.01 (48 h aging) (*3*), 0.05 (*4*), 0.05 (48 h aging) (*5*), 0.1 M (*6*).

- steep rise of absorption in UV range ~ 200–240 nm. The rise is observed both at M = 0 and M ≠ 0. When M = 0 the above-mentioned zone of absorption can be divided into two. The 200–220 spectrum region has a higher average absorption level and typical peaks at 200, 205, 210–212 nm, that are most dramatic at 5 min exposition (Figure 17.6, curve *1*), and the 220–240 nm region with the absorption level 20–25% lower;
- zone of absorption rise (5–7%) within the region of 240–350 nm with peaks at 270, 300, 325 and 350 nm (Figure 17.6, curves *1, 3, 5*; Figure 17.7, curves *1, 3, 5, 6*);
- absorption dips ("bleaching") in narrow spectrum interval at the beginning (245–255 nm) and at the end (~355 nm) of the mentioned region (Figure 17.6, curve *1, 4, 5*; Figure 17.7, curves *3, 5, 6*);
- increase (3–4%) of absorption within the interval of 440–480 nm with maxima at 445, 460–470 nm (Figure 17.6, curves *1, 4, 5*).

The increase of ablation time and SDS concentration in solution gives a considerable rise of absorption level, especially in the long wave spectrum range, due to particle concentration growth. At that it does not change much within the wave range of 350–600 nm, that is typical of colloids with formed fractal structures of the solid phase [25, 61]. Colloid aging for 24 h and more does not practically change spectrum shape in the initial region of 200–260 nm (Figure 17.6, Figure 17.7 curve *3, 5*). However, beyond this boundary the absorption decrease towards longer wave lengths is more dramatic, and the final level is considerably lower. That speaks to sedimentation of large units of the colloid solid phase.

The analysis of the mentioned features of the absorption spectrum with consideration of the phase mixture (cubic, monoclinic and tetragonal phases) and dimensional composition of ZrO_2 nanoparticles (40–100 nm) leads to the following conclusions. The steep rise of absorption in UV range from 200 to 240 nm seems to be associated to $O^{2-}(2p) \rightarrow Zr^{4+}(4d)$ transitions. Similar phenomenon was observed in a number of other researches [56, 57]. The spectrum region at 200–220 nm is determined by the presence of cubic and tetragonal phases, and of metal zirconium, as well. It is shown by typical absorption peaks at 200, 205, 210–212 nm [62, 63]. The appearance of monoclinic phase means the transformation of crystalline lattice structure, its symmetry. Coordination number of zirconium ions at transition from tetragonal to monoclinic phase changes from 8 to 7. This change leads to a split of 4d levels of Zr^{4+} ion and the appearance of long wave branch in $O^{2-}(2p) \rightarrow Zr^{4+}(4d)$ transition spectrum, that is, the boundary of absorption band shifts to the red region up to 240–250 nm [56].

The band gap of zirconia E_g is calculated from the dependence of the absorption coefficient (α) on the photon energy (hv) in the band-edge spectral region for a direct transition by a known formula:

$$\alpha h v = \text{const } (hv - E_g^{bulk})^{1/2} \tag{1}$$

where, h is Plank constant, E_g^{bulk} is the band gap of the solid [64]. It gives values for samples obtained at ablation on deionized water $\sim$ 5.27 eV and $\sim$ 5.76 eV that is close to the mentioned above. The increase of SDS concentration in solution up to M = 0.1 reduces E_g to the level of $\sim$ 4.54 and 4.78 eV. This fact is determined by growth of lattice defects at M $\neq$ 0.

Zirconia, representing photoresistant metal oxides, possesses a typical property to capture charge carriers (electrons and holes) by crystalline lattice defects [65] with the formation of centers of absorption (F, V color centers) or luminescence [65, 66]. Under the influence of radiation of a certain wavelength range the captured electrons or holes get free and recombine, in particular, by radiative emission, that is, color (absorption) centers disappear. In that case gaps ("bleaching" zones) appear in absorption spectrum within rather narrow wavelength range. We should note that the process of nanostructure synthesis at metal ablation in liquid by nanosecond pulses is far from equilibrium. The presence of SDS molecules and ions in the solution also increases potential generation of defects inside and on the surface of crystals. With this point of view, gaps in spectra presented in curves *1*, *4*, *5* (Figure 17.6) and *3*, *5*, *6* (Figure 17.7) at wavelengths of 245–255 nm and ~ 355 nm, enhanced absorption in regions of 240–350 nm, 440–480 nm and 540–600 nm can be attributed to the discussed above character. Similar facts were observed and discussed in Ref. [59].

17.3.5   PHOTOLUMINESCENCE

Figures 17.8 and 17.9 give colloid spectra obtained at different excitation wavelengths and SDS molar fractions in the initial solution. The spectra

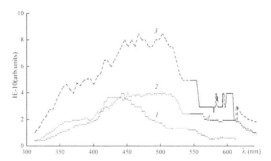

FIGURE 17.8   Photoluminescence spectra of colloid depending on the concentration (M) SDS λ_{exc} = 271 nm: 0 (*1*), 0.05 (*2*), 0.1 (*3*).

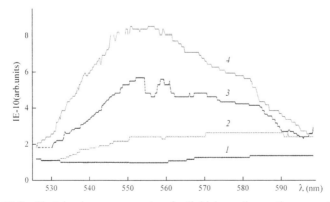

FIGURE 17.9   Photoluminescence spectra of colloid depending on the concentration (M) SDS λ_{exc} = 510 nm: 0 (*1*), 0.01 (*2*), 0.05 (*3*), 0.1 (*4*).

(Figure 17.7, curves 1, 3) excited by UV radiation ($\lambda_{exc} = 271$ nm) in colloids with M = 0.05 and 0.1 take up a rather continuous region in wavelengths of 340–610 nm. In both cases two parts are obviously specific: 310–400 nm and 400–535 nm. Maximal level of the signal in the second region is about 1.5–2.0 times higher. Moreover, luminescence intensity for the spectrum in general at M = 0.1 is higher than at M = 0.05. In the absence of SDS, firstly, the general level of luminescence goes down by several times, secondly, the spectrum starts from 330 nm. In this case we can see several regions of enhanced luminescence: 360, 420–425, 435–450 nm. Spectra maxima at M $\neq$ 0 are shifted into the red region by $\sim$ 50–60 nm and have flatter wide top. In the spectrum (Figure 17.8) the luminescence region lies at 530 up to 575–580 nm at M = 0.01, while at M = 0 there is no luminescence. When analyzing causes of luminescence in colloids, it seems appropriate to consider a chain of processes mentioned in Ref. [59]:

$$Z \div hv_{ex} \rightarrow Z^* \tag{2}$$

$$Z^* \rightarrow Z \div hv_{lum} \tag{3}$$

$$Z \div hv_{ex} \rightarrow Z^+ + e \text{ (photoionization of defects, Z)} \tag{4}$$

$$Z^* \rightarrow Z^+ + e \text{ (thermal ionization of defects, Z*)} \tag{5}$$

$$Z^+ + e \rightarrow (Z^*) \rightarrow Z + hv_{lum} \tag{6}$$

where Z and Z* are basic and excited states of luminescence centers, process (6) is recovery of Z state by means of electron capture. With consideration of conditions of ZrO_2 nanostructure synthesis processes (2) – (6) seem very probable in the present experiment. Introduction of SDS into the initial solution may considerably increase the number of intrinsic and, especially, surface defects, different vacancies, precondition appearance of new centers of absorption and luminescence in nanocrystals obtained after ablation. The indicated circumstances at M $\neq$ 0 seem to determine a considerable growth of luminescence, enlargement of its spectrum to UV and red regions at $\lambda_{exc} = 271$ nm, appearance of luminescence at $\lambda_{exc} = 510$ nm. It should be noted that the obtained data both in the wavelength range and in luminescence intensity coincide with the results of a number of researches, where synthesis of zirconia was performed by different chemical methods [60, 67–69].

17.4   CONCLUSIONS

In the present research ZrO_2 nanostructure synthesis was performed by laser ablation under conditions of increased pulse repetition rate onto zirconium target both in deionized water and aqueous solutions of SDS. That mode provides optimal conditions for productive generation of zirconia and permits a relatively flexible transformation of properties in the obtained product. XRD data and RAMAN spectra point out an interesting feature of the experiment, that is, prevalence of dioxide amorphous phase in the product of synthesis and the presence of all tree ZrO_2 phases: cubic, tetragonal and monoclinic in the crystallized part of colloid. Application of surfactant SDS in the process of zirconia producing can be treated as an additional factor of stabilization for high temperature cubic and tetragonal phases. Besides, it seems the fact of synthesizing organic-inorganic nanocomposites based on zirconium oxides or hydroxides and SDS alkyl chains was fixed in the experiment for the first time. The metal zirconium ablation in deionized water for the first time produced hollow nano and microspheres from zirconia. The authors assume that a basic mechanism for their formation is the adsorption of particles on the surface of vapor-gas bubbles and Kirkendal effect cannot be excluded, as well. There are certain references [70] that similar structures are rather promising for practical application in biomedicine, pharmacology, for developing new materials, etc.

KEYWORDS

- laser ablation
- metal nanocompositions
- RAMAN Spectra
- ZrO_2

REFERENCES

1. Shukla, S., Seal, S., Vanfleet, R. *J. Sol-Gel Sci. Technol.* 2003. v. 27. I. 2. p. 119.
2. Salavati-Niasarim, M., Dadkhah, M., Davar, F. *Inorg Chim Acta.* 2009. v. 362. p. 3969.
3. *Feng,* X., *Bai,* Y. J., Lu, B., *Zhao,* Y. R., Yang, J., Chi, J. R. *J. Cryst. Growth. 2004. v. 262. p. 420.*
4. Ray, J. C., Pramanik, P., Ram, S. *Mater. Lett.* 2001. v. 48. I. 5. p. 281.
5. Sliem, M. A., Schmidt, D. A., Bétard, A., Kalidindi, S. B., Gross, S., Havenith-Newen, M., Devi, A., Fischer, R. A. *Chem. Mater.* 2012. v. 24. p. 4274.
6. Tok, A. I. Y., Boey, F. Y. C., Du, S. W., Wong, B. K. *Mater. Sci. Eng. B.* 2006. v. 130. p. 114.
7. Meskin, P. E., Ivanov, V. K., Barantchikov, A. E., Churagulov, B. R., Tretyakov Yu. D. *Ultrasonics Sonochem.* 2006. v. 13. p. 47.
8. Chen, L., Mashimo, T., Omurzak, E., Okudera, H., Iwamoto Ch., Yoshiasa, A. *J. Phys. Chem. C.* 2011. v. 115. p. 9370.
9. Cao, G. Nanostructures and nanomaterials: Synthesis, Properties and Applications. Imperial College Press, 2004. 433 pp.
10. Botta, S. G., Navio, J. A., Hidalgo, M. C., Restrepo, G. M., Litter, M. I. *J. Photochem. Photobiol. A: Chemistry.* 1999. v. 129. p. 89.
11. *Subbarao,* E. C., *Maiti,* H. S. *Adv. Ceram.* 1988. v. 24. p. 731.
12. Latha Kumari, Du, G. H., Li, W. Z., Selva Vennila, R., Saxena, S. K., Wang, D. Z. *Ceramics International.* 2009. v. 35. I. 6. p. 2401.
13. Varaksin, A. Yu, Protasov, M. V. Teplitsky Yu. S. *High Temperature.* 2014. T. 52. № 4. p. 581.
14. Kumar, B., Thareja, R. K. *J. Appl. Phys.* 2010. v. 108. p. 064906.
15. Stratakis, E., Zorba, V., Barberoglou, M., Fotakis, C., Shafeev, G. A. *Appl. Surf. Sci.* 2009. v. 255. p. 5346.
16. Liu, P., Cai, W., Fang, M., Li Zh., Zeng, H., Hu, J., Luo, X., Jing, W. *Nanotechnology.* 2009. v. 20. p. 285707.
17. Dezhi Tan, Geng Lin, Yin Liu, Yu Teng, Yixi Zhuang, Bin Zhu, Quanzhong Zhao, Jianrong Qiu. *J. Nanopart. Res.* 2011.V.13, P. 1183.
18. Dezhi Tan, Yu Teng, Yin Liu, Yixi Zhuang, Jianrong Qiu. *Chem. Lett.* 2009. v. 38. p. 1102.
19. Mahmoud, A. K., Fadhill, Z., Ibrahim Al-Nasser, S., Ibrahim Hussein, F., Akman, E., Demir, A. *J. Mat. Sci. and Eng. B3.* 2013. V.6. P.364.
20. Chao-Hsien Wu, Chang-Ning Huang, Pouyan Shen, Shuei-Yuan Chen. *J. Nanopart. Res.* 2011.V.13, P. 6633.
21. Golightly, J. S., Castleman, A. W. *Zeitschrift für Physikalische Chemie.* 2010. v. 221. p. 1455.
22. Simakin, A. V., Voronov, V. V., Shafeev, G. A. *Phys. of Wave Phenomena.* 2007. v. 15. p. 218.
23. Bozon-Verduraz, F., Brayner, R., Voronov, V. V., Kirichenko, N. A., Simakin, A. V., Shafeev, G. A. Quantum Electron. 2003. v. 33. p. 714.
24. Yang, G. W. *Progress in Material Science.* 2007. v. 52. p. 648.

25. Karpuhin, V. T., Malikov, M. M., Val'yano, G. E., Borodina, T. I., Gololobova, O. A. *High Temperature*, 49, 681 (2011).
26. Batenin, V. M., Bokhan, P. A., Buchanov, V. V., Evtushenko, G. S., Kazaryan, M. A., Karpukhin, V. T., Klimovskii, I. I., Malikov, M. M. Lazery na samoogranichennykh perekhodakh atomov metallov-2 (Lasers on Self-Terminating Transitions in Metal Atoms: II). *Publishing House of Physica, Mathematical Literature (in Rus)*, Moscow. 2011. Vol. 2.
27. Stefanic, G. Music, S. *Croatica Chem. Acta.* 2002. v. 75. p. 727.
28. Li, C., Li, M. *J. Raman Spectrosc.* 2002. v. 32. p. 301.
29. Pesika, N. S., Hu, Z., Stebe, K. J., Searson, P. C. *J. Phys. Chem. B.* 2002. v. 106. p. 6985.
30. Kandare, E., Chigwada, G., Wang, D., Wilkie, C. A., Hossenlopp, J. M. *Polymer Degradation and Stability.* 2006.V. 91. p. 1781.
31. Q. Huo, D. I. Margolese, U. Ciesla, Demuth, D. G., Feng, P., Gier, T. E., Sieger, P., Firouzi, A., Chmelka, B. F., Schüth, F., Stucky, G. D. *Chemistry of Materials.* 1994. v. 6. № 8. p. 1176.
32. Karpukhin, V. T., Malikov, M. M., Val'yano, G. E., Borodina, T. I., Gololobova, O. A. *J. Nanotechnol.*, 2012, Article ID 910761 (2012); doi: 10.1155/2012/910761.
33. Kandare, E., Chigwada, G., Wang, D., Wilkie, C. A., Hossenlopp, J. M. *Polymer Degradation and Stability*, 91, 1781 (2006).
34. Borodina, T. I., Val'yano, G. E., Gololobova, O. A., Karpukhin, V. T., Malikov, M. M., Strikanov, D. A. *Quantum Electron.* 2013. v. 43. p. 563.
35. Smith, L. A., Duncan, A., Thomson, G. B., Roberts, K. J., Machin, D., McLeod, G. *Journal of Crystal Growth.* 2004. v. 263. I. 1–4. p. 480.
36. Yan, Z., Bao, R., Wright, R. N., Chrisey, D. B. *Appl. Phys. Lett.* 2010. v. 97. p. 124106.
37. Yan, Z., Bao. R., Huang, Y., Caruso, A. N., Qadri, S. B., Dinu, C. Z., Chrisey, D. B. *J. Phys. Chem. C.* 2010. v. 114. p. 3869.
38. Yan, Z., Bao, R., Huang, Y., Chrisey, D. B. *J. Phys. Chem. C.* 2010. v. 114. p. 11370.
39. Yan, Z., Bao, R., Chrisey, D. B. *Nanotechnology.* 2010. v. 21. p. 145609.
40. Lim, K. Y., Quinto-Su, P. A., Klaseboer, E. A., Khoo, B. C., Venugopalan, V. C., Ohl, C. *Phys. Rev. E.* 2010. v. 81. p. 016308.
41. Yavas, O., Leiderer, P., Park, H. K., Grigoropoulos, C. P., Poon, C. C., Leung, W. P., Do, N., Tam, A. C. *Phys Rev Lett.* 1993. v. 70. p. 1830.
42. Ohl, C. D., Lindau, O., Lauterborn, W. *Phys. Rev. Lett.* 1998. v. 80. p. 393.
43. Brenner, M. *Rev. of Modern Phys.* 2002. v. 74. p. 425.
44. Li, X., Shimizu, Y., Pyatenko, A., Wang, H., Koshizaki, N. *Nanotechnology.* 2012. v. 23. p. 115602.
45. Takeda, S., Ikuta, Y., Hirano, M., Hosono, H. *J. Mater. Res.* 2001. v. 16. p. 1003.
46. Pyatenko, A., Yamaguchi, M., Suzuki, M. *J. Phys. Chem. C.* 2007. v. 111. p. 7910.
47. Binks, B. P. *Current Opinion in Colloids and Interface Sci.* 2002. v. 7. p. 21.
48. Ostwald, W. Lehrbuch der Allgemeinen Chemie. v. 2. Leipzig, Germany, 1896. p. 1163.
49. Ratke, L., Voorhees, P. W. Growth and Coarsening: Ostwald Ripening in Material Processing. Springer, 2002.

50. Orrù, R., Licheri, R., Locci, A. M., Cincotti, A., Cao, G. *Mater. Sci. Eng. R.* 2009. v. 63. p. 127.
51. Kang Suk-Joong, L. Sintering: Densification, Grain Growth, and Microstructure. Elsevier Ltd., 2005.
52. Smigelskas, A. D., Kirkendall, E. O. *Trans. AIME.* 1947. v. 171. p. 130.
53. Niu, K. Y., Park, J., Zheng, H., Alivisatos, A. P. *Nano Lett.* 2013. v. 13. p. 5715.
54. Niu, K. Y., Yang, J., Kulinich, S. A., Sun, J., Du, X. W. *Langmuir.* 2010. v. 26. p. 16652.
55. Yang, J., Hou, J., Du, X. School of Materials Science and Engineering. Tianjin: Tianjin University, 2013. p. 300072.
56. Bluvshtein, Z. M., Nizhnikova, G. P., Farberovich, U. V., *Sov. Phys. Solid State* (in Rus), 32, 548 (1990).
57. López, E. F., Escribano, V. S., Panizza, M., Carnasciali, M. M., Busca, G. *J. Mater. Chem.*, 11, 1891 (2001).
58. Sutton, D. Electronic spectra of transition metal complexes (McGraw-Hill, New York, 1968)
59. Emeline, A., Kataeva, G. V., Litke, A. S., Rudakova, A. V., Ryabchuk, V. K., Serpone, N. *Langmuir*, 14, 5011 (1998).
60. Cong, Y., Li, B., Yue, S., Fan, D., Wang, X.-J. *J. Phys. Chem. C.* 2009. v. 113. p. 13974.
61. Karpov, S. V. and Slabko, V. V. Optical and Photophysical Properties of Fractal-Structured Metal Sols, Russian Academy of Sciences, Siberian Branch, Novosibirsk, Russia, 2003. (in Rus)
62. Sahu, H. R., Rao, G. R. *Bull. Mater. Sci.* 2000. v. *23.* p. *349.*
63. Geethalakshmi, K., Prabhakaran, T., Hema, J. *World Academy of Sci. Eng. Tech.* 2012. v. 64. p. 150.
64. Pankove, J. I. *Optical properties in semiconductors (Prentice Hall, Englewood Cliffs, NJ, 1971).*
65. Lushchik Ch.B., Lushchik, A. C. Electronic excitations with the formation of defects in solids (Nauka, Moscow, 1989). (in Rus)
66. Strekalovsky, V. N., Polezhaev Yu. M., Palguev, S. F. Oxides with impurity disordering. Composition, structure, phase transformations, (Moscow: Nauka, 1987). (in Rus)
67. Sliem, M. A., Schmidt, D. A., Bétard, A., Kalidindi, S. B., Gross, S., Havenith-Newen, M., Devi, A., Fischer, R. A. *Chem. Mater.* 2012. v. 24. p. 4274.
68. Reddy Channu, V. S., Kalluru, R. R., Schlesinger, M., Mehring, M., Holze, R. *Coll. Surf. A: Physicochem. Eng. Aspects.* 2011. v. 386. p. 151.
69. Neppolian, B., Wang, Q., Yamashita, H., Choi, H. *Appl. Catal. A: General.* 2007. v. 333. p. 264.
70. Zhou, J., Wu, W., Caruntu, D., Yu, M. H., Martin, A., Chen, J. F., O'Connor, C. J., Zhou, W. L. *J. Phys. Chem. C. 2007. v. 111 (47).* p. 17473.

CHAPTER 18

CALCIUM SOAP LUBRICANTS

ALAZ IZER, TUGCE NEFISE KAHYAOGLU and DEVRIM BALKÖSE

Izmir Institute of Technology Department of Chemical Engineering Gulbahce Urla Izmir Turkey,E-mail: devrimbalkose@gmail.com

CONTENTS

ABSTRACT

The preparation and characterization of calcium stearate ($CaSt_2$) and a lubricant by using calcium stearate were aimed at in this study. Calcium stearate powder was prepared from sodium stearate and calcium chloride by precipitation from aqueous solutions. $CaSt_2$ and the Light Neutral Base oil were mixed together to obtain lubricating oil. It was found that $CaSt_2$ had a melting temperature of 142.8°C and in base oil it had a lower melting point, above 128°C. It was dispersed as lamellar micelles as the optical micrographs had shown. From rate of settling the size of dispersed particles were found to be 1.88 μm and 0.11 μm for lubricants having 1% and 2% CaSt$_2$, respectively. The friction coefficient and wear scar diameter of base oil 0.099 and 1402 nm were reduced to 0.0730 and 627.61 nm, respectively, for the lubricant having 1% CaSt$_2$. Lower wear scar diameter (540 nm) was obtained for lubricant with 2% CaSt$_2$. CaSt$_2$ improved the lubricating property of the base oil but did not improve its oxidative and thermal stability.

18.1   INTRODUCTION

A lubricant provides a protective film, which allows for two touching surfaces to be separated, thus lessening the friction between them. Lubricating oil is a liquid lubricant that reduces friction, protects against corrosion, reduce electric currents and cool machinery temperature. It is most often used in the automobile industry and is applied to bearings, dies, chains, cables, spindles, pumps, rails and gears to make them run smoother and more reliably. Lubricating oil is a substance introduced between two moving surfaces to reduce the friction and wear between them. Lubricating oils consist of a liquid paraffinic or vegetable oil and surface-active agents, antioxidants and anticorrosive additives. Metal soaps in pure form or dispersed in paraffinic oils are used as lubricants. Felder et al. [1] used sodium and calcium soap coatings on steel wires for drawing the wires. Calcium stearate had good lubricating efficiency at low wire drawing rates [1]. The possibility for the production of a motor oil with improved operating characteristics and a higher stability by applying of composite additives has been studied by Palichev et al. [2]. For this purpose two multifunctional additives, synthesized by them have been used. They used additives containing calcium stearate and calcium salts of nitrated polypropylene and oxidized paraffin, urea, ethylene diammine, stearic acid. The additives improved the anticorrosion, viscosity-temperature, antiwear and antisludge properties of the lubricant [2]. The optimum concentration of the additive, which enables the production of a high-quality motor lubricant, has been found to be 5% [2]. Cutting oils were obtained by adding CaSt$_2$ to dry paraffin oil up to 5% together with other additives [3]. Thus the gelation was prevented and an easily flowing cutting oil was obtained. Savrik et al. [4] prepared lubricants using base oil, surface-active agent Span 60 and zinc borate particles. They used 1% Span 60 and 1% zinc borate. Surface-active agent Span 60 was found to be very effective in reducing the friction coefficient and wear scar diameter in four ball tests. As surface-active agents metal soaps are also used. Metal soaps are transition metal salts of the fatty acids and the alkaline earth elements. Although, the alkali salts of the fatty acids such as sodium and potassium are water soluble, metal soap is water insoluble but more soluble in nonpolar organic solvents. Calcium stearate Ca $(C_{17}H_{35}COO)_2$, in short form $CaSt_2$ is the one of the important ionic surfactants of metal soaps. Calcium Stearate, is a nontoxic, white powdery substance. It is a calcium salt derived from stearic acid and is widely used in cosmetics, plastics, pharmaceuticals and lubricants [5].

Metal soaps can be obtained by neutralization of long chain organic acids with bases or by precipitation process. Moreria et al. [6] investigated formation of $CaSt_2$ from stearic acid and calcium hydroxide in different solvents and a complete conversion to $CaSt_2$ was obtained in ethanol medium [6]. The precipitation process generally produces metal soap in powder form by the reaction of aqueous solutions of a water-soluble metal salt and a fatty acid alkali metal salt at a temperature below the boiling point of water at atmospheric pressure. Filtering, washing, drying are the important steps in this method. Calcium stearate is produced in pure form by using this process [5].

Production of a lubricant by using a neutral base oil and calcium stearate is the aim of this study. The lubricating effects were tested by a four-ball tester for this purpose.

18.2 MATERIALS AND METHODS

18.2.1 MATERIALS

Calcium chloride, $CaCl_2 \cdot 2H_2O$ (98%, Aldrich), and sodium stearate, (NaSt) C17H35COONa (commercial product, Dalan Kimya A. S., Turkey), were used in the synthesis of $CaSt_2$. The acid value of stearic acid, used in the NaSt synthesis, was 208.2 mg of KOH/g of stearic acid and it consists of a C16–C18 alkyl chain and with 47.7% and 52.3% by weight, respectively [5].

Spindle Oil from TUPRAS Izmir was used as base oil in the preparation of the lubricants.

18.2.2 PREPARATION OF CALCIUM STEARATE POWDER

Calcium stearate powder was prepared from sodium stearate and calcium chloride by precipitation from aqueous solutions according to reaction (1).

$$2C_{17}H_{35}COO - {}^{+}Na_{(aq)} + Ca^{2+}{}_{(aq)} \rightarrow (C_{17}H_{35}COO)_2Ca_{(s)} + 2Na^{+}{}_{(aq)} \quad (1)$$

About 5.000 g (0.016 mol) of sodium stearate, (NaSt) was dissolved in 200 cm³ of deionized water in a stainless steel reactor at 75°C. 1.7984 g (0.012 mol) of calcium chloride (50% excess) was dissolved in 100 cm³ of deionized water at 30°C and added to sodium soap solution at 75°C. The mixture was stirred at a rate of 500 rpm at 75°C by a mechanical stirrer for 30 min. Since the by-product, NaCl, is soluble in water the reaction media was filtered by using Büchner funnel and flask under 600-mmHg vacuum level. To remove the NaCl completely, wet $CaSt_2$ was washed by 200 cm³ deionized water once and then, wet $CaSt_2$ cake was dried in a vacuum oven under 2×10^4 Pa pressure. The KBr disc spectrum of the powder was taken with Shimadzu FTIR spectrophotometer. The SEM micrograph of the dried powder was taken with Scanning electron microscopy (Philips XL30 SFEG).

18.2.3 LUBRICANT PREPARATION

About 1 g of $CaSt_2$ and 100 cm³ spindle oil were mixed together at 160°C at 880 revolution min⁻¹ rate for 30 min and then cooled to 25°C by continuously stirring. At the mixing experiments, a heater and magnetic stirrer (Ika Rh Digital KT/C) and a thermocouple (IKA Werke) were used. The experiment was repeated with 2 g of $CaSt_2$ in 100 cm³ oil.

18.2.4 LUBRICANT CHARACTERIZATION

The dispersion of $CaSt_2$ in base oil was observed by optical microscopy. The phase change behavior of $CaSt_2$ and lubricants with increasing temperature was observed with an optical microscope equipped with a hot plate. The stabilities of the lubricants having different calcium stearate contents were determined by measuring the rate of settling of calcium stearate particles in base oils. The chemical structures of calcium stearate and the prepared lubricants were investigated by FTIR spectroscopy. The tribologic behavior of the lubricants was tested with a four-ball tester. Four ball tests were done using the four-ball tester from DUCOM Corporation (Figure 18.1) to determine the friction coefficient and wear scar diameter of the lubricants. The test was performed according to ASTM D 4172–94 at 392 N and 1200 rpm and the test duration was 1 h. The wear scar diameter was reported as the average of the wear scar diameter of the three fixed balls.

The visible spectrum of base oil separated by centrifugation from base oil was taken by using Perkin Elmer UV-Vis spectrophotometer by using base oil without any additive as the reference.

18.2.5 OPTICAL MICROSCOPE

Melting behavior of CaSt$_2$ in powder form and in dispersed form in the base oil on a microscope slide was observed by using the transmission optical microscope (Olympus, CH40) with a heated hot stage controlled by a temperature controller (Instec, STC 200C). The samples were heated at 5°C/min rate from room temperature up to 190°C. The photographs were taken with Camedia Master Olympus Digital camera.

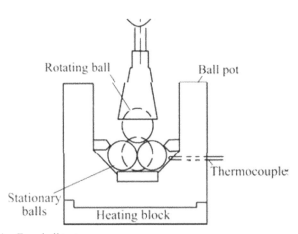

FIGURE 18.1 Four ball tester.

18.3   RESULTS AND DISCUSSION

18.3.1   CaSt₂ POWDERS

FTIR spectrum of calcium stearate powder obtained by precipitation process is shown in Figure 18.2. The characteristic peaks of calcium stearate at 1542 cm⁻¹ and 1575 cm⁻¹ were observed. These bands are due to anti-symmetric stretching bands for unidendate and bidendate association of carboxylate groups with calcium ions [5, 7]. Antisymmetric and symmetric methylene stretching, and methylene scissoring bands ($v_a CH_2$, $v_s CH_2$, and $\delta_s CH_2$) were observed at about 2914 cm⁻¹, 2850 cm⁻¹ and 1472 cm⁻¹, respectively. These bands are due to the alkyl chain in the calcium stearate structure [5, 7].

The SEM micrograph of the CaSt₂ powder shown in Figure 18.3 indicated that the particles were flat in in shape and had a broad size distribution ranging from 200 nm to 1 μm. The average diameter of particles was 600 nm.

18.3.2   FTIR SPECTRA OF LUBRICANTS

The prepared lubricants were also examined by FTIR spectroscopy. Their FTIR spectra are shown in Figure 18.4. The peaks at 2918 and 2848 cm⁻¹, 1454 cm⁻¹ are due to stretching and bending vibrations of the methylene groups in base oil structure. The stretching and bending vibrations of the

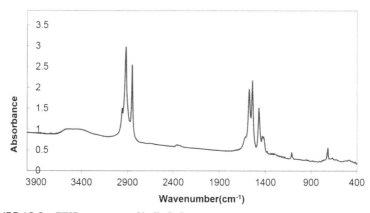

FIGURE 18.2   FTIR spectrum of bulk CaSt₂.

FIGURE 18.3 SEM micrograph of CaSt$_2$ powder.

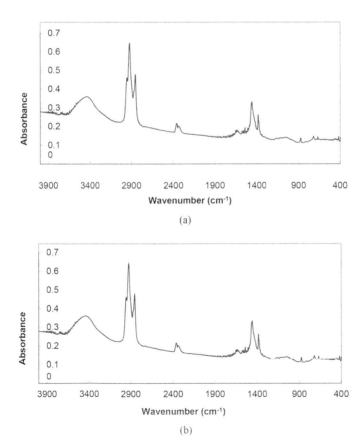

(a)

(b)

FIGURE 18.4 FTIR spectra of lubricants with a. 1% CaSt$_2$ b. 2% CaSt$_2$.

methyl group are observed at 2951 and 1385 cm^{-1}. At 3414 cm^{-1} a broad peak related to hydrogen bonded OH groups are present. The antisymmetric stretching bands for unidendate and bidendate association of carboxylate groups with calcium ions at 1542 cm^{-1} and 1575 cm^{-1} are observed as small peaks in the spectra.

18.3.3   STABILITY OF LUBRICANTS AND THE PARTICLE SIZE OF THE CaSt$_2$ DISPERSED IN BASE OIL

The stability of the lubricant suspensions was determined by recording the height of the line separating the oil phase and the suspension phase. Due to gravity settling of the particles the level of this line decreases continuously with time as seen in Figure 18.5. The settling velocity is directly proportional to the radius of the particle as shown in Eq. (2) [8].

$$dx/dt = 2r^2(\rho-\rho_o) \, g/ \, 9\eta \qquad\qquad (2)$$

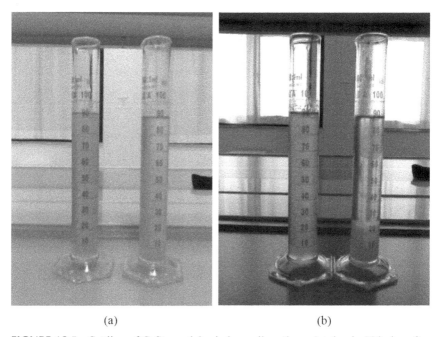

(a)   (b)

FIGURE 18.5   Settling of CaSt$_2$ particles in base oil on the a. 1st day b. 15th day after mixing.

where: dx/dt is rate of settling (cm/s); ρ_o is the density of medium (g/cm³), ρ is the density of particle (g/cm³), η is viscosity of medium (g/cm.s), r is radius of particle (cm), g is 981 cm/s². The radius of particles was calculated from the slopes of the lines in Figure 18.6. The results were evaluated for the settling of particles within 15 days. The oil density and viscosity used for the calculations are 0.86 g/cm³ and 0.35 (g/cm.s). The density of $CaSt_2$ is 1.12 g/cm³. The initial rate of settling was calculated as 0.188×10^{-7} cm/s and 0.635×10^{-12} cm/s for oils with 1% and 2% $CaSt_2$, respectively, from Figure 18.6. Apparent radius of the $CaSt_2$ particles dispersed in base oil was 1.88 μm and 0.11 μm, respectively, for 1% and 2% $CaSt_2$ added samples, respectively. The $CaSt_2$ particles were molten and recrystallized in base oil during preparation of the lubricant. Thus they have a different particle size than the original powder. At higher $CaSt_2$ content the formed $CaSt_2$ crystals were in smaller size due to fast nucleation and slow growth of crystals. The gelation of $CaSt_2$ and base oil system is also another possibility affecting apparent size of particles.

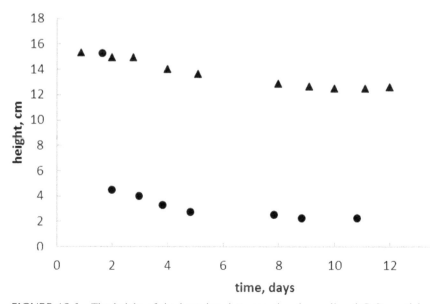

FIGURE 18.6 The height of the boundary between clear base oil and $CaSt_2$ particles settling in base oil.

18.3.4 MELTING BEHAVIOR OF PURE CaSt₂ AND CaSt₂ IN MINERAL OIL

CaSt₂ powder melts at 120°C as determined by DSC and at 148°C by optical microscopy in a previous study [3]. In Figure 18.7 micrographs of the CaSt₂ powders before and after melting are seen. Before melting CaSt₂ appears as a white powder and on melting it is transformed into a transparent liquid. It was found that CaSt₂ had a melting temperature of 142.8°C by optical microscopy in the present study. CaSt₂ particles in base oil also showed a similar phase transition behavior as bulk CaSt₂. They were dispersed as particles in base oil at room temperature. The particles kept their shape up to 113°C and they melted and mixed with mineral oil homogeneously at 128°C as seen in Figure 18.8.

18.3.5 FRICTION AND WEAR BEHAVIOR OF THE LUBRICANTS

The lubricants with CaSt₂ efficiently decreased the friction and wear between metal surfaces. The four ball test results are shown in Table 18.1, wear scar's optical micrographs are seen in Figure 18.9 and the change of friction coefficient during 1 h test duration is seen in Figure 18.10. The friction coefficient and wear scar diameter of base oil 0.099 and 1402 nm were reduced to 0.0730 and 627.61 nm, respectively, for the lubricant

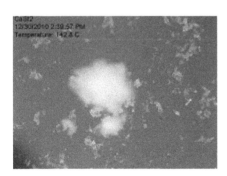
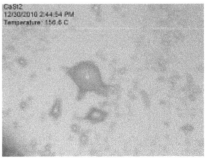

(a)   (b)

FIGURE 18.7 Optical micrographs of CaSt₂ powder at: (a) 142.8°C, and (b) 156.6°C.

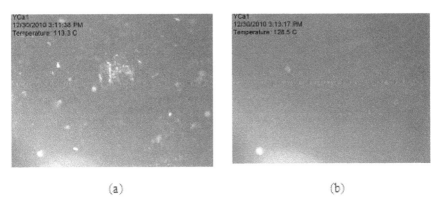

(a)   (b)

FIGURE 18.8   Optical micrographs of CaSt$_2$ (1%) dispersed in mineral oil at: (a) 113°C, (b) 128°C.

TABLE 18.1   Friction Coefficient and Wear Scar Diameter of Base Oil and Lubricants With 1% and 2% CaSt$_2$

Property	Base oil [2]	Base oil with 1% CaSt$_2$	Base oil with 2% CaSt$_2$
Friction coefficient	0.099	0.0730	0.8150
Wear Scar Diameter, nm	1402	627.61	540.88

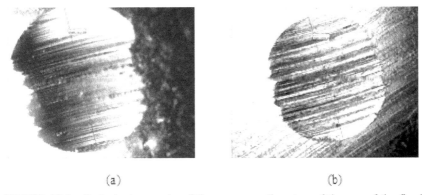

(a)   (b)

FIGURE 18.9   Optical micrographs of the wear scar diameters of the one of the fixed balls of four ball tests for: (a) 1% CaSt$_2$, (b) 2% CaSt$_2$ containing lubricant.

having 1% CaSt$_2$. For 2% CaSt$_2$ containing lubricant the friction coefficient and the wear scar diameter were 0.815 and 0.540, respectively. As the

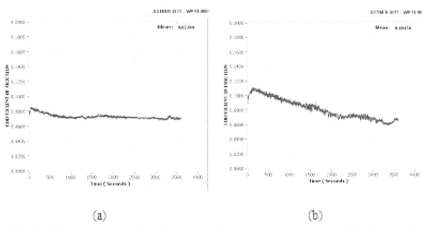

FIGURE 18.10 Change of the friction coefficient with time for: (a) 1% CaSt$_2$, (b) 2% CaSt$_2$ containing lubricants.

CaSt$_2$ content increased better lubricating efficiency were observed. The four ball tests are done at 75°C. At this temperature CaSt$_2$ is in solid form in base oil. However, by the kinetic energy of the rotating ball over fixed balls the temperature of the oil should have been increased to melt the CaSt$_2$ crystals in base oil and to cover the surface of the balls by a smooth lubricating layer. The solid CaSt$_2$ particles similar to other nano particles can also fill the crevices and holes on the steel surface reducing the friction and wear.

18.3.6   THE EFFECT OF FOUR BALL TESTS ON THE COLOR OF THE BASE OIL

The lubricants change their color due to oxidation, hydrolysis and thermal degradation during its use. Contaminants from the eroding surfaces also change the color of the oil. The solid colorants in the lubricating oil can be filtered and the filter surface color can be measured [9]. In the present study the color change of the lubricating oils during four ball tests were investigated by visible spectroscopy. The visible spectra of the lubricants shown in Figure 18.11 were taken using the base oil as the reference. The base oil with 1% CaSt$_2$ were lighter in color than the reference base oil as indicated by the negative absorbance values in Figure 18.11. CaSt$_2$

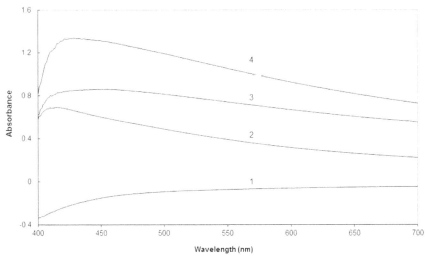

FIGURE 18.11 Visible spectra of the base oil: (1) before four ball tests, (2) after four ball tests for 1% CaSt$_2$ and (3) before four ball tests, and (4) after four ball tests for 2% CaSt$_2$.

adsorbed the coloring material initially existing in the base oil. After four ball test the base oil become dark yellow due to oxidation and crosslinking reactions in base oil. The base oil having 2% CaSt$_2$ had higher absorbance values at all wavelengths and the absorbance was maximum at 420 nm. It also had a darker color after the test. Thus CaSt$_2$ improves the lubricating efficiency of the base oil, but it does not increase the oxidative and thermal stability. Adding antioxidants to the system would help the thermal and oxidative stability, which could be the subject of further investigations.

18.4 CONCLUSION

Calcium stearate powder prepared from sodium stearate and calcium chloride by precipitation from aqueous solutions and Light Neutral Base oil were mixed together to obtain lubricating oils. It was found that CaSt$_2$ powder had a melting temperature of 142.8°C and in the base oil it melted above 128°C. From rate of settling of the particles in base oil the size of dispersed particles were found to be 1.88 μm and 0.11 μm, respectively, for lubricants having 1% and 2% CaSt$_2$. The friction coefficient (0.099) and wear scar diameter of base oil (1402 nm) were reduced to 0.0730 and 627.61 nm, respectively, for the lubricant having 1% CaSt$_2$. Lower wear scar diameter (540 nm) was obtained for lubricant with 2% CaSt$_2$. Calcium stearate when added to base oil reduces the friction and wear of metal surfaces sliding on each other. It covers the cracks and groves of the metal surface with a smooth film. Thus it is an efficient lubricant additive. However, CaSt$_2$ did not increased oxidative and thermal stability of the base oil. Thus further studies for the antioxidant selection should be made.

ACKNOWLEDGEMENTS

The authors thank to Opet Fuchs Turkey for the four ball tests.

KEYWORDS

- calcium stearate
- oxidative and thermal stability
- melting temperature
- optical micrographs
- lubricants
- friction coefficient

REFERENCES

1. Felder, E., Levrau, C., Mantel, M., Dinh, N. G. T. (2011) "Experimental study of the lubrication by soaps in stainless steel wire drawing" Proceedings Of The Institution Of Mechanical Engineers Part J-Journal Of Engineering Tribology, 225(19, J9), 915–923.
2. Palichev, T., Kramis, F. S., Petkov, P. (2008) "Operating-conservation motor oil with composite additives" Oxidation Communications. 31 (1) 223–230.
3. Hughes, E. C., Harrison, M. S. (1956) US2995516 A, 1956.
4. Savrık, A. S., Balköse D, Ülkü, S. (2011) "Synthesis of zinc borate by inverse emulsion technique for lubrication" J. Therm. Anal. Calorim. 104, 605–612.
5. Gönen, M., Öztürk, S., Balköse, D., Ülkü, S. (2010) "Preparation and Characterization of Calcium Stearate Powders and Films Prepared by Precipitation and Langmuir-Blodgett Techniques", Ind. and Eng. Chem. Res. 49, 1732–1736.

6. Moreira, A. P. D., Souza, B. S., Teixeira, A. M. R. F. (2009) "Monitoring of calcium stearate formation by thermogravimetry" J Therm Anal Calorim. 97, 647–652.
7. Lu, Y., Miller, J. D. (2002) "Carboxyl stretching vibrations of spontaneously adsorbed and LB transferred calcium carboxylates as determined by FTIR internal reflection spectroscopy" J. Colloid and Interface Sci. 256, 41–52.
8. Alberty, R. (1997) Physical chemistry, McGraw Hill, New York.
9. Akira Sasaki, Hideo Aoyama, Tomomi Honda, Yoshiro Iwai, Yong, C. K. (2014) A Study of the Colors of Contamination in Used Oils, Tribology Transactions, 57(1), 1–10.

CHAPTER 19

A NEW APPROACH TO THE CREATION CARBON-POLYMER NANOCOMPOSITES WITH POLYETHYLENE AS A BINDER

SERGEI V. KOLESOV,[1] MARINA V. BAZUNOVA,[2] ELENA I. KULISH,[3] DENIS R. VALIEV,[4] and GENNADY E. ZAIKOV[5]

[1]Professor of the Department of High-Molecular Connections and General Chemical Technology of the Chemistry Faculty of the Bashkir State University, 450076, Ufa, Zaks Validi Street, 32, Russia, Tel.: (347) 229–96–86; E-mail: kolesovservic@mail.ru

[2]Docent of the Department of High-Molecular Connections and General Chemical Technology of the Chemistry Faculty of the Bashkir State University, 450076, Ufa, Zaks Validi Street, 32, Russia, Tel.: (347) 229–96–86; Mobile: 89276388192; E-mail: mbazunova@mail.ru

[3]Professor of the Department of High-Molecular Connections and General Chemical Technology of the Chemistry Faculty of the Bashkir State University, 450076, Ufa, Zaks Validi Street, 32, Russia, Tel.: (347) 229–96–86; E-mail: onlyalena@mail.ru

[4]Student of the Department of High-Molecular Connections and General Chemical Technology of the Chemistry Faculty of the Bashkir State University, 450076, Ufa, Zaks Validi Street, 32, Russia, Tel.: (347) 229–96–86; F-mail: valief@mail.ru

[5]Institute of Biochemical Physics Named N.M. Emanuel of Russian Academy of Sciences, 4 Kosygina Street, 119334, Moscow, Russia, E -mail: chembio@sky.chph.ras.ru

CONTENTS

ABSTRACT

A new approach of obtaining of the molded composites on the basis of the mixtures of the powders of nano-dispersed polyethylene, cellulose and the ultra-dispersed carbonic materials is developed. These materials possess the assigned sorption properties and the physic-mechanical characteristics. They are suitable for the usage at the process of cleaning and separation of gas mixture.

19.1   INTRODUCTION

In solving problems of environmental protection, medicine, cleaning and drying processes of hydrocarbon gases are indispensable effective sorbents, including polymer nanocomposites derived from readily available raw materials.

The nature of the binder and the active components, and molding conditions are especially important at the process of sorption-active composites creating. These factors ultimately exert influence on the development of the porous structure of the sorbent particles and its performance. In this regard, it is promising to use powders of various functional materials having nanoscale particle sizes at the process of such composites creating. First, high degree of homogenization of the components facilitates their treatment process. Secondly, the high dispersibility of the particles allows them to provide a regular distribution in the matrix, whereby it is possible to achieve improved physical and mechanical properties. Third, it is possible to create the composites with necessary sorption, magnetic, dielectric and other special properties combining volumetric content of components [1].

Powders of low density polyethylene (LDPE) prepared by high temperature shearing (HTS) used as one of prospective components of the developing functional composite materials [2, 3].

Development of the preparation process and study of physicochemical and mechanical properties of sorbents based on powder mixtures of LDPE, cellulose (CS) and carbon materials are conducted. As the basic sorbent material new – ultrafine nanocarbon (NC) obtained by the oxidative condensation of methane at a treatment time of 50 min (NC1) and 40 min (NC2) having a specific surface area of 200 m^2/g and a particle size of 30–50 nm is selected [4]. Ultrafine form of NC may give rise to technological difficulties, for example, during regeneration of NC after using in gaseous environments, as well as during effective separation of the filtrate from the carbon dust particles. This imposes restrictions on the using of NC as an independent sorbent. In this connection, it should be included in a material that has a high porosity. LDPE and CS powders have great interest for the production of such material. It is known that a mixture of LDPE and CS powders have certain absorption properties, particularly, they were tested as sorbents for purification of water surface from petroleum and other hydrocarbons [5].

Thus, the choice of developing sorbents components is explained by the following reasons:

1. LDPE has a low softening point, allowing to conduct blanks molding at low temperatures. The very small size of the LDPE particles (60–150 nm) ensures regular distribution of the binder in the matrix. It is also important that the presence of binder in the composition is necessary for maintaining of the material's shape, size, and mechanical strength.

2. Usage of cellulose in the composite material is determined by features of its chemical structure and properties. CS has developed capillary-porous structure, that's why it has well-known sorption properties [5] towards polar liquids, gases and vapors.

3. Ultrafine carbon components (nanocarbon, activated carbon (AC)) are used as functionalizing addends due to their high specific surface area.

19.2   EXPERIMENTAL PART

Ultrafine powders of LDPE, CS and a mixture of LDPE/CS are obtained by high temperature shearing under simultaneous impact of high pressure and shear deformation in an extrusion type apparatus with a screw diameter of 32 mm [3].

Initial press-powders obtained by two ways. The first method is based on the mechanical mixing of ready LDPE, CS and carbon materials' powders. The second method is based on a preliminary high-shear joint grinding of LDPE pellets and sawdust in a specific ratio and mixing the resulting powder with the powdered activated carbon (БАУ-А mark) and the nanocarbon after it.

Composites molding held by thermobaric compression at the pressure of 127 kPa.

Measuring of the tablets strength was carried out on the automatic catalysts strength measurer ПК-1.

The adsorption capacity (A) of the samples under static conditions for condensed water vapor, benzene, n-heptane determined by method of complete saturation of the sorbent by adsorbate vapor in standard conditions at 20°C [6] and calculated by the formula: A=m/(M·d), wherein m – mass of the adsorbed benzene (acetone, n-heptane), g; M – mass of the dried sample, g; d – density of the adsorbate, g/cm^3.

Water absorption coefficient of polymeric carbon sorbents is defined by the formula: $K = \dfrac{m_{absorbed..water}}{m_{sample}} \times 100\%$, wherein $m_{absorbed\ water}$ is mass of the water, retained by the sorbent sample, m_{sample} is mass of the sample.

Experimental error does not exceed 5% in all weight methods at P = 0.95 and the number of repeated experiments n=3.

19.3   RESULTS AND DISCUSSION

Powder components are used as raw materials for functional composite molding (including the binder LDPE), because molding of melt polymer mixtures with the active components has significant disadvantages. For example, the melt at high degrees of filling loses its fluidity, at low degrees of filling flow rate is maintained, but it is impossible to achieve the required material functionalization.

It is known that amorphous-crystalline polymers, which are typical heterogeneous systems, well exposed to high-temperature shear grinding process. For example, the process of HTS of LDPE almost always achieves a significant results [3]. Disperse composition is the most important feature of powders, obtained as result of high-temperature shear milling. Previously, on the basis of the conventional microscopic measurement, it was believed that sizes of LDPE powder particles obtained by HTS are within 6–30 micrometers. Electron microscopy gives the sizes of 60 to 150 nm. The active powder has a fairly high specific surface area (up to 2.2 m^2/g).

The results of measurement of the water absorption coefficient and of the static capacitance of LDPE powder by n-heptane vapor are equal to 12% and 0.26 cm^3/g, respectively. Therefore, the surface properties of LDPE powder more developed than the other polyethylene materials'.

19.3.1   SELECTION OF MOLDING CONDITIONS OF SORBENTS BASED ON MIXTURES OF LDPE, CS AND ULTRAFINE CARBON MATERIALS' POWDERS

Initial press-powders obtained by two ways. The first method is based on the mechanical mixing of ready LDPE, CS and carbon materials' powders. The second method is based on a preliminary high-shear joint grinding of LDPE pellets and sawdust in a specific ratio and mixing the resulting powder with the powdered activated carbon and the nanocarbon after it. The method of molding – thermobaric pressing at a pressure of 127 kPa.

The mixture of LDPE/CS compacted into cylindrical pellets at a temperature of 115–145°C was used as a model mixture for selection of composites molding conditions. Pressing temperature should be such that the LDPE softens but not melts, and at the same time forms a matrix to prevent loss of specific surface area in the ready molded sorbent due to fusion of pores with the binder. The composites molded at a higher temperature, have a lower coefficient of water absorption than the tablets produced at a lower temperature, that's why the lowest pressing temperature (120°C) is selected. At a higher content of LDPE the water absorption coefficient markedly decreases with temperature.

Cellulose has a high degree of swelling in water (450%) [5], this may lead to the destruction of the pellets. Its contents in samples of composites, as it has been observed by the sorption of water, should not exceed 30 wt. %. There is a slight change of geometric dimensions of the pellets in aqueous medium at an optimal value of the water absorption coefficient when the LDPE content is 20 wt. %.

Samples of LDPE/CS with AC, which sorption properties are well studied, are tested for selecting of optimal content of ultrafine carbon. The samples containing more than 50 wt. % of AC have less water absorption coefficient values. Therefore, the total content of ultrafine carbon materials in all samples must be equal to 50 wt. %.

Static capacitance measurement of samples, obtained from mechanical mixtures of powders of PE, CS and AC, conducted on vapors of n-heptane and benzene, to determine the effect of the polymer matrix on the sorption properties of functionalizing additives. With a decrease of the content of AC in the samples with a fixed (20 wt. %) amount of the binder, reduction of vapor sorption occurs. It indicates that the AC does not lose its adsorption activity in the composition of investigated sorbents.

Strength of samples of sorbents (Figure 19.1) is in the range of 620–750 N. The value of strength is achieved in the following molding conditions: t = 120°C and a pressure of 127 kPa.

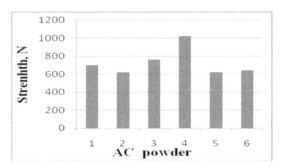

FIGURE 19.1 Comparison of strength of pellets, based on LDPE, CS (different species of wood) and AC powders.1 – sorbent of LDPE/AC/CS = 20/50/30 wt. % based on the powders of jointly dispersed pellets of LDPE and softwood sawdust with subsequently addition of AC; 2 – sorbent of LDPE/AC/CS = 20/50/30 wt. % based on the powders of jointly dispersed pellets of LDPE and hardwood sawdust with subsequently addition of AC; 3 – sorbent of LDPE/AC/CS = 20/50/30 wt. % based on the mechanical mixtures of the individual powders of LDPE, CS from softwood and AC; 4 –AC tablet; 5 – sorbent of LDPE/CS = 20/80 wt. %; 6 – sorbent of LDPE/AC = 20/80 wt. %

Thus, optimal weight composition of the matrix of LDPE/CS composition – 20/30 wt.% with 50 wt. % containing of carbon materials.

19.3.2 SORPTION PROPERTIES OF CARBON – POLYMER COMPOSITES BY CONDENSED VAPORS OF VOLATILE LIQUIDS

For a number of samples of sorbents static capacitance values by benzene vapor is identified (Figure 19.2). They indicate that the molded mechanical mixture of 20/25/25/30 wt.% LDPE/AC/NC1/CS has a maximum adsorption capacity that greatly exceeds the capacity of activated carbon. High sorption capacity values by benzene vapor appears to be determined by weak specific interaction of π-electron system of the aromatic ring with carbocyclic carbon skeleton of the nanocarbon [7].

Static capacitance of obtained sorbents by heptane vapors significantly inferiors to capacity of activated carbon (Figure 19.3), probably it is determined by the low polarizability of the molecules of low-molecular alkanes. Consequently, the investigated composites selectively absorb benzene and can be used for separation and purification of mixtures of hydrocarbons.

Molded composite based on a mechanical mixture of LDPE/AC/NC1/CS = 20/25/25/30 wt. % has a sorption capacity by acetone vapor comparable with the capacity of activated carbon (0.36 cm^3/g) (Figure 19.4).

Sorbents' samples containing NC2 have a low values of static capacity by benzene, heptanes and acetone vapor. It can be probably associated

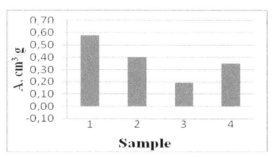

FIGURE 19.2 Static capacitance of sorbents, A (cm^3/g) by benzene vapor (20°C). 1 –molded mechanical mixture of LDPE/AC/NC1/CS= 20/25/25/30wt. %; 2 – molded mechanical mixture of LDPE/AC/NC2/CS = 20/25/25/30 wt. %; 3 – molded mechanical mixture of LDPE/AC/CS=20/50/30 wt. %; 4 – AC medical tablet (controlling)

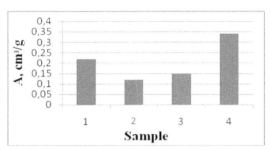

FIGURE 19.3   Static capacitance of sorbents, A (cm^3/g) by n-heptane vapor (20°C). 1 – molded mechanical mixture of LDPE/AC/NC1/CS= 20/25/25/30wt. %; 2 – molded mechanical mixture of LDPE/AC/NC2/CS = 20/25/25/30 wt. %; 3 – molded mechanical mixture of PE/AC/CS=20/50/30 wt. %; 4 – AC medical tablet (controlling).

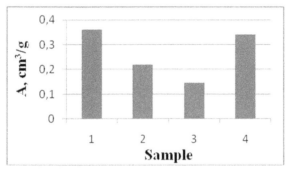

FIGURE 19.4   Static capacitance of sorbents, A (cm^3/g) acetone vapor (20°C). 1 – molded mechanical mixture of LDPE/AC/NC1/CS= 20/25/25/30wt. %; 2 – molded mechanical mixture of LDPE/AC/NC2/CS = 20/25/25/30 wt. %; 3 – molded mechanical mixture of LDPE/AC/CS=20/50/30 wt. %; 4 – AC medical tablet (controlling).

with partial occlusion of carbon material pores by remnants of resinous substances – by products of oxidative condensation of methane, and insufficiently formed porous structure.

The residual benzene content measuring data (Table 19.1) shows that the minimal residual benzene content after its desorption from the pores at t = 70°C for 120 min observes in case of sorbent LDPE/AC/NC1/CS composition = 20/25/25/30 wt. %. It allows to conclude that developed

TABLE 19.1   Sorbents' Characteristics: Total Pore Volume V_{tot}: Static Capacitance (A) by Benzene Vapors at the Sorption Time of 2 days; Residual Weight of the Absorbed Benzene After Drying at t = 70°C for 120 min

LDPE/AC/NC/CS sorbent composition, wt. %	V_{tot}, cm³/g	A, cm³/g	Residual benzene content as a result of desorption, %
20/25/25/30	1.54	0.5914	2.9
20/50/ – /30	1.21	0.1921	10.3
- /100/ – / -	1.60	0.3523	32.0

sorbents have better ability to regenerate under these conditions in comparison with activated carbon.

19.4   CONCLUSIONS

Thus, the usage of nanosized LDPE as a binder gives a possibility to get the molded composite materials with acceptable absorption properties. Optimal conditions for molding of sorbents on the basis of mixtures of powdered LDPE, cellulose and ultrafine carbon materials were determined: temperature 120°C and pressure of 127 kPa, content of the binder (polyethylene) is 20 wt. %.

Varying the ratio of the components of the compositions on the basis of ternary and quaternary mixtures of powdered LDPE, cellulose and ultrafine carbon materials it is possible to achieve the selectivity of sorption properties by vapors of certain volatile liquids. Established that molded mechanical mixture of LDPE/AC/NC1/CS 20/25/25/30wt. % has a static capacity by condensed vapors of benzene and acetone 0.6 cm³/g and 0.36 cm³/g respectively, what exceeds the capacity of activated carbon. The static capacitance of the compositions by the n-heptane vapors is 0.21 cm³/g, therefore, the proposed composites are useful for separation and purification of gaseous and steam mixtures of different nature.

Developed production method of molded sorption-active composites based on ternary and quaternary mixtures of powdered LDPE, cellulose and ultrafine carbon materials can be easily designed by equipment and can be used for industrial production without significant changes.

KEYWORDS

- cellulose
- high-temperature shift crushing
- nano-carbon
- polyethylene
- sorbents

REFERENCES

1. Akbasheva, E. F., Bazunova, M. V. Tableted sorbents based on cellulose powder mixtures, polyethylene and ultra-dispersed carbon. *Materials Open School Conference of the CIS "Ultrafine and Nanostructured Materials" (11–15 October 2010),* Ufa: Bashkir State University, p.106. (2010)
2. Enikolopyan, N. S., Fridman, M. L., Karmilov, A. Yu. Elastic-deformation grinding of thermo-plastic polymers. *Reports AS USSR*, 296, №1, pp. 134–138. (1987)
3. Akhmetkhanov, R. M., Minsker, K. S., Zaikov, G. E., On the mechanism of fine dispersion of polymer products at elastic deformation effects. *Plasticheskie Massi*, №8, pp.6–9. (2006).
4. Aleskovskiy, V. B., Galitseisky, K. B., Russian patent "Method of Ultrafine Carbon" № 2287543 from 20.11.2006.
5. Raspopov, L. N., Russiyan, L. N., Zlobinsky, Y. I., Waterproof composites comprising wood and polyethylene dispersion. *Russian Polymer Science Journal*. 2007. Б, 50, №3. pp. 547–552.
6. Keltsev, N. V., *Fundamentals of Adsorption Technology*. Moscow: Chemistry, 1984, 595 p.
7. Valinurova, E. R., Kadyrov, A. D., Kudasheva, F. H., Adsorption properties of carbon rayon// *Vestn. Bashkirs. Univer.*, V. 13, № 4. pp. 907–910 (2008).

CHAPTER 20

THERMOOXIDATION OF THE BLENDS LOW DENSITY POLYETHYLENE AND BUTYL RUBBER

T. V. MONAKHOVA,[1] L. S. SHIBRYAEVA,[1] N. N. KOLESNIKOVA,[1] A. I. SERGEEV,[2] S. G. KARPOVA,[1] and A. A. POPOV[1]

[1]*Institute of the Russian Academy of Sciences N.M. Emanuel Institute of Biochemical Physics, Russian Academy of Sciences, Moscow, Russia*

[2]*Institute of the Russian Academy of Sciences N.N. Semenov Institute of Chemical Physics, Russian Academy of Sciences, Moscow, Russia*

CONTENTS

20.1   INTRODUCTION

Studies aimed at developing new polymeric composites consisting of plastic and rubber, which do not require vulcanization and reinforcement, represent a significant interest at present. These studies are mainly based on the fact that the introduction of polyethylene in elastomers (butyl rubber, ethylene propylene diene rubbers and others) makes it possible to receive the systems with a sufficiently high cohesive strength [1–3]. Method for creating these materials – mechanical melt mixing, makes their resistance to thermal and thermo-oxidative destruction an important problem. The same property is needed for processing and use of the products made of polymer composites. The main challenge facing the researchers and manufacturers occupied in the field of developing polymeric materials is to increase their thermo-oxidative stability. The study was aimed at establishing a relationship between the structure, molecular dynamics and thermo-oxidative stability of low density polyethylene and butyl rubber blends.

20.2   EXPERIMENTAL PART

Binary blends of butyl rubber and low density polyethylene were investigated. We used butyl rubber 1675N (hereinafter BR) and polymer LDPE 273–76. The blends contained 0, 10, 20, 30, 40, 60, 70, 80 and 100 wt % of PE were prepared in a Brabender type mixer at a temperature of 170°C. The rotor speed was 60 r / min. The mixing time was 15 min. Film samples were obtained by compressing on a laboratory press at a temperature of 170°C, followed by rapid cooling to room temperature.

Thermophysical parameters were determined by differential scanning calorimetry, using the microcalorimeter DCM-10. As a standard was used Indium (Tm = 156.5°C; specific enthalpy is 28.4419 J/g). The thermal melting effect of the samples was determined by the peak area between the DSC curve and the baseline. We calculated the enthalpy of melting, based on the obtained thermograms. To determine the degree of crystallinity of LDPE, the value of the specific melting heat of PE crystallites was assumed to be 288 J/g. Error in the determination of melting point did not exceed 1°C, the melting heat – 10%.

The structure of PE and BR polymer chains was determined by IR spectroscopy. Error in the determination of structural parameters did not exceed 15%

To determine segmental mobility of the chains in the amorphous regions of blends we used the paramagnetic probe method. A stable nitroxyl radical 2,2,6,6-tetramethylpyperidyn-1-oxyl was used as a paramagnetic probe, which was injected in the polymeric film of saturated vapor. Spectra of nitroxyl radicals introduced in the sample were obtained by EPR. The radical probe correlation time (τ_c) was calculated from the spectrum by the formula [4]:

$$\tau = \Delta H + (\sqrt{I+/I_-} - 1)\, 6.65 \times 10^{-10},$$

where $\Delta H+$ – width of the spectral components, located in the weak field, $(\sqrt{I+/ I_-} -1)$ – intensity component in the weak and strong field, respectively.

Error in the determination of τ was within 5%.

Proton transverse magnetic relaxation PE-BK samples was performed on Bruker Minispek PC-120 spectrometer. This spectrometer operates at proton resonance frequency of 20 MHZ. The length of 90° pulse – 2.7 μs and dead time 7 μs. Two different pulse sequences were used for measurement of T2 relaxation time and the amount of a rigid (crystalline) and soft (amorfous) phase components. To evaluate the spin-spin relaxation time T2 and the fraction of protons with different degrees of mobility, we used techniques for studying induced signal decay after a 90° pulse (FID) and (CPMG).

Error in the determination of parameters did not exceed 5%.

Kinetic oxidation curves of the mixtures were obtained using a nanometric device with a circulating pump and by freezing volatile oxidation products at a temperature of 180°C and an oxygen pressure of 300 mm Hg (40 kPa). Error in the determination of kinetic parameters did not exceed 10%.

20.3   RESULTS AND DISCUSSION

The purpose of the research was to study binary blends of butyl rubber and low density polyethylene, containing 0, 10, 20, 30, 40, 60, 70, 80 and 100 wt % of PE. DSC method was used to determine the structural parameters for the crystalline regions of the sample of blends. Thermal and thermophysical parameters were determined by the same method. The results of research are presented in Table 20.1 for all the mixtures. The

TABLE 20.1 Thermal and Thermo-Physical Parameters of BR – PE Blends

Composition of BR/PE sample, wt %	0/100	20/80	40/60	60/40	70/30	80/20
$T_{m\,max,}$ °C	129.0	129.0	128.0	128.0	127.0	127.0
ΔH_m of the blends, J/g	180.0	143.0	111.0	72.0	48.0	30.0
$\Delta H_{m\,PE}$, J/g	180.0	179.0	185.0	180.0	160.0	150.0
χ_{PE}, %	61.0	61.0	63.0	61.0	54.0	51.0

melting endotherms of the sample of mixtures have a single melting peak of PE. As seen from the table data, the temperature at the maximums of PE melting peaks is insignificantly shifted towards low temperatures with a decrease in the PE content of the mixture, which may be associated with the formation of defective or smaller crystalline structures in the LDPE phase. At the same time, the degree of crystallinity of polyethylene (χ) decreases with the increase in its content only in the samples with a high rubber content (70 wt. %). This is most likely to be associated with the changes in the phase structure of the mixture, phase inversion. Rubber forms a dispersion medium in the samples of this composition, where PE is distributed in the form of small particles of the dispersed phase.

The changes in the structure of amorphous regions were determined by IR spectroscopy method. The content of straightened and coiled conformers in the amorphous regions of the PE component was defined by the changes in the intensity of the bands responsible for deformation vibrations of the chains in straightened and folded conformations. The concentration of the former in PE was determined by the intensity of the band at 720 cm^{-1} and the D_{720}/D_{730} ratio. The band at 720 cm^{-1} is responsible for the fluctuations of methylene sequences – (CH2) n, n > 5 in a trans zigzag conformation in the amorphous regions of PE. The content of folded conformers of the TGT and GG type was determined according to the intensities of the bands at 1080 and 1306 cm^{-1}, correspondingly. The obtained data on the changes in the structure of PE chains depending on the composition of the PE-BR blends are shown in Table 20.2. As seen from the table, the content of straightened conformers T-T in the PE component increases, while the content of the coiled GG-conformers decreases with the increased addition of rubber in PE.

TABLE 20.2   The Data on Changes in the Structure of PE and BR Chains Depending on the Composition of PE – BR Blends

PE (mas.%)	$\dfrac{D_{1305}}{D_{2740}}$	$\dfrac{D_{1080}}{D_{2740}}$	$\dfrac{D_{720}}{D_{730}}$	$\dfrac{D_{1230}}{D_{2740}}$	$\dfrac{D_{853}}{D_{2740}}$
0	-	-	-	26.4	0.764
30	-	-	1.263	42.0	0.979
40	0.062	-	1.22	48.7	1.254
60	0.750	0.520	1.19	65.4	1.838
80	0.875	0.417	1.18	96.6	2.720
100	1.229	0.312	1.11	-	-

The nature of changes in the intensities of the above-mentioned bands in the PE component suggests the presence of structural rearrangements by the type of conformational transitions GG→ TGT → TT; the mechanism of this transition was previously established for the PE subjected to orientation drawing. Tensions generated in the transduction chains under the influence of deformations as a result of mixing components may lead to similar changes in the conformational composition of chains in the volume of a polymer. Another reason for the enrichment of PE transduction chains with straightened conformers is implementation of the interphase phenomenon, in other words, the emergence of tension as a consequence of intermolecular interactions at the phase separation border.

The character of the changes in the BR structure can be established from the changes in the intensity of the bands at 1230 and 853 cm^{-1}. They grow monotonously with the increased content of BR, which apparently points to an increase in the size of its particles, and, hence, the length of the PE-BR phase separation border. This leads to a speculation that the enrichment of PE matrix chains by straightened conformers with a decrease of its content in the mixture occurs on the boundary with the surface of the rubber particles. Changes in the structure of components and their chains must lead to a change of molecular dynamics of polymers.

Changes in the structure of components and their chains should result in the alterations of the molecular dynamics of polymers. To determine segmental mobility of the chains in the amorphous regions of mixtures, we used the paramagnetic probe method. A stable nitroxyl radical was used as a paramagnetic probe, which was injected in the polymeric film of saturated vapor. Spectra of nitroxyl radicals introduced in the sample were obtained by EPR. The radical probe correlation time (τ_c) was calculated from the spectrum. Segmental mobility is characterized by the value reciprocal to τ_c. The curves showing the changes in τ_c of the mixture are shown in Figure 20.1. As can be seen, the curve of the relationship between τ_c and the BR content in the mixture has a complex shape. With the increase of the content of BR in the mixture up to 40%, the increase in the correlation time of the radical probe, that is, a drop of the segmental mobility, is observed in the sample with 40% BR. The correlation time of the probe decreases with the growth of the rubber content in the mixture, thus increasing the mobility of the chains. Analysis of the data on the solubility of radical probe in mixtures shows that the highest concentration of the nitroxyl radical probe at the highest correlation time is observed in the region of the phase inversion (Figure 20.2). It is speculated that the rearrangement of the conformational structure of the polymer chains in the PE matrix provides an increase in the free volume and simultaneously reduces the segmental mobility of the chains. Rubber fills the elements in the free volume of the mixture and forms its own phase with the growth

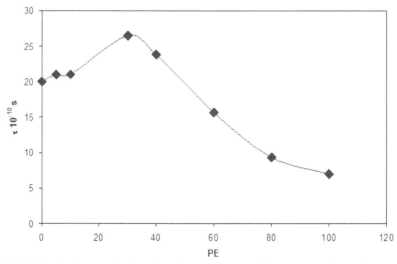

FIGURE 20.1 Curves of changing the mobility of chains PE-matrix in the sample mixtures with a high content of BR

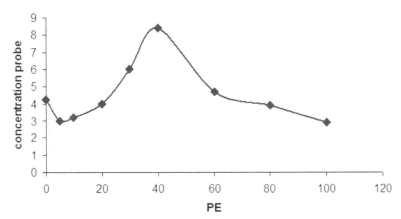

FIGURE 20.2   The curves of the change in the correlation time of the radical probe separately in each phase of PE and BR components.

of its content. This leads to the reduction of the solubility of the nitroxyl radical, but decreases the radical probe correlation time. Given that the intermolecular interactions are absent, the radical probe correlation time characterizing the chain mobility will be determined by the flexibility of PE chains and its changes under the influence of the rubber introduced in polyethylene. Obviously, the maximum concentration of the radical probe corresponds to the phase inversion.

For definition of segmental mobility of chains in the amorphous regions of blends we determined the correlation time (τ_c) paramagnetic probe. Segmental mobility is characterized by the value inverse to the τ_c. Curves demonstrating changing the mobility of chains PE-matrix in the sample mixtures with a high content of BR in the mixture are shown in Figure 20.1.

As you can see, the curve of dependence of the correlation time τ_c of content BR in a mixture is a complex shape. With an increase of BR in the PE to 40%, increasing the correlation time of the radical of the probe, that is, point structural mobility is observed in a sample of 40% BR for which the value of τ_c, exceeds the time that is typical for pure PE and BR. With further increase in the content of the rubber component in a mixture of time correlation probe is reduced, the same time the mobility of the chains increasing.

Analysis of data on the solubility of low-molecular radical probe in the mix shows that the highest concentration is in the longest time observed in the region of phase inversion. Suggesting that a change in the conformational structure of a polymer chain in the PE- matrix provides the increase of the free volume and simultaneously reduces the segmental mobility chains.

Rubber particles fill the items in the free volume PE-matrix and lead to the formation of its phases. This reduces the solubility nitroxyl radical, but reduces the correlation time.

Believing that no intermolecular forces, the correlation time of the radical probe, which characterizes the mobility chain will determine the flexibility PE chain and its change under the influence of rubber particles entered into polyethylene. Apparently, the maximum concentration of radical probe corresponds to phase inversion.

The curves describing the change in the correlation time of the radical probe separately in each phase of PE and BR components are presented in Figure 20.2. As seen, the calculated data, obtained from the ratio of the mixture composition and solubility radical in each component in the range 30–80% BR PE differ from the experimental data. Experimental values significantly above typical for pure BR and PE, shows that the reduction caused by segmental motion mobility chains in both polymers. The fact that mobility polyethylene chain in mixtures with a high content of rubber lower than in mixtures with phase inversion, despite having the highest value of the free volume and lower content of the straightened chains PE, indicates the presence of intermolecular interactions in amorphous regions.

Molecular dynamics samples mixtures PE-BR studied by the method of proton magnetic relaxation on NMR was performed on Bruker Minispek PC-120 spectrometer. This spectrometer operates in the proton resonance frequency of 20 MHZ. Length 90 degree pulse – 2.7 ISS and dead time 7 ISS. Two different pulse sequences used to measure the relaxation time T2 and amount of firm (crystalline) and soft (amorfous) phase components:

1. free induction decay (FID) after a single pulse (90 degrees) excitation;
2. Carr-Purcell-Meiboom-Gill – (CPMG) – multiple-echo pulse sequence $90° - (\tau - 180° - \tau)_n$.

The FID pulse sequence [5, 6] was used to register quickly decaying part of the slow-moving protons (solid phase). The time of relaxation of these protons is a few microseconds and the duration of the collapse of heavily depends on the inhomogeneity of the magnetic field Bo in the sample. It is not possible to use this method for the precise determination of T2, which lasted for more than 100 μs. Method CPMG removes the effect of the inhomogeneity of the magnetic field, but it can be used only for the registration of the slow part of collapse.

Our experiments were performed at the following conditions:

FID
The temperature (t) = 40°C
The recycle delay between scans (RD) = 1 sec
The beginning of the decay measurement (BD) = 10 μs
The end of the decay measurement (ED) = 100 μs
The number of experimental points (n) = 50
The number of scans (NS) = 25

CPMG
t = 40°C, RD = 1 sec, n = 30, BD = 48 μs, ED = 1440 μs
NS–(25–144)
Time between 90° and 180° pulses (τ) – 12 μs

Our results to FID and CPMG experiments presented in Table 20.3.

FID experiment for pure PE and PE-BR compositions (80 and 20 wt% PE) demonstrated low mobile proton fraction for all samples (T2=4–5 ms).

The relative amount of these protons was 90–95%. When the part of BK in composition increased up to 80 wt%, the amount of mobile protons decreased to 47%. This fact is likely to indicate the increase in the segmental mobility of the chains in the mixtures as compared with pure PE and PE matrix. Since PE crystallites can serve clamps for transition chains and, hence, inhibit the relaxation of protons, one of the reasons for the growing mobility of the chains in the mixture may be reduction in the crystalline regions of polyethylene due to dilution of the polymer mass by an amorphous rubber.

The cristallinity of PE for the samples with its different content (from 40 wt% up to 100 wt%) did not change, but the volume of the crystalline phase in PE-BK compositions decreased more than twice. The molecular

TABLE 20.3 Magnetic – Relaxation Characteristics of PE-BR Blends

Composition of BR / PE sample, wt.%	χ of PE, χ of the blend %	SIS		CPMG	
		Proton spin-spin relaxation time T2, mc.sec	Relative content, %	Proton spin-spin relaxation time T2, mc.sec	Relative content, %
0/100	61/61	5.2 194	92.0 8.0	1357 147	49.0 51.0
20/80	61/50	3.3 128	95.0 5.0	1010 150	60.0 40.0
40/60	63/38	–	– –	689 151	74.0 26.0
60/40	61/25	–	– –	636 240	61.0 39.0
70/30	54/16	5.2 214	–	651 239	80.0 20
80/20	51/10	–	47.0 53.0	649 230	85.0 15.0
100/0	0	–	–	653	100

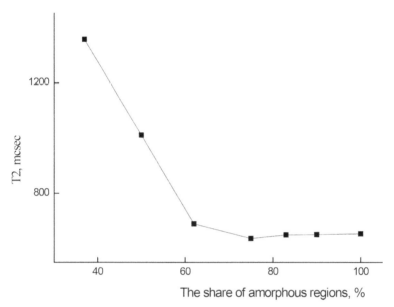

FIGURE 20.3 Proton relaxation time T2 (microseconds) of the slow (S) component depending on the content of the amorphous phase in the PE-BR mixture

mobility of polymer chains also depends on the structure amorphous phase of composition and on a crystal – amorphous interface, which can be detected either as crystalline or as amorphorous fraction depending on the method used. Proton transverse relaxation curves deconvolution for different polymers (polyethylene, polypropylene and their composition) reveals several components, which can be attributed to the crystalline, amorphous phases and crystal-amorphous interface [7, 8]. It is possible to suggest that the less mobile proton fraction in our experiment (T2 = 5 μs) is, the fraction of crystalline phase in PE-BK composition. CPMG investigations of PE-BR composition demonstrated two exponential components of the experimental relaxation curves – slow (S) and fast (F) decay components. S-component could characterize the amorphous phase and F component (as a slow part of decay in FID experiment) – intermediate phase. When BR content in samples increased the contribution of S component in experimental curves of decay increased too (Figure 20.3). It's possible that S component and F component reflect the rubber and polyethylene proton mobility, respectively. S component relaxation times T2 for BR samples

(40, 60, 70, 80 wt% BR) are close to T2 of pure BR (T2 =650 μs) and it confirms our assumption. The growth BR fraction in the composition gave increase T2 of F component and it suggested proton mobility increase in this intermediate phase. Obviously, this fact points to the formation of the amorphous regions in the mixtures enriched by rubber, in which intermolecular interactions between PE and BR are manifested. The regions such as these represent interphase layers. The decline in the crystallinity degree of the polyethylene component also says about the formation of interphase layers. PE crystallites in these samples are surrounded by the amorphous regions containing rubber macrochains, which prevents formation of crystallites and reduces the degree of crystallinity of the polymer. Changes in the molecular mobility of chains have an impact on the thermo-oxidative resistance of PE – BR sample mixtures.

Figure 20.4 shows kinetic curves of oxygen absorption by the tested samples. As seen from the figure, mixing polyolefin and rubber leads to the changes in the kinetics of oxygen absorption. Moreover, the reaction rates vary depending not only on the composition of the mixture, but also

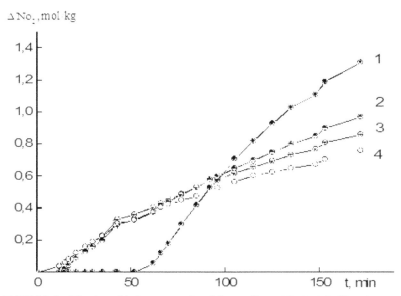

FIGURE 20.4   Kinetic oxidation curves for different films: PE (1), BR (2), mixtures of BR / PE (70/30) (3) and BR / PE (60/40) (4) at 180°C, the oxygen pressure is 300 mm Hg

correlate with the changes in the segmental mobility of the chains. For example, the nature of the curve showing the relationship between the radical probe correlation time τ_c and the composition of the mixture corresponds to the changes in the rate of polymer oxidation. Introduction of rubber in PE leads to a slower oxidation process of the PE component in all the mixtures and of the BR component in the mixtures containing 30 and 40 wt % of PE. The oxidation reaction for PE starts with a delay – the induction period was about 50 min, while for a mixture containing 80% of PE – 20 min, 60% – 13 min. Other mixtures, as well as BR, are oxidized without an induction period. The dependence of the maximum rates of oxidation (Wo$_2$) of the sample mixtures on their composition has a complex nature (Table 20.4). Figure 20.5 (curve 1) shows the dependence of the maximum oxidation rate on the composition of the sample mixtures, which was obtained from the experiment, taking into consideration the oxidation of amorphous regions of the oxygen absorption. The two regions can be distinguished on the curve (compare with the correlation time of the radical probe). The first region corresponds to the change of the composition from 5 to 40 wt % of PE. Moreover, a clear drop in the maximum rate of oxygen absorption is observed with the increase in the content of PE. In the second region of the curve, Wo$_2$ monotonously increases with the increase of the PE content up to 40 wt % or more. We used three models, to describe the kinetics of oxygen absorption by the PE-BR sample mixtures. The first model was based on the following approach: since the rate of BR oxidation is lower than the rate of PE oxidation, the first value was neglected. We assumed that the kinetic curves describing the absorption of oxygen by sample mixtures of different composition could be considered the curves of PE component oxidation. A kinetic curve for each mixture

TABLE 20.4 Dependence of the Rate of Oxidation at 180°C Blends PE and BR on the Composition

Composition of BR / PE sample, wt.%	0/100	20/80	40/60	60/40	70/30	80/20	90/10	95/5	100/0
The rate of oxidation $\times 10^4$, mol/kg*s	2.0	1.5	1.4	0.5	0.6	0.9	1.2	1.1	0.9

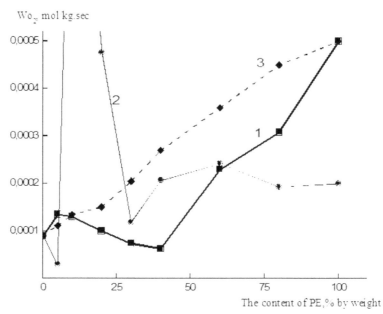

FIGURE 20.5 Relationships of the oxidation maximum rate and the composition of sample mixtures obtained from the experiment, taking into account oxidation of the amorphous regions (1), and theoretical ones, obtained from PE oxidation conditions (2), and from participation in oxidation of both components according to the additive scheme (3)

was transformed into a curve, which took into account oxidation of PE, that is, was calculated per 100% of its content. The dependence of the values of oxidation rates obtained from the transformed curves on the content of PE is shown in Figure 20.5 (curve 2). Another approach took into consideration independent oxidation of PE and BR. At the same time, the oxidation rate of the mixture was determined on the basis of the additive dependence as the sum of the rates of oxygen consumption by its components. Theoretical curves of oxygen absorption by pure polymers taking into account the reactivity of the amorphous phase of PE and BR and the composition of the mixture were obtained in the study. Based on the curves, gross values of the rates were defined and the dependence of these rates on the composition of the samples was obtained (Figure 20.5, curve 3). Comparison of the theoretical curves (2 and 3) and the experimental curve (1) shows significant differences of their shapes. The curve describing the localization of oxidation in the PE component has the form of the exponent. Moreover, a sharp increase in the rate of oxygen absorption is observed for the samples with a low content of the polyethylene component. All this fundamentally distinguishes curve 2 from the experimental curve. In accordance with curve 3, which takes into account the additive contribution of the components in the oxidation of the mixture, the oxidation rate of the samples should increase monotonously with the increase of the PE content. This also contradicts the observed experimental dependence.

Data presented in Figure 20.5 show that a model, which takes into account cross-radical reactions between components is needed to describe the kinetics of oxidation of PE-BR mixtures. In this case, the oxidation of PE-BR mixtures can be described, taking into account the oxidation peculiarities of the rubbers, by the following kinetic scheme:

Initiation of kinetic oxidation chain:

$$R_{pe}H \longrightarrow R_{pe}*$$

$$R_{br}H \longrightarrow R_{br}*$$

Development of the kinetic chain:
For PE:
k_{11}

$$R_{pe}* + O_2 \longrightarrow R_{pe}O_2* \tag{1}$$

k_{p11}

$$R_{pe}O_2* + R_{pe}H \longrightarrow R_{pe}OOH + R_{pe}* \tag{2}$$

For BR:
k_{22}

$$R_{br}* + O_2 \longrightarrow R_{br}O_2* \tag{3}$$

k_{p22}

$$R_{br}O_2* + R_{br}H \longrightarrow R_{br}OOH + R_{br}* \tag{4}$$

$$R_{br}O_2* \longrightarrow R_{1br}CHO + R_{2br}O* \tag{5}$$

$$R_{2br}O* + RH \longrightarrow R_{2br}OH + R* \qquad (6)$$

BR chain transfer reactions:

$$R_{2br}O* + -CH=CH- \longrightarrow -CH-C*H- \qquad (7)$$
$$|$$
$$OR_{2br}$$

Secondary reactions of BR:

$$R_{2br}OH + O_2 \longrightarrow aldehydes \qquad (8)$$

$$R_{1br}CHO + O_2 \longrightarrow acids \qquad (9)$$

Cross reactions of the chain development:

$$k_{p12}$$
$$R_{pe}O_2* + R_{br}H \longrightarrow R_{pe}OOH + R_{br}* \qquad (10)$$

$$k_{p21}$$
$$R_{br}O_2* + R_{pe}H \longrightarrow R_{br}OOH + R_{pe}* \qquad (11)$$

Chain termination (a quadratic break on peroxide radicals takes place at high oxygen pressure, alkyl and alkoxide radicals may participate in the break in the rubber):

$$k_{t11}$$
$$2R_{pe}O_2* \longrightarrow products \qquad (12)$$

$$k_{t22}$$
$$2R_{br}O_2*(R*_{BR}, RO*_{br}) \longrightarrow inactive\ products\ or cross-links \qquad (13)$$

Cross-termination reaction:

$$k_{t12}$$
$$R_{pe}O_2* + R_{br}O_2*(R*_{br,}\ RO*_{br}) \longrightarrow inactive\ products\ or cross-links$$
$$(14)$$

Free valence, which emerged in the PE component of the mixture, was released from the "cell" and localized in the interphase layer or at the interphase boundary, can participate in the BR component in the processes which are competitive to oxidation of hydrocarbon radicals, for example, in the destruction by reaction (5) or cross-linking of polymer chains by reaction (14). Moreover, this chain process of hydrocarbon oxidation may be blocked, which can be represented as a linear break:

$$k'_{tl2}$$

$$R_{pe}OO* \left(R_{pe}* \right) + RbrH \longrightarrow \text{inactive products} \qquad (15)$$

or

$$k'_{tl2}$$

$$R_{pe}OO* + R_{br}H \longrightarrow \left[R_{br}* + R_{pe}OOH \right] \longrightarrow R_{br} - O - R_{pe} \qquad (16)$$

Intermolecular interactions are necessary for cross-reactions. They can be realized through the contact of different polymer chains at the interphase boundary or interphase layer. One might speculate that the complex (bimodal) dependence of the oxidation rate on the mixture composition, which we observed in the study, is due to the peculiar features of the structure of PE – BR sample mixtures. Slowdown of the oxidation rate of the sample mixtures is associated with the formation of interphase boundaries and layers, that is, transition of molecules between the polyethylene phase and elastomeric matrix, thus creating a structure to ensure the development of cross-reactions.

The authors speculate that the radicals arising from cross-reactions lead to the transfer of the free valence of macromolecules from rigid-chain PE to flexible-chain BR; for this reason, they can pass into inactive form, or die in the reaction of rubber destruction or cross-linking. Obviously, the greatest contribution of cross-reactions leading to a decrease in the oxidation activity of the mixtures is made in the samples enriched with rubber. In mixtures with a high content of PE, the contribution of cross-reactions decreases, chain reactions of polyethylene oxidation in its phase dominate, and the total rate of the mixture oxidation starts to increase up to the oxidation rate of a pure polymer. Thus, the oxidation ability of PE-BR mixtures depends on the contribution of cross-reactions, the mechanism for the reaction of the free valence transfer from one component to another, whose implementation is determined by intermolecular contacts of PE and BR macromolecules in the interphase layers, or at the interphase boundaries.

KEYWORDS

- blends
- butyl rubber
- low density polyethylene
- molecular dynamics
- thermal oxidation
- thermophysical parameters

REFERENCES

1. Zakharchenko, P. I., Yashunskaya, F. I., Evstratov, V. F., Orlovsky, P. N. in *Spravochnik rezinshchika* Moskow: Chemistry, 1971. p. 342–395.
2. Schwartz, A. G., Dinzburg, B. N. in *The Combination of Rubbers with Plastics*. Moskow: Khimiya, 1972. 224 p.
3. Xakimulin, Volfson, S. I., Kimel'blat, B. I. *Caoutchouc and Rubber*, 200. № 3. P.32.
4. Piotrovsky, K. B., Tarasova, Z. N. in *Aging and Stabilization of Synthetic Rubbers and of Vulcanizates*. Moscow: Khimiya, 1980.
5. Carr, H. Y., Purcell, E. M. *Phys. Rev.* 1954. V 94. p. 630.
6. Meiboom, S., Gill, D. *Rev. Sci. Instrum.* 1958. V. 29. p. 688.
7. Hedesiu, C., Dan, E. Demco, D. E., Kleppinger, R., Buda, A. A., Blümich, B., Remerie, K., Victor, M. Litvinov, V. M. *Polymer,* 2007. V48. №3. p. 763.
8. Blom H. P., Teh, J. W., Bremner, T., Rudin, A. *Polymer*. 1998. V.39. № 17. p. 4011.

CHAPTER 21

NANOFIBROUS WEB FOR REMOVAL OF BACTERIAL

MOTAHAREH KANAFCHIAN, MOHAMMAD KANAFCHIAN, and A. K. HAGHI

University of Guilan, Rasht, Iran

CONTENTS

ABSTRACT

The Chitosan (CS) based nanofibers web is a biocompatible, biodegradable, antimicrobial and nontoxic structure, which has both physical and chemical properties to effectively capture and neutralize toxic pollutants from air and liquid media. The purpose of this study is to characterize CS-based nanofibers web for filtration. Antibacterial experiments were performed to examine the amount of bacteria reduction. Nanofibers analyzed with FTIR and DSC instruments. In antibacterial test, Turbidimetric method was more suitable than agar diffusion method and the web with 60/40 weight ratio was demonstrated most bacteria reduction. FTIR analysis demonstrated that there were strong intermolecular hydrogen bonds between CS and PVA molecular. In DSC analysis, it was known that filters made of CS/PVA nanofibers are not suitable and applicable for high temperatures.

21.1   INTRODUCTION

Recently, membrane filtration in water treatment and air cleaning has been used worldwide [1]. Filters have been widely used in both households and industry for removing substances from air or liquid. Filters for environment protection are used to remove pollutants from air or water. In military, they are used in uniform garments and isolating bags to decontaminate aerosol dusts, bacteria and even virus, while maintaining permeability to moisture vapor for comfort. Respirator is another example that requires an efficient filtration function. Similar function is also needed for some fabrics used in the medical area [2]. Central to this application is also the ability of the various membrane filtration processes to remove pathogenic microorganisms such as protozoa, bacteria and viruses [3]. Among the membrane processes, nanofiltration is the most recent technology, having many applications, especially for drinking water and wastewater treatment and air filtration [4]. Fibrous media in the form of nonwovens have been widely used for filtration applications. Non-woven filters are made of randomly laid micron-sized fibers, which provide a physical, sized-based separation mechanism for the filtration of air and water borne contaminants [5]. Non-woven nanofibrous filter media (nanofiber is defined as having diameter <0.5 μm by nonwovens industry [6]) would offer a unique advantage as they have high specific surface area, good interconnectivity of pores, and ease of incorporation of specific functionality on the surface effectively filtering out contaminants by both physical and chemical mechanisms [6]. Nonwoven nanofibrous media have low basis weight, high permeability, and small pore size that make them appropriate for a wide range of filtration applications. In addition, nanofibers web offers unique properties like high specific surface area (ranging from 1 to 35 m^2/g depending on the diameter of fibers) and the potential to incorporate active chemistry or functionality on nanoscale [7].

21.1.1 CHITOSAN

Over the recent years, interest in the application of naturally occurring polymers such as polysaccharides and proteins, owing to their abundance in the environment, has grown considerably [8, 9]. CS is a modified natural amino polysaccharide derived from chitin, known as one of the most abundant organic materials in nature. Chitin is the major structural component in the exoskeleton of arthropods and cell walls of fungi and yeast [10]. Commercial chitin is mainly prepared from crab, lobster and shrimp shells, which are the massive waste products of seafood industries [11]. Applications for chitin are very limited because of its poor solubility in common solvents resulting mainly from its highly extended hydrogen-bonded semicrystalline structure [12]. Chitin, the second most polysaccharide found on earth next to cellulose, is a major component of the shells of crustaceans such as crab, shrimp and crawfish. The structural characteristics of chitin are similar to those of glycosaminoglycans. When chitin is deacetylated over about 60% it becomes soluble in dilute acidic solutions and is referred to CS or poly (N-acetyl-D-glucosamine). CS and its derivatives have attracted much research because of their unique biological properties such as antibacterial activity, low toxicity, and biodegradability [8, 9]. Thus, chitin is often converted to its more deacetylated derivative called CS. Chitin is very similar to cellulose, except for the hydroxyl group at C_2 position that is replaced by the acetylamino group. Depending on the chitin source and the methods of hydrolysis, CS varies greatly in its molecular weight (MW) and degree of deacetylation (DDA). The MW of CS can vary from 30 kDa to well above 1000 kDa and its typical DDA is over 70%, making it soluble in acidic aqueous solutions. At a pH of about 6–7, the biopolymer is a polycation and at a pH of 4.5 and below, it is completely protonated. The fraction of repeat units, which are positively charged is a function of the degree of deacetylation and solution pH. A higher degree of deacetylation would lead to a larger number of positively charged groups on the CS backbone [13]. As mentioned above, CS has several unique properties such as the ability to chelate ions from solution and to inhibit the growth of a wide variety of fungi, yeasts and bacteria. Although the exact mechanism with which CS exerts these properties is currently unknown, it has been suggested that the polycationic nature of this biopolymer that forms from acidic solutions below pH 6.5

is a crucial factor. Thus, it has been proposed that the positively charged amino groups of the glucosamine units interact with negatively charged components in microbial cell membranes altering their barrier properties, thereby preventing the entry of nutrients or causing the leakage of intracellular contents. Another reported mechanism involves the penetration of low MW CS in the cell, the binding to DNA, and the subsequent inhibition of RNA and protein synthesis. CS has been shown also to activate several defense processes in plant tissues, and it inhibits the production of toxins and microbial growth because of its ability to chelate metal ions [14, 15]. To date, the most successful method of producing nanofibers is electrospinning.

21.1.2 ELECTROSPINNING OF CS

CS is insoluble in water, alkali, and most mineral acidic systems. However, though its solubility in inorganic acids is quite limited, CS is in fact soluble in organic acids, such as dilute aqueous acetic, formic, and lactic acids. CS also has free amino groups, which makes it a positively charged polyelectrolyte. This property makes CS solutions highly viscous and complicates its electrospinning [16]. Furthermore, the formation of strong hydrogen bonds in a 3-D network prevents the movement of polymeric chains exposed to the electrical field [17]. Different strategies were used for bringing CS in nanofiber form. The three top most abundant techniques include blending of favorite polymers for electrospinning process with CS matrix [18, 19] alkali treatment of CS backbone to improve electrospinnability through reducing viscosity [20] and employment of concentrated organic acid solution to produce nanofibers by decreasing of surface tension [21]. Electrospinning of polyethylene oxide (PEO)/CS [18] and polyvinyl alcohol (PVA)/CS [19] blended nanofiber are two recent studies based on first strategy. In the second protocol, the MW of CS decreases through alkali treatment. Solutions of the treated CS in aqueous 70–90% acetic acid produce nanofibers with appropriate quality and processing stability [20]. Using concentrated organic acids such as acetic acid [21] and triflouroacetic acid (TFA) with and without dichloromethane (DCM) [22] has been reported exclusively for producing neat CS nanofibers. They similarly reported the decreasing of surface tension and at the same time enhancement of charge density of CS solution without significant effect on viscosity. This new method suggests significant influence of the concentrated acid solution on the reducing of the applied field required for electrospinning. The electrospinning process uses high voltage to create an electric field between a droplet of polymer solution at the tip of a needle and a collector plate. When the electrostatic force overcomes the surface tension of the drop, a charged, continuous jet of polymer solution is ejected. As the solution moves away from the needle and toward the collector, the solvent evaporates and jet rapidly thins and dries. On the surface of the collector, a nonwoven web of randomly oriented solid nanofibers is deposited [7].

21.2 EXPERIMENTAL PART

21.2.1 MATERIALS

CS (degree of deacetylation 0.85) and medium molecular weight was supplied by SIGMA-ALDRICH. PVA (degree of hydrolysis, 98%) and acetic acid (AA) purchased from MERK. Nutrient Broth and Nutrient Agar was supplied from LIOFILCHEM COMPANY. *Staphylococcus aureus* bacteria used for this research.

21.2.2 PREPARATION OF CS/PVA SOLUTIONS

CS/PVA solution was prepared by blending of CS and PVA solution with concentration 20 wt% and 3 wt%, respectively. PVA solution was prepared by dissolving PVA polymer in warm water (80°C) with magnetic stirring apparatus until a clear solution be made. Also, CS solutions were prepared by dissolving CS in aqueous 2%v/v acetic acid under magnetic stirring overnight at room temperature to obtain homogeneous solutions. The weight ratios of CS to PVA were selected as ranging from 10/90 to 70/30, respectively. These blends were stirred for 6 h.

21.2.3 ELECTROSPINNING

After the preparation of spinning solution, it was imported in a 2 ml syringe with a stainless steel needle (Inner diameter 0.4 mm) and then the syringe was placed in a metering pump from WORLD PRECISION INSTRUMENTS (Florida, USA). The electrospinning instrument is shown in Figure 21.1. The needle was connected with a high voltage power supply, which could generate positive DC voltages. A piece of aluminum foil was selected as a collector. The electrospun webs were obtained at 10 kV Voltage, 10 cm tip to collector distance and 0.1 mL/hr Feed rate.

21.2.4 SCANNING ELECTRON MICROSCOPY (SEM)

The morphology of the electrospun fibers of CS/PVA was observed under a LEO 1455 VP scanning electron microscope after gold coating. 100 fibers randomly selected in SEM micrographs and their diameter measured by IMAGE J software. Finally, the average fiber diameter and diameter distribution were reported.

21.2.5 FOURIER-TRANSFORM INFRARED SPECTROSCOPY (FTIR)

A sample of electrospun fibers were prepared by electrospinning of CS/PVA solutions at 10 kV, 10 cm collection distance. FT-IR measurements

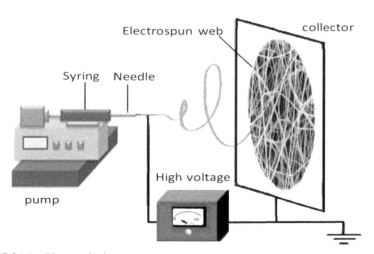

FIGURE 21.1   Electrospinning setup.

were performed in a FT-IR Spectrometer IR 560 (Nickolet Magan) to obtain functional groups and the formed chemical bonds between PVA and CS in the fiber.

21.2.6 DIFFERENTIAL SCANNING CALORIMETRY (DSC)

To investigate the melting point and the shift of endothermic peaks of electrospun web, it was placed in a *BAHR Thermo Analyze* DSC 302 at 10°C/min heating rate from room temperature to 250°C. The sample was stored in a desiccator prior to analysis.

21.2.7 ANTIBACTERIAL ASSESSMENT

21.2.7.1 Nutrient Broth (NB) Solution Preparation

NB solution was prepared by adding NB powder to distilled water and agitating this mixture to reach a clear solution. Then, the solution was sterilized in a steam autoclave under these conditions, steam at a pressure of about 15 pounds per square inch (121°C) in about 15 min.

21.2.7.2 Nutrient Agar (NA) Preparation

Preparation of NA solution was performed in the same way as described for NB. To make homogeneous solution, NA added to heated distilled water and then sterilized. Then, it was pour into a plate to cool the solution and to achieve a solid Agar medium.

21.2.7.3 Antimicrobial activity

To evaluate antibacterial activity, CS/PVA webs with different weight ratio were tested. Before bacterial testing, the NA and NB solutions were sterilized under UV laminar flow and autoclave apparatus. At this research, *Staphylococcus aureus* bacteria is used. Two method used for antimicrobial test: Agar plate and Turbidimetric method. In Agar plate, First,

CS/PVA webs were cut in 1×1 cm. Then, the bacteria suspension was filled in Agar plate and the samples placed on the agar medium. Finally, the Agar plate was tacked into incubator with 37°C for 4 h.

For Turbidimetric method, bacteria culture was performed into liquid medium. In this method, 5 ml solution which it had included NB and *Staphylococcus aureus* of bacteria and salt solution was prepared for each electrospun web. Then, the webs imported into this solution and then absorbance of solution was read by spectrophotometric UV-visible at wavelength of 600 nm after 3 h.

21.3 RESULTS AND DISCUSSIONS

21.3.1 EFFECT OF BLEND WEIGHT RATIO

Table 21.1 shows SEM images of CS/PVA webs with different weight ratio of CS to PVA and its fiber diameter distribution. In this study, we prepared the electrospun web of CS/ PVA using acetic acid-water solution as a spinning solvent. As we know, CS is a cationic polysaccharide with amino groups at the C_2 position, which are ionizable under acidic or neutral pH conditions. Therefore, the morphology and diameter of electrospun fibers will be seriously influenced by the weight ratio of CS/PVA. As seen at Table 21.1, the fiber diameter gradually decreased with increasing CS content in the blend. When CS content was more than 60%, an electrospun web was created with a lot of beads. These behaviors can be explained as the following. CS is ionic polyelectrolytes, therefore a higher charge density on the surface of ejected jet forms during electrospinning. This aggregation of charge causes a higher elongation force imposed to the jet under the electrical field and the diameter of final fibers becomes smaller.

For filtration application, we needed a nanofiber web that it have smaller diameter, uniform and beadles fibers. Thus, the nanofiber web of CS/PVA with weight ratio 40/60,50/50 and 60/40 selected, because, these weight ratios have smaller diameters in comparison with others. The average fiber diameter of these webs was 97.70, 129.53 and 176.79 nm, respectively. Finally after evaluation of antibacterial activity, the best weight ratio of nanofiber web can determine for filter application.

TABLE 21.1 SEM Photographs of Nanofiber Web in Different Weight Ratios of CS to PVA

Weight ratio (CS /PVA)	SEM image	Average Diameter(nm)	Distribution of nanofibers diameter
70/30		54.13±14	
60/40		97.70 ±35	

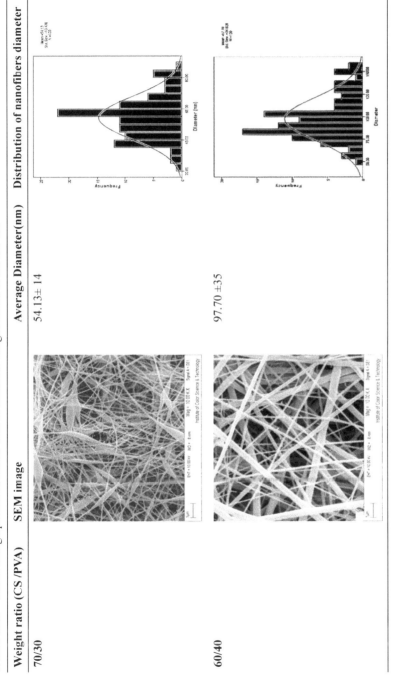

TABLE 21.1 Continued

Weight ratio (CS /PVA)	SEM image	Average Diameter(nm)	Distribution of nanofibers diameter
50/50		129.53 ± 24	
40/60		176.79 ± 28	

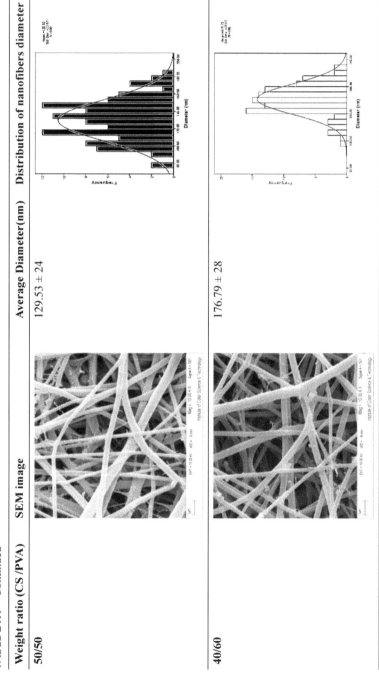

TABLE 21.1 Continued

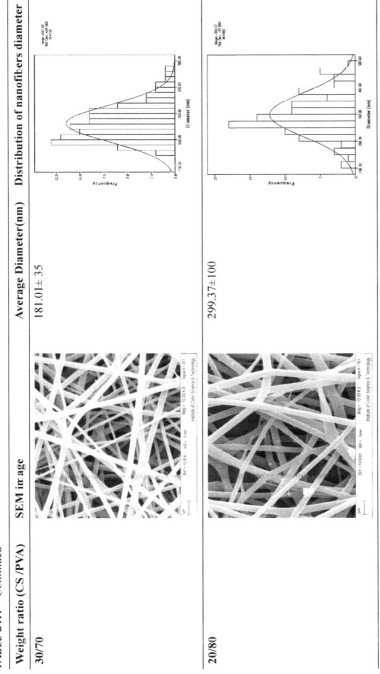

Weight ratio (CS/PVA)	SEM image	Average Diameter(nm)	Distribution of nanofibers diameter
30/70		181.01 ± 35	
20/80		299.37 ± 100	

Weight ratio (CS /PVA)	SEM image	Average Diameter(nm)	Distribution of nanofibers diameter
10/90	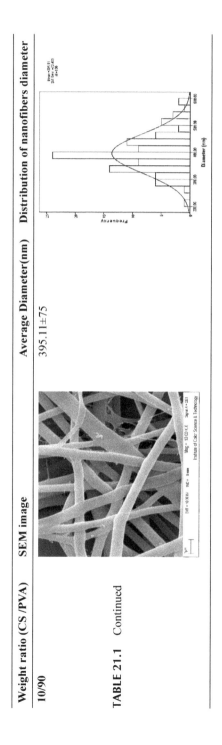	395.11±75	

TABLE 21.1 Continued

21.3.2 ANTIBACTERIAL ACTIVITY

In antibacterial assessment, 3 wt% CS/20 wt% PVA electrospun fibers with different weight ratio prepared from electrospinning process at 10 kV applied electrical potential, 10 cm collection distance and 0.1 mL/hr feed rate. These electrospun fibers exposed to *Staphylococcus aureus* (*S. aureus*) bacteria. Before exposing, all samples had been sterilized with Ultraviolet (UV) because the electrospun fibers can be deformed by alcohol or autoclave. There are many methods to determine the antibacterial activity Such as Agar plate diffusion, Agar tube diffusion assay for Agar medium, and Turbidimetric, pH change and broth dilution assay for broth medium. In this research, the antimicrobial activity of CS in CS/PVA fibers was tested by an Agar plate diffusion method and Turbidimetric method. Antibacterial activity determine with *inhibition zone* around the samples in Agar plate diffusion. Figure 21.2 shows that there is no *inhibition zone* around electrospun fiber with different weight ratio but bacteria is grew on sample 10/90 (when amount of CS was minimum) completely. It is possible that CS in the electrospun web dissolves in bacteria suspension or has no diffusion. In Agar plate diffusion need a chemical active material to be able to diffuse out of the matrix into the plate in order to prohibit microorganism growth. This experiment indicates that the Agar plate diffusion method was not suitable for testing antibacterial assessment for CS/PVA fiber system because, it is able to show antibacterial

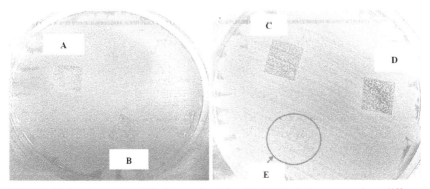

FIGURE 21.2 Agar plate diffusion method for CS /PVA electrospun web at different weight ratio (A: 50/50, B:40/60, C:30/70, D: 20/80, E:10/90).

activity and is not a quantitative method and do not exhibit *inhibition zone* easily.

Therefore, Turbidimetric method was used for antibacterial assay. In this method, the absorbency of solution which it is included of bacteria, brine and webs measured by spectrophotometer UV-visible. Figure 21.3 shows the relationship of bacteria absorbance and CS content at 600 nm wavelength. This diagram presents that solution absorbency decreased by increasing of CS content to 60/40 weight ratio. In other words, maximum reduction of bacteria was occurred at 60/40 weight ratio, because, bacteria can prevent of light transmission from the cell. Thus, the percent of bacteria reduction increases by increasing of CS content and electrospun web of CS/PVA with 60/40 weight ratio has better bacteria reduction.

21.3.3 CHARACTERIZATION OF ELECTROSPUN FIBERS

To confirm the existence of PVA and CS in electrospun fibers, an electrospun web was prepared from spinning solution 20wt% PVA /3wt% CS and evaluated by FTIR and DSC experiments.

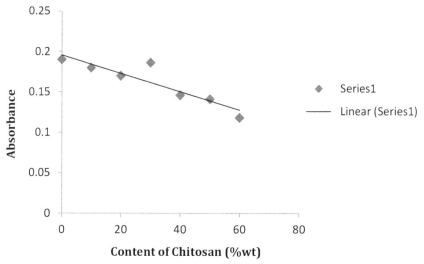

FIGURE 21.3 The relationship of bacteria absorbance and CS content at 600 nm wavelength.

21.3.4 FTIR ANALYSIS

The FTIR test of composites was carried out in order to characterize the participated functional groups in formation of CS/PVA nanofibers web. Figure 21.4 presents the FTIR spectra of electrospun web CS/PVA with weight ratio 60/40 in comparison with pure PVA and CS.

The FTIR spectra of PVA reveals peaks at 1430 cm^{-1} (CH$_2$ bending) and 3354 cm^{-1}(-OH stretching). It represents the characteristic broad band at 2900–3000 cm^{-1} for CH$_2$ group and CH$_3$ group, respectively. The CS exhibited characteristic broad band of OH group at 3400–3500 cm^{-1} [23]. The bands of NH$_2$ group (1560–1640) and O-C-NH$_2$ (1600–1640) group can be observed at1634 cm^{-1}. The broad bands of CH$_3$ group and CH$_3$-O group can be observed at 1000- 1200 cm^{-1} [24]. It can be found the peaks over the wave number range of 3345–3356 cm^{-1} that represent to OH stretching and -NH stretching. FTIR Spectroscopic Measurement exhibited the existence of relevant functional groups of both PVA and CS in CS /PVA nanofibers web.

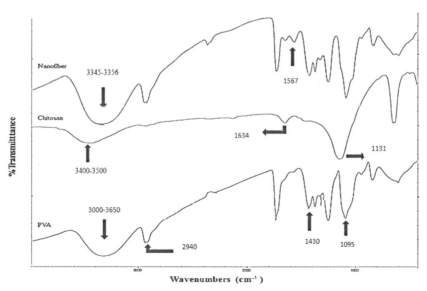

FIGURE 21.4   FTIR spectra of nanofiber CS /PVA: 60/40, PVA and CS powder.

21.3.5 DSC ANALYSIS

DSC thermograms of the electrospun fibers CS/PVA, pure PVA and CS are shown in Figure 21.5. The pure PVA showed a relatively large and sharp endothermic curve with a peak at 200°C. More polysaccharides do not melt but they, because of associations through hydrogen bonding, degrade under heating above a certain temperature. Below degradation temperature of polymer, its thermogram shows a very broad endothermic peak that is associated with the water evaporation. In CS thermogram, a broad peak at approximately 100°C was seen that it was corresponding to the water evaporation process. However, for CS/PVA blend nanofibers, endothermic curve became broad and obtuse, and the peak shifted toward the low temperature. This indicated that the crystalline microstructure of electrospun fibers did not develop well. The reason of this phenomenon is that the majority of chains are in noncrystalline state due to the rapid solidification process of stretched chains during electrospinning. This demonstrated that CS content in the blend caused to decreasing thermal stability of PVA/CS in comparison with pure PVA. Thus, the filters made of CS/PVA nanofibers are not suitable and applicable for high temperatures.

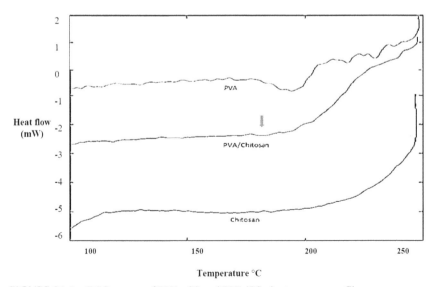

FIGURE 21.5   DSC curves of PVA, CS and PVA/CS electrospun nanofibers.

21.4   CONCLUSIONS

The electrospun nanofibrous web of PVA/CS blends was fabricated. The effects of the blend weight ratio on structure and morphology of the fibers were investigated. The result indicated that the average diameter of the fiber gradually decreased with increasing CS content from 10% to 60%. Above 60% CS, the nanofibers had a lot of beads. In antibacterial test, Turbidimetric method was more suitable than agar diffusion method and the web with 60/ 40weight ratio was demonstrated most bacteria reduction. FTIR and DSC analysis demonstrated that there were strong intermolecular hydrogen bonds between CS and PVA molecular. It was known the filter made nanofibers CS/PVA can be able to inhibit bacteria growth so can use as antibacterial filter. In DSC analysis, it was known that filters made of CS/PVA nanofibers are not suitable and applicable for high temperatures.

KEYWORDS

- antibacterial
- chitosan
- filtration
- nanofibers web

REFERENCES

1. Song, J., Fallgren, H., M. Morria and Ch. Qi, Removal of bacteria and viruses from waters using layered double hydroxide nanocomposites, Sci. Tech. Adv. Mater, 8, 67–70 (2007).
2. Kosmider, K., Scott, J., Polymeric nanofibers exhibit an enhanced air filtration performance. Filtr Separat, 39, 20–22 (2002).
3. M. Bohonak and Zydney, L., Compaction and permeability effects with virus filtration membranes, J. Membr. Sci, 254, 71–79 (2005).
4. Tahaikta, M., R. El Habbania, A. Ait Haddoua, Acharya, I., Amora, Z., Takya, M., Alamib, A., Boughribab, A., M. Hafsib and Elmidaouia, A., Fluoride removal from groundwater by nanofiltration, Desalination, 212, 46–53 (2007).
5. Kaur, S., Gopal, R., WNg, J., Ramakrishna, S., Matsuura, T., Next-generation fibrous media for water treatment, MRS Bulletin 33(1), 21–26 (2008).
6. Wang, J., Kim, S. C., Pui, D. Y. H., Investigation of the figure of merit for filters with a single nanofiber layer on a substrate, J. Aerosol Sci, 39 (4), 323–334 (2008).

7. J.Kim, S., D.Reneker, H., Polybenzimidazole nanofiber produced by electrospinning, Polymer Engineering & Science, 39 (5), 849–854 (1999).

8. Fang, S. W., Li, C. F., Shih, D. Y. C., Antifungal activity of Chitosan and its preservative effect on low-sugar candied kumquat, J. Food Prot, 57, 136–140 (1994).

9. Ignatova, M., Manolova, N., Markova, N., Rashkov, I., Electrospun nonwoven nanofibrous hybrid mats based on Chitosan and PLA for wound-dressing applications, Macromol. Biosci, 9, 102–111 (2009).

10. Pillai, C. K. S., Paul, W., et al., Chitin and Chitosan polymers: Chemistry, solubility and fiber formation, Progress in Polymer Science, 34(7), 641–678 (2009).

11. Kumar, M., A review of chitin and Chitosan applications, Reactive & Functional Polymers 46(1), 1–27 (2000).

12. Kumar, M., Muzzarelli, R. A. A., et al., Chitosan chemistry and pharmaceutical perspectives, Chemical Reviews 104(12), 6017–6084 (2004).

13. Ki Myong Kim, Jeong Hwa Son, Sung-Koo Kim, Curtis L. Weller and Milford A, Properties of Chitosan Films as a Function of pH and Solvent Type, Hanna Journal of Food Science, 71(3), 119–124 (2006).

14. I.Helander, M., E.Nurmiaho-Lassila, L., Ahvenainen, R., Rhoades, J., Roller, S., Chitosan disrupts the barrier properties of the outer membrane of Gram-negative bacteria, International Journal of Food Microbiology, 71(2–3), 235–244 (2001).

15. Devlieghere, F., Vermeulen, A., Chitosan: antimicrobial activity, interactions with food components and applicability as a coating on fruit and vegetables, Debevere, J., Food Microbiol, 21, 703−714 (2004).

16. Aranaz, I., Mengíbar, M., Harris, R., Paños, I., Miralles, B., Acosta, N., Galed, G., Heras, Á., Functional characterization of chitin and chitosan, Curr. Chem. Biol., 3(2), 203–230 (2009).

17. Neamnark, A., Rujiravanit, R., Supaphol, P., Electrospinning of hexanoyl Chitosan, Carbohydr. Polym, 66, 298–305 (2006).

18. Duan, B., Dong, C., Yuan, X., Yao, K., Electrospinning of Chitosan solutions in acetic acid with poly(ethylene oxide), J. Biomater. Sci. Polym. Ed,15(6), 797–811 (2004).

19. Jia, Y. T., Gong, J., Gu, X. H., Kim, H. Y., Dong, J., Shen, X. Y., Fabrication and characterization of poly (vinyl alcohol)/chitosan blend nanofibers produced by electrospinning method, Carbohydr. Polym., 67(3), 403–409 (2007).

20. Homayoni, H., Ravandi, S. A. H., Valizadeh, M., Electrospinning of Chitosan nanofibers: Processing optimization, Carbohydr. Polym, 77(3), 656–661 (2009).

21. Geng, X., O.-Kwon, H., Jang, J., Electrospinning of Chitosan dissolved in concentrated acetic acid solution, Biomaterials, 26(27), 5427–5432 (2005).

22. Vrieze, S. D., Westbroek, P., T.Camp, V., L.Langenhove, V., Electrospinning of chitosan nanofibrous structures: Feasibility study, J. Mater. Sci, 42, 8029−8034 (2007).

23. Boonsongrit, Y., Mueller, B. W., Mitrevej, A., Characterization of drug–Chitosan interaction by H NMR, FTIR and isothermal titration calorimetry Eur. J. Pharm. Biopharm., 69, 388–395 (2008).

24. Mincheva, R., Manolova, N., Sabov, R., Kjurkchiev, G., Rashkov, L., Hydrogels from Chitosan cross-linked with polyethylene glycol di-acid as bone regeneration materials, E-polymer, 58, 1–11 (2004).

CHAPTER 22

MODEL REPRESENTATIONS OF THE EFFECT OF TEMPERATURE ON RESISTANCE POLYPROPYLENE FILLED WITH CARBON BLACK

N. N. KOMOVA[1] and G. E. ZAIKOV[2]

[1]Moscow State University of Fine Chemical Technology, 86 Vernadskii prospekt, Moscow 119571, Russia, E-mail: Komova_@mail.ru

[2]N.M. Emanuel Institute of Biochemical Physics, Russian Academy of Sciences, 4 Kosygin str., Moscow 119334, Russia, E-mail: Chembio@sky.chph.ras.ru

CONTENTS

ABSTRACT

The analysis of the relative resistivity changing, crystallization and kinetics of crystallization was carried out using the difference scanning calorimetry and ohmmeter for the carbon black filled samples of polypropylene. Two stages of crystallization were observed. They were described by means of the generalized equation of Erofeev. Obtained by modeling representations the kinetic parameters are in satisfactory agreement with the experimental data.

22.1 INTRODUCTION

Addition of various kinds of fillers is used to make composite materials based on polymers required for engineering properties. One of the fillers applied in high-molecular materials is carbon used in a form of natural carbon black or graphite. The carbon black is an active strengthening material. Its strengthening action is the stronger the smaller are the dimensions and specific surface area and the higher is the surface energy of the particles. Black particles are composed of a large number of agglomerated crystalline elements called micro crystallites. In general, the system of microcrystallites have the shape of spheres or sintered spheres which may form spatial branched chains [1]. Forming such chains, aggregates exhibit the properties of fibrous fillers. As a result, the resistance of the composite material filled with carbon black particles is reduced even at a small content of carbon black [2].

Polypropylene (PP) filled with carbon black with content ranging from 10 wt. %, is a conductive material. On the other hand, the mechanical properties of the polypropylene filled with carbon black such as strength and impact resistance are satisfactory when the contents of carbon black is less than 20% [2, 3]. Properties of electrical conductivity in the composites are based on polypropylene and the carbon black is determined by the formation of chains of particles conglomerates in the amorphous phase [3–5]. Along with this, in ref. [6] on the basis of molecular dynamic simulation it was shown that the adsorption energy and the portion of the polymer on the surface of amorphous carbon black are determined by the area of the hydrophobic contact (CH_2 and CH_3 groups) and the presence of electronegative atoms in the structure of the polymer. The conformation of the polymer fragment, or complementarity of its geometry on the surface of carbon black also affects the interaction between the polymer molecules and a solid particle surface. Calculation by quantum mechanical simulation showed that the enthalpy of adsorption to particles of carbon black PP is amount of 24.57 kJ/mol, which is 1.5 higher than for PE [6]. So, we can conclude that the formation of the morphological structure of the polymer matrix is affected by the filler both displaced into amorphous phase and contacted with the filler particles. The temperature effect on each of the factors determining the morphology of the polymer matrix is not unique. Since the morphology of the system significantly affects the electrical conductivity the effect of temperature on the electrical conductivity also has a complex character.

In this chapter, we analyze the relative resistivity change and the degree of crystallinity of the system (PP filled with carbon black) affected by temperature. A description of processes kinetics and estimation of experimental results have been made.

22.2   EXPERIMENTAL PART

The object of study is a composition based on the brand PP 01050 (TU 2211–015–00203521–99) (density of 900 kg/m3. Melt index 4.0–7.0 g/10 min), with the same content of conductive carbon black (CB) PA-76 (TU 38–10001–94) – about 11.7% (20% weight fraction).

Preparation of the compositions was performed in a closed rotary mixer "Brabender" volume of the working chamber with 30 mL of liquid heated for 10 min at a rotor speed of 50 rev/min. Mixing temperature was 190°C. The process of sample making for tests was carried out in a hydraulic press at a temperature of 200°C. Samples were cooled between the steel plates in the mold for 20 min from 200 to 70°C after they were finally cooled in air. Measurement of volume resistivity samples (ρ) at elevated temperatures (T) was carried out in a heat chamber as much as 3°C/min. Electrical resistance of the samples was measured at a voltage ohmmeter DT9208A 9 V. Melting process electroconductive samples were examined by differential scanning calorimetry (DSC) by TA instrument Pyris 6 DSC Perkin Elmer with heating from 25 to 250°C at 3°C/min (as in the study of changes in electrical resistance when heated). The degree of crystallinity was calculated using the equation: $\alpha_{cryst} = \Delta H_{melt} / \Delta H_{melt0}$, where ΔH_{melt} – enthalpy of polymer fusion phase of the sample calculated from the results of DSC based on the mass fraction of carbon, ΔH_{melt0}-melting enthalpy of the polymer with a 100% crystallinity [8].

22.3 RESULTS AND DISCUSSION

Results of temperature effect study on the resistance of PP filled with 20% wf. carbon black are presented as relative change in electrical resistivity at ρ_T / ρ_{20} where ρ_T – electrical resistivity at temperature T, ρ_{20}-resistivity at room temperature (Figure 22.1). This dependence has a clear extremum, and the maximum (peak) electric resistance is in the temperature range of the polymer matrix melting.

Represented dependence can be divided into four characteristic temperature range: 1 – weak growth resistance with increasing temperature, 2 – a sharp increase of resistance, and 3 – a sharp drop of resistance, 4 – slight decrease of resistance. Define ranges of these sites allows linearization of dependencies logarithm of relative resistivity versus the reciprocal of temperature. Throughout the temperature range investigated

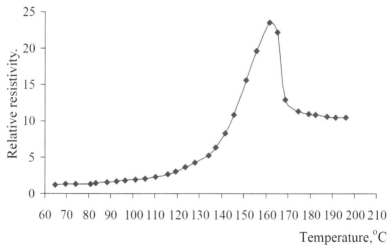

FIGURE 22.1   Dependence of the relative resistivity of PP-20% wf. on crystallization temperature.

characteristic curves have a satisfactory linear relationship (R^2 value of reliable approximation from 0.8 to 0.99).

The boundaries of these temperature ranges between the values: 1 range: 60°C–120°C; 2 range: 120°–160°C; 3 range 160°–171.44°C; 4 range above 171.44°C.

The increase of resistance occurs as the temperature rises at the first and second ranges. Temperature coefficient of resistance change is positive for systems in these areas. The activation energy of changes in resistivity when heated PP filled up in the first region is amount to 12 kJ/mol, and the second – to 83 kJ/mol. At 3 and 4 ranges of resistance decreases with increasing temperature, the activation energy is as follows: 120 kJ/mol 5.54 kJ/mol, respectively. The final resistance of the PP – CB filler reaches a constant value and is greater than the initial approximately 5 times.

To explain the extreme dependence of the resistance of PP filled with carbon black it is advisable to raise enough developed apparatus of model representations of heterogeneous processes. The process of changing the conductivity of PP-black with an increase in temperature is associated with changes in the morphology of the system, which largely forms the filler.

Changes in the degree of crystallinity with temperature for unfilled and filled with carbon black PP are shown in Figure 22.2. The Figure 22.2 shows that the degree of crystallinity of PP increases twice by adding black.

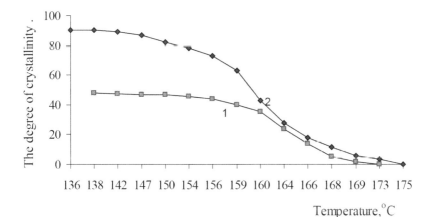

FIGURE 22.2 Dependence of the degree of crystallinity of PP (1) and PP filled 20% wf. carbon black (2) on the temperature.

Formally kinetic analysis of processes in heterophase surfaces can be carried out in terms of topochemical approach, which is carried out on the basis of the theoretical curves (power, exponential) and is based on assumptions about the various laws of the formation and growth of nuclei – nucleation [9]. Most often Erofeev equation is used for the analysis of such processes

$$\alpha = 1 - \exp(-kt^n) \qquad (1)$$

where α is the proportion of reacted material; k is rate constant; t is time expansion, n – kinetic parameter. Value of the exponent n gives an indication of the conditions of formation and growth of nuclei and the reaction mechanism. If the velocity is proportional the weight of the unreacted substance, n is equal 1. When the reaction is located in the kinetic region n > 1. The reaction rate depends only weakly on the nucleation rate and the growth of existing nuclei is determined when n >> 1. If the reaction is limited by diffusion, the greater the deviation from the n units, the greater the effect of diffusion processes (n < 1). Differential form of Eq. (1)

$$\frac{d\alpha}{dt} = nk^{1/n}[-\ln(1-\alpha)^{1-1/n}](1-\alpha) \qquad (2)$$

From Eq. (2) can obtain an approximate equation

$$\frac{d\alpha}{dt} = k\alpha^a (1-\alpha)^b \qquad (3)$$

where a and b are constants corresponding to certain values of n [9].
   Equation (1) in the form

$$-\ln(1-\alpha) = kt^n \qquad (4)$$

is known as the equation Iohansona – Mele – Avrami – Erofeev – Kolmogorov [6] (in the literature it is referred to as the Avrami equation – or Kolmogorov Avrami – Erofeev). In Eq. (4) k is a generalized rate constant, n is exponent. In Ref. [11] k and n are considered as parameters of the Avrami, depending on the geometry of the growing crystals and the nucleation process. These parameters are also a convenient way to represent empirical data crystallization. The equation describing the crystal growth rate is presented in Ref. [11]

$$v = \frac{d\alpha}{dt} = v_o e^{-\Delta G/RT} \tag{5}$$

where α is a degree of crystallinity of the polymer matrix at due time in an isothermal process, ΔG is the total free energy of activation, which is equal to the enthalpy change during crystal growth. $\Delta G = \Delta G^* + \Delta G_\eta$. ΔG^* is the free enthalpy of nucleation of a nucleus of critical size, ΔG_η the free enthalpy of activation of molecular diffusion across the phase boundary. Equating the right hand side of Eqs. (5) and (3), we obtain the expression

$$k\alpha^a (1-\alpha)^b = v_o e^{-\Delta G/RT} \tag{6}$$

Studies in [11], allowed to determine the exponent n Avrami equation – Erofeev-Kolmogorov crystallization of PP filled with carbon black. In the initial stage of crystallization according to [11], it is 2, and the second stage is 3. Isothermal melt crystallization process is reversed so that at low temperatures must take place a decrease of the crystal phase, which was formed on the last (second) stage. Therefore, by solving equation for $n = 3$ and taking the values of constants for this type of a heterophasic processes a= 2/3, b = 0,7 [9] we obtain the expression

$$2/3\ln\alpha + 0.7\ln(1-\alpha) = \ln\frac{v_o}{k} - \frac{\Delta G}{RT_c} \tag{7}$$

where T_c – crystallization temperature for different weight contents of carbon black.

Depending defined for isothermal processes with a certain degree of approximation can be used for processes occurring in the quasi-isothermal conditions (at a relatively slow heating, which is implemented in the research process). Dependence of the degree of crystallinity of PP filled in the considered coordinates is presented in Figure 22.3.

Dependence in Figure 22.3 is well approximated by two straight lines intersecting at a temperature of 162°C. The degree of crystallinity at this temperature is 30%. Analysis of temperature effect on the degree of

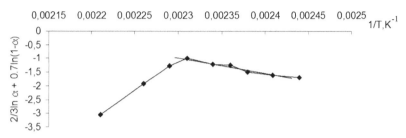

FIGURE 22.3   The crystallinity degree dependence of PP filled 20% wf. carbon black on the temperature in the coordinates of the Avrami-Erofeev-Kolmogorov equation.

crystallinity according to Eq. (7) allows to determine ΔG for two characteristic temperature regions, which are determined by the intersection of the lines in Figure 22.3. Plot for low temperatures (up to 160°C) ΔG is 47 kJ/mol. The activation energy of the process of increasing resistance in this area is 83 kJ/mol.

The difference in the values of defined energies gives reason to assume that the change in conductivity is involved not only a change in the structure of the system, but the result affects some another processes. Is possible these are the fluctuation processes along with the redistribution of the filler particles in increasing the amorphous phase and the agglomerates are destroyed due to the interaction with the polymer segments sites. The activation energy of the process of changing the degree of crystallinity in the area of high temperature (when n = 3) is amount to 122 kJ/mol. The activation energy for decreasing the relative resistivity is 120 kJ/mol. Energy values of the two processes are very similar, which gives grounds to consider these as interrelated processes. Thus, not only low-temperature stages, but also in high temperature change affects the degree of crystallinity of the PP-conductivity specification. Since as shown in Ref. [6] molecular segments PP and PP molecules generally have sufficient adsorption energy to the particles carbon black, it can be assumed that the carbon black particles are displaced into the amorphous phase and collected in agglomerates as a result of this change occurs in crystallinity, and electrical resistance at range 1 and 2. However, a certain part of the carbon black particles served as centers of crystallization filled system remains bound to the polymer molecules.

Having destroyed such polymer crystals relieve certain portion of the carbon black particles, which either form additional conduction channels or enhance existing ones. As a result, the resistance of the entire system

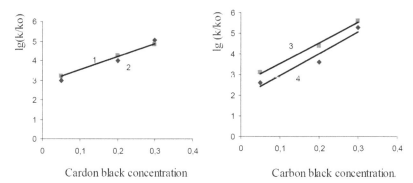

FIGURE 22.4 Dependence of the relative values of the constants in the Eq. (4) on the concentration carbon black at 128°C (1) and 129°C (2) (a) 130°C (3) and 131°C (4) (b).

decreases. The results of studies carried out in Ref. [11], allow us to determine the effect of concentration of filler (carbon black) forming the morphology of the polymer matrix composite PP-W (carbon black) and temperature on the characteristic parameter k (Avrami parameter) reflecting the growth of crystals and the process of nucleation.

A generalization of explored dependences in the range of investigated concentrations of carbon black and at 128–131°C gives the dependence of the relative value of the constant k (k/ko) of the defining parameters in Eq. (8)

$$\ln (k/k_o) = A(T){\cdot}C + B(T) \tag{8}$$

where C is the concentration of carbon black in polypropylene, A and B-parameters depending on the temperature at which the crystallization process proceeds, k_o – characteristic parameter for pure PP.

Dependence of the parameters A and B have a complex temperature dependence of the third-degree polynomial, which in some approximation can be approximated by a linear dependence with the value of R^2 reliable approximation not less than 0.8. As a result, the values for the Eq. (8)

$$\ln (k/k_o) = (3T{-}1191){\cdot}C - 0{,}7 {\cdot}T +265 \tag{9}$$

Comparison of the relative values of the parameters calculated by the proposed correlation and obtained experimentally, is presented in Figure 22.5. Values coincide satisfactorily, especially 129° C and 130°C. By substituting the value of (ln k) of Eq. (9) into Eq. (7), we obtain the expression

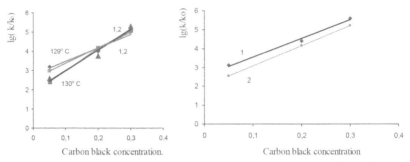

FIGURE 22.5 Comparison of the relative values of the parameter depending on k, obtained from model calculations (1) and experimental (2) at temperatures of 129°C and 130°C (a) and 131°C (b).

$$2/3 \ln \alpha + 0.7\ln(1-\alpha) = \ln v_0 - \ln k_0 - (3T-1191)C + 0.7T - 265 - \Delta G/RT \tag{10}$$

or

$$2/3 \ln \alpha + 0.7\ln(1-\alpha) = A - B(T) - \Delta G/RT \tag{11}$$

where $A = \ln v_0 - \ln k_0 - 265$; $B(T) = (3T-1191)C + 0.7T$.

Dependence in Figure 22.6 has an extreme character with a peak at 162°C and is approximated by two straight lines as in Figure 22.3. Determined from this dependence activation energy for temperatures up to 160°C is −48 kJ/mol, which corresponds to the value found from the Eq. (7). For temperatures above 160°C energy is 373 kJ/mol. Found the energy characteristics of the process changes the crystallinity of PP filled with soot have a value higher than the energy of the relative resistivity change for this system, but the signs of these parameters are the same.

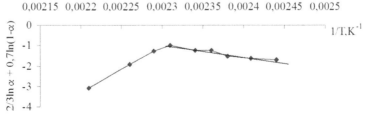

FIGURE 22.6 Dependence of crystallinity degree of carbon black-filled PP on the temperature in the coordinates of Eq. (11).

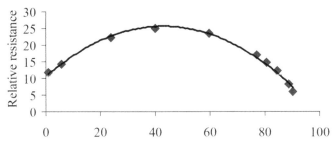

FIGURE 22.7   Relative resistivity dependence on the degree of crystallinity obtained in the process of heating the PP filled with 20% wf. carbon black.

Dependence of resistivity on the relative degree of crystallinity obtained in the heating speed of 3 deg/min PP, filled with 20% wf. carbon black is shown in Figure 22.7. The dependence has an extreme character: The highest relative resistivity appears at the degree of crystallinity of 30% ÷ 60%, which corresponds to a temperature of 160–162°C. Depending on the nature of the polynomial describing the dependence

$$\rho/\rho_{20} = -0.008\alpha^2 + 0.7\alpha + 10.6$$

22.4   CONCLUSIONS

During the heating process there is a change in crystallinity and resistance of the composite material based on polypropylene and carbon black. As a result, there are two mechanisms, one of which is realized to 160° C, and the other is above this temperature. Change the kinetic characteristics both for the crystallization process in the proposed coordinates and for the process of change resistance according to experimental data occurs at the same temperature. For both processes, these dependencies are approximated by linear and have the same sign temperature coefficients. The activation energy of changes in the degree of crystallinity of model representation and the activation energy of resistance change have the same values for the temperature range above 160°C. It was determined that the resistance value takes the maximum at a temperature of energy change for both processes (at 162°C). The value of crystallinity degree is equal to 30% under such conditions. The proposed equations to describe the process satisfactorily coincide with the values obtained from the experiment.

KEYWORDS

- carbon black
- polypropylene
- resistance
- temperature

REFERENCES

1. Sichel, E. K., editor. Carbon black polymer composites. New York: Marcel Dekker, 1989.
2. Dafu, W., Tiejun, Z., Yi, X.-S. Resistivity-Volume Expansion Characteristics of Carbon Black- Loaded Polyethylene. J. Appl. Polym. Sci. 2000. Vol. 77. P. 53–58.
3. Yamaki, J., Maeda, O., Katayama, Y. Electrical conductivity of carbon fiber composites.. Kobunshi Ronbunshi, 1975. V.32. N1, P.42.
4. Moonen, J. T., van der Putter, D., Brom, H. B. Physical properties of carbonblack/polymer compounds. Synthetic Metals. 1991. V.41–43. P.969–972.
5. Carbon black polymer composites: the physics of electrically conducting composites. Ed. Siohel, E. K.- N. Y., Basel: Marcel Dekker. 1982, 214 c.
6. Yanovsky, Y. G., Nikitina, E. A., Karnet, Y. N. Molecular modeling of mesoscopic composite systems. Structure and micromechanical properties.. Physical Mesomechanics. (in Rus.) V.8, №5, 2005. P. 61–75.
7. Shevchenko, V. G., Ponomarenko, A. T., Carl Klason, Tchmutin, I. A., Ryvkina, N. G. Electromagnetic properties of synthetic dielectrics from insulator-coated conducting fibers in polymeric matrix. Electromagnetics, J., 1997, V17, N.2, P. 157–170.
8. Jauffres, D., Lame, O., Vigier, G., Dore, F. Microstructural origin of physical and mechanical properties of ultra high molecular weight polyethylene processed by high velocity compaction. Polymer. 2007. Vol. 48, no. 21. P. 6374–6383.
9. Haines, P. J., Thermal methods of analysis: principles, applications and problems, Blackie Academic & Professional, Chapman & Hall, 1995.
10. Tretyakov, Y. D. Solid-state reactions.. "Khimiya" ("Chemistry", in Rus.) Publishing House. 1978. 192 p.
11. Mucha, M., Marszalek, J., Fidrych, A. Crystallization of isotactic polypropylene containing carbon black as a filler. Polymer. V. 41. 2000. P. 4137–4142.

CHAPTER 23

ENVIRONMENTAL DURABILITY OF POWDER POLYESTER PAINT COATINGS

T. N. KUKHTA,[1] N. R. PROKOPCHUK,[2] and B. A. HOWELL[3]

[1]Scientific-research Institute BelNIIS RUE, 15b Frantsisk Skorina St., 220114, Minsk, Belarus, E-mail: kuhta_tatiana@mail.ru

[2]Belarusian State Technological University, 13a Sverdlov St., 220006, Minsk, Belarus, E-mail: prok_nr@mail.by

[3]Central Michigan University, Chemical Faculty, Mount Pleasant, MI, USA, E-mail: bob.a.howell@cmich.edu

CONTENTS

ABSTRACT

For the first time ever durability of powder polyester paint coatings was determined by an express method. The method is based on empirical exponential dependence of coating durability on activation energy of thermal-oxidative degradation of paint filming agent. Quantitative evaluation of the impact of key destructive factors on coatings durability was made. The express method noticeably shortens time for certification of powder paints by such factor as "durability", and their manufacturing application in obtaining coatings for metal, concrete, asbestos boards.

23.1 INTRODUCTION

Technological advance in the field of paintwork materials (PWM) that requires enhancement of coating protective properties under severe service conditions, as well as solution of a number of ecological and economical problems, has resulted in the development of brand new PWM – powder ones. Within relatively short period of time these materials proved themselves to be quite promising, with their formulation being one of the highest priorities of the current material science development. At the present time powder paints in terms of coatings manufacture technology, durability, as well as ecology and economics, have practically no alternatives. Absence of solvents in powder paints dramatically reduces environmental pollution, with absence of expenditures connected with organic solvents (30–70% of liquid PWM makeup), treatment of air and sewage waters. On top of that, the technology of powder paint coating fabrication is nonwaste (production waste is fully recyclable), less energy-consuming (no power is required for solvent evaporation, the costs of production premises ventilation drop), more automated (maintenance personnel and production floor are reduced), more manufacture efficient (several times).

Now the most wide-spread are powder paints on the basis of the following filming agents: epoxy resins, epoxy- polyester oligomers (combination of epoxy and polyether resins), hybrid filming agents; not-saturated polyester resins [1].

In selection of powder paints one of the most important properties is resistance to weather conditions. The influence of ambient environment leads to energy absorption in ultraviolet band of electromagnetic spectrum. This energy has negative effect both on film-forming polymer, and the pigment resulting in loss of glitter and change of color. Due to tendency to chalking epoxy and hybrid paint coatings are usually not recommended for use in the open air.

Polyether materials form coatings that provide good resistance to the influence of ultraviolet rays.

Composition of polyether powder paints contain curing agents that cross–link oligomer macromolecules and form low-molecular polymer spatial cross-linked structure. Powder paints with extensively used curing agents were selected for the research.

One of them is triglycidyl isocyanurate (TGIC), which was used for rather long period of time, but lately regarded as a harmful reagent in some European countries.

Another one is hydroxyalkylamide considered as harmless and known under the trade name "primide." Therefore, study of environmentally resistant polyether powder paint coatings cured with TGIC and primide, as compared to coatings of hybrid powder makeup is a crucial task.

Work objective is to quantitatively evaluate destructive factors that influence polymer protective coatings made of powder paints when in service; propose empiric exponential relationship between durability of metal tiles and asbestos cement boards coatings and activation energy of thermal oxidative breakdown of filming agent for polyether powder paints; develop express method for evaluation of these coatings durability with regard to influence of environmental factors.

23.2 EXPERIMENTAL PART

Subjects of the research were films of 0.3–0.4 mm in thickness, (10 ± 2) mm in width and 100 mm in length, and coatings of 0.1 mm in thickness fabricated of powder paints, samples of:

- 1, 2, 3, 4 – polyether paint, curing agent – primide, colors – red, white, green and black, correspondingly;
- 5, 6 – polyether paint, curing agent – TGIC, colors – green, black, correspondingly;
- 7, 8 – hybrid paint, colors – blue and black, correspondingly.

The films were formed on fluoroplastic sheet. Powder paint was deposited through the sieve.

The tests were performed in the climatic chamber "Feutron", type 3826/16 (Germany) according to the following cycle:

- moisture treatment of samples at temperature (40 ± 2) °C and relative humidity $(97 \pm 3)\%$ for 2 h;
- moisture treatment free of heating at relative humidity $(97 \pm 3)\%$ for 2 h;
- freezing at temperature minus (30 ± 3)°C for 6 h;
- irradiation of samples by creating light flux with surface density of total radiation energy $(730 + 140)$ W/m^2 with surface density of UV radiation flux $(30 + 5)$ W/m^2 and periodic sprinkling with water for 3 min in each 17 min for 5 h;
- freezing at temperature minus (60 ± 3)°C for 3 h;
- conditioning at temperature 15–30°C and relative air humidity of 80% for 6 h.

Samples were taken each 25; 50; 75; 100 cycles.

The films were artificially aged under influence of UV and IR radiation by means of dummy emitter of sunlight SOL 1200S (Germany).

Artificial aging mode:

- temperature in the climatic chamber – 50°C;
- relative air humidity – 60%;
- UV radiation mode – 57.7 W/m^2; IR – 730 W/m^2;
- visible range – 320 W/m^2.

Total optical radiation flux from dummy emitter HSA 1200S at a distance of 60 cm from radiation source was 1,107.7 W/m^2.

Magnitude of samples radiation energy from dummy emitter for 600 h amounted to 2,393 MJ/m^2; 1,200 h – 4,786 MJ/m^2; 2,400 h – 9,572 MJ/m^2.

Films porosity was determined by their specific area values calculated by BET method (Brunour, Emmet, Teller). Nitrogen adsorption isotherms were read out on instrument NOVA 2200. Gaseous nitrogen with operation temperature of 77 K was obtained by evaporation of liquid nitrogen. Measurements error did not exceed 10% of specific area values.

Mechanical tests were conducted on tensile testing machine T 2020 DC 10 SH (Alpha Technologies UK, USA).

Ambient air temperature – 18°C, speed of top grip motion – 10 mm/min, clamping length of samples – 54 mm, number of test samples – 10.

By diagrams "tensile strain σ (MPa) – tensile deformation ε (%)" using the instrument computer program, the rupture resistance (σ, MPa), relative elongation (ε, %), Young elasticity modulus (E, MPa) were computed as mean arithmetic of ten measurements.

The value of activation energy E_a was determined by Broido computational method on the basis of dynamic thermogravimetric data [2].

Morphology of films surface (in research film surface being in contact with air and not with fluoroplastic substrate was used in order to exclude influence of substrate) was studied on scanning electronic microscope JSM 5610LV (Jeol, Japan).

23.3   RESULTS AND DISCUSSION

Results of films mechanical tests before influence of artificial climatic factors are given in Figure 23.1.

Data analysis shows that more durable and elastic films are formed from polyether films cured with primide in particular. Durability of 1–4 samples on the average is 24.6 MPa, whereas durability of 7, 8 samples is only 17.4 MPa. Relative elongation at rupture – 2.35% and 1.66%, elasticity modulus 1,610 and 1,865 MPa correspondingly.

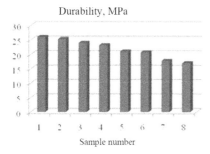

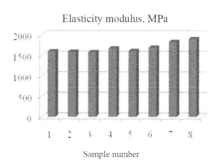

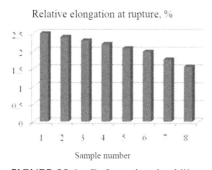

FIGURE 23.1   Deformation-durability and elastic properties of films made of powder paints.

Therefore, films of polyether powder paints are more durable than films of hybrid powder compositions by 41%, and more elastic by 42%. Less than by 16% elasticity modulus of polyether films as against hybrid ones gives ground to believe that smaller inherent stresses will be developed in coatings of polyether powder paints due to their greater relaxation in film that is easier deformed. In its turn, small inherent stresses in the film provide its longer service life (service life of coatings without loss of their protective properties).

Comparing compositions of polyether powder, containing different curing agents, one may note that primide is more preferable than TGIC. The films cured by primide are more durable than films cured by TGIC by 14% on the average, and more elastic by 16%.

This is explained by differences in molecular structure of primide and TGIC. Chemical reaction of cross-linking of unsaturated polyether oligomeric molecules with primide molecules proceeds slower and requires higher temperatures and more time consuming. However, it forms more uniform, durable and elastic polymer network.

As a result, durability and relative elongation at rupture is higher, and elasticity modulus is lower for primide cured films. Coatings made of polyether powder compositions containing primide have objectively to be more durable than TGIC cured coatings because of smaller inherent stresses that develop in them.

Since films made of powder paints represent chemically cross-linked spatial network patterns, their disintegration has fragile nature. For each sample the following ratio is observed:

$$\sigma = K \bullet E \bullet \varepsilon,$$

where K= 0.0066–0.0063 for 1–4 samples; 0.0062–0.0063 for 5,6 samples; and 0.0057–0.0055 for 7,8 samples.

In spatially cross-linked (network) polymers their mechanical properties are greatly influenced by ratio between molecular mass of section between network nodes and molecular mass of kinematic segment [3].

If molecular weight of kinetic segment (MWk.s) << molecular weight of an interval between crosslinks (MWc) (kinematic linkage is flexible and network is sufficiently wide), then variation of network density practically affects neither highly elastic deformation nor temperature of polymer glass transition. But if MWc > MWk.s, then increase of network density (reduction of Mc) results in reduction of highly elastic deformation and rise of glass transition temperature [4].

At very high density of three-dimensional network the highly elastic deformation is impossible, and at room temperature material is in glass state. Variation of durability and increase of network density is, as a rule, expressed by a curve with maximum. Small amount of cross linkages does not hamper straightening of chains in deformation, resulting in the enhancement of durability. Still increase of density over optimal values hinders orientation processes during stretching of films, and their durability starts to degrade the more the larger density of formed three-dimensional network is.

According to the data obtained by DSC (differential scanning calorimetry) (Figure 23.2), the lowest glass state temperature have nonadhesive films obtained from primide cured polyether filming agent (glass-transition temperature of sample No 2 is 62°C). During solidification of this filming agent with TGIC curing agent the glass state temperature goes up by 6°C (samples No 5 and 6 in Figure 23.4).

Finally, with the highest glass transition temperature (77–82°C) is characterized by samples of films No. 7 and 8 obtained from hybrid filming agent (combination of epoxy and polyester resins). Consequently, logical

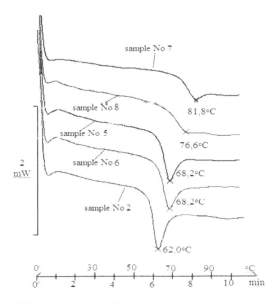

FIGURE 23.2   DSC of powder paint films.

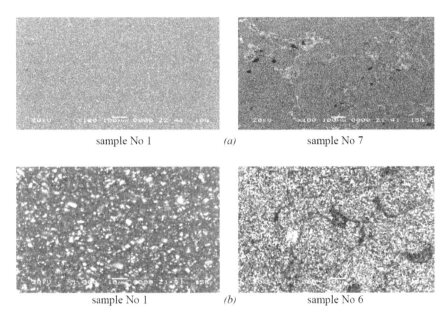

sample No 1   *(a)*   sample No 7

sample No 1   *(b)*   sample No 6

FIGURE 23.3   Electron micrographs of polymer films with magnification: a – x100; b – x1000.

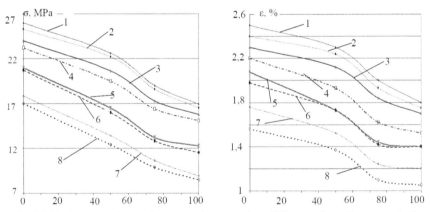

FIGURE 23.4   Dependence of stress-strain properties of films on time of exposure in the climatic chamber.

interrelationship is observed between values of glass transition temperature, durability, elasticity modulus, relative elongation upon rupture of films.

Electron micrographs (Figure 23.3) give evidence of different nature of films surface, along with various size and shape of coloring agents particles.

Surface of films of primide cured polyether powder paint (sample No 1) is smooth and uniform.

Network structure is formed free of considerable inherent stresses.

Surface of films of hybrid paint (sample No 7), and TGIC cured polyether paint (sample No 6) is texturized, as it was formed under conditions of great inherent stresses.

Upon the influence of weather factor on films in climatic chamber rigidness of the three-dimensional network has small variation, so elasticity modulus varies within several percent (10% maximum). However, under impact of heat, UV-radiation and humidity sections of macromolecules between network nodes disintegrate and its elasticity drops. With practically linear rise of film stress upon its extension, and smaller relative elongation of upon rupture lower durability is achieved. Due to this with the increase of the number of cycles affecting films, both durability and relative elongation diminish. Dependences of durability and relative elongation upon rupture on exposure time have similar S-shaped nature (Figure 23.4).

As kinetics of radical reactions in solid polymers has strong dependence on the degree of macromolecules cross-linking and mechanical stress on their chemical bonds [5], the films under study differ in aging rate.

So films obtained from primide cured polyester resin (samples No. 1–4) retain 65% of initial durability and 72% of relative elongation upon rupture after impact of 100 cycles in the climatic chamber.

Invariance of values of these indices for TGIC cured polyester films is equal to 60 and 70% correspondingly, and for hybrid epoxy-polyester resin films – 50 and 67%.

The data we obtained for powder paint films comply with provision of the kinetic theory of strength [6] stating that durability of a solid, including that of polymer, decreases with the increase of mechanical stress σ that affects chemical carbon-to-carbon bonds in the backbone chain.

On trials in the climatic chamber polymer films were periodically subjected to 100% moistening. An important role in the reduction of durability and deformation characteristics of stressed composite materials is played by liquid media affecting them on a long-term basis [7].

Cross-linking reaction of polyester binder oligomeric molecules when cured by primide is exercised through chemical interaction of resin molecules terminal carboxyl with hydroxyl groups of curing agent (Figure 23.5a).

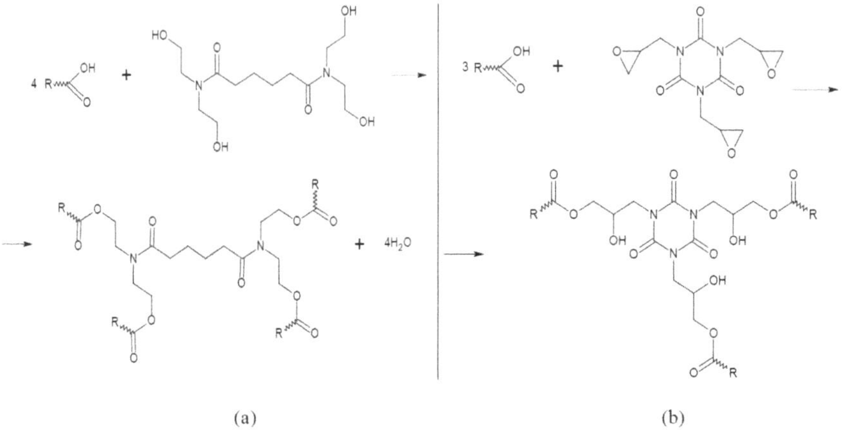

(a)   (b)

FIGURE 23.5   Diagrams of polyester resins cured by primide (a) and TGIC (b).

Alongside with that, molecular structure of primide (aliphatic architecture of molecule matrix) provides comparably flexible mobile cross linkage of cured film three-dimensional network and, as a consequence, good mechanical properties and high weathering resistance.

With the same TGIC-cured powder polyester paint, terminal carboxyl of the binder interacts with three epoxy groups of the curing agent (Figure 23.5b).

Furthermore, ester links are formed and water is exuded which, with large thickness of films, may form defects in the form of punctures. However, this negative effect can be brought to a minimum by using degassing additives in make up of powder paints, along with selection of optimal temperature-time mode of coatings formation.

Evaluation of films porosity, made by BET method showed that it practically does not depend on the chemical nature of curing agent used. Values of specific surface area of film samples No. 1–6 varied from 10 to 11 m^2/g, i.e., being within the measuring error. For this reason one may say that porosity has identical effect on all film samples during their tests. Whereas differences in mechanical properties and weathering resistance of powder polyester paint films are conditional on different nature of chemical cross-links in the three-dimensional network, i.e., on differences in chemical architecture of curing agent molecules.

As a result, individual molecules of polyester resin are cross-linked into the three-dimensional network without emission of volatile low-molecular compound. Enhanced reacting capacity of TGIC epoxy groups provides more easy and quick cross-linking at lower temperatures and within shorter time of coatings cure. Meanwhile cyclic TGIC structure specifies higher rigidity of network cross links, with inherent stress slow relaxation. Due to this complex of mechanical properties of films and their weathering resistance worsen.

Analysis of Figure 23.4 shows that after impact of 100 cycles in the climatic chamber the films still retain on the average 65% of initial durability and 72% of relative elongation upon rupture. However, with regard to impact of other operational factors on protective films, like inherent stresses, external mechanical forces, it is reputed that 100 exposure cycles result in complete loss of films service life, as it is common knowledge that durability corresponds to time within which durability and/or elasticity drops twofold.

We have established that nonadhesive films of primide-cured powder polyester paint of different colors disintegrate with energy practically identical to activation energy of thermal-oxidative degradation E_d equal to 140±2 kJ/mol.

The key prevailing destructive factors that lower potential barrier of filming agent bond opening E_d are:

ΔE^{clim} – lowering of E_d due to exposure in the climatic chamber;

ΔE^{UV} – lowering of E_d under influence of sunlight-simulating radiation;

$\Delta E^{in.str}$ – lowering of E_d by inherent stresses that occur due to differences in coefficients of thermal expansion of coating and protected surface;

$\Delta E^{mech.in.}$ – lowering of E_d due to static and dynamic loads on coatings. The following values of destruction factors were obtained:

$\Delta E^{clim} = 50$ kJ/mol, $\Delta E^{UV} = 10$ kJ/mol; $\Delta E^{in.str} = 7$ kJ/mol; $\Delta E^{mech.in} = 3$ kJ/mol.

Due to this calculated value of thermal-oxidative degradation activation energy of coatings that determines their durability equals to:

$E_{calc} = 140–50–10–7–3=70$ kJ/mol.

Earlier we [8–10] have proposed and implemented in the system of certification tests the express method for determination of rubber and thermoplastic articles durability, which is based on the interrelation between durability of polymer material τ and value of activation energy of its thermal-oxidative degradation E_d. It has been repeatedly demonstrated that E_d determines quality of polymer material and is reduced under influence of service factors.

In this work for the first time ever the method is proposed for evaluation of durability of powder paint coatings having tridimensional cross-linked structure.

Durability of coatings τ_{T_s} at set value of service temperature T_s is calculated by empirical formula we have established:

$$\tau_{T_s} = K \cdot (10^{-\alpha \cdot E_{calc.} - \beta} \cdot e^{E calc./R \cdot T_s})$$

where $K = 2.74 \times 10^{-3}$ years, $\alpha = -0.1167$ mol/kJ, $\beta = 0.090$.

Values of coefficients are established by mathematical treatment of experimental datasets obtained at prolonged (six months) aging of powder polyester paint films from different manufacturers differing considerably in values of E_d.

Calculated durability of polymer coating in years (τ_{total}) at variable values of article service temperature is determined by the formula:

$$\tau_{total} = [\sum_{i=1}^{n} \frac{m_i}{\sum m_i}] \cdot \tau_{T_s}$$

where m_1 – number of hours of impact at particular values of service temperature; Σm_1 – total number of hours of impact at variable values of service temperature; τ_{Ts} – durability of a polymer article in years at particular value of article service temperature.

For climatic conditions of Eastern Europe duration in hours was determined for service within one year and at temperatures that develop within materials upon impact on them of direct sun beams (Table 23.1).

Example of durability calculation for on-metal coatings from polyester powder paint produced by MAV PUPE is given below.

TABLE 23.1   climatic conditions of Eastern Europe duration in hours

Index name	Value				
Number of hours of temperature impact, h	2,250	1,400	440	200	100
Temperature, °C	20	30	40	50	60

Calculated durability of coating in years at temperatures 20,30,40,50 и 60°C is:

$$\tau_{20°C} = 2.74 \cdot 10^{-3} \cdot [10^{-0.1167 \cdot 70 - 0.090} \cdot e^{70/2.435}] = 46$$

$$\tau_{30°C} = 2.74 \cdot 10^{-3} \cdot [10^{-0.1167 \cdot 70 - 0.090} \cdot e^{70/2.518}] = 17.9$$

$$\tau_{40°C} = 2.74 \cdot 10^{-3} \cdot [10^{-0.1167 \cdot 70 - 0.090} \cdot e^{70/2.601}] = 7.4$$

$$\tau_{50°C} = 2.74 \cdot 10^{-3} \cdot [10^{-0.1167 \cdot 70 - 0.090} \cdot e^{70/2.684}] = 3.2$$

$$\tau_{60°C} = 2.74 \cdot 10^{-3} \cdot [10^{-0.1167 \cdot 70 - 0.090} \cdot e^{70/2.767}] = 1.5$$

Calculated durability of coating in years when in service in climatic conditions of Eastern Europe is:

$$\tau_{total} = \left[\frac{2250}{4392} \cdot \frac{1}{46} + \frac{1400}{4392} \cdot \frac{1}{17.9} + \frac{440}{4392} \cdot \frac{1}{7.4} + \frac{200}{4392} \cdot \frac{1}{3.2} + \frac{100}{4392} \cdot \frac{1}{1.5} \right]^{-1} = 14$$

23.4 CONCLUSIONS

Express method was developed for evaluation of durability of protective coatings made of powder paints of polyester class. The method is based on empirical exponential dependence of coatings durability on activation energy of thermal-oxidative degradation of powder paint filming agent. Evaluation of the impact of destructive factors acting on macromolecules of protective coatings when in service, and lowering potential barrier of empirical bonds opening of polymer macromolecules was made and, as a consequence, the coating deformation-durability properties and service life. For the first time data was obtained on real service time of powder paint coatings, notably made of local powder paint. Determination of coating durability by express method is performed in one working day, whereas the method established by standard takes four months. This makes it possible to promptly evaluate quality level of powder polyester paints from different manufacturing companies available on the market of paintwork materials of the Republic of Belarus. Besides, producers of powder paints of polyester class are permitted to noticeably shorten time for the development of new formulations of durable protective coatings.

KEYWORDS

- activation energy of thermal-oxidative degradation
- climatic factors
- coating
- durability
- powder polyester paint

REFERENCES

1. Yakovlev, A. D. Powder paints. A. D. Yakovlev. L.: Chemistry, 1987. – 216 p.
2. Broido, A. Sensitive Graphical Method of Treating Thermogravimetrie Analysis Data. A. Broido, A. Simple. J. Polymer. Sci. Part A-2. 1969. Vol.7, No. 10. pp. 1761–1773.
3. Tager, A. A., Physics-chemistry of polymer. A. A. Tager; edited by prof. A. A. Askersky. M.: Nauchniy mir, 2007. 573 p.
4. Encyclopedia of polymers: in V. A. Kabanov (editor-in-chief) [et.al.]. M.: "Soviet Encyclopedia", 1977. V. 3: P-Y 1152 p.
5. Emanuel, N. M. M. Chemical physics of aging and stabilization of polymers. N. M. Emanuel, A. L. Bugachenko. M.: Nauka, 1982. – 360p.
6. Regel, V. R. Kinetic nature of solid durability. V. R. Regel, A. I. Slutsker, E. M. Tomashevsky. M.: Nauka, 1974. 560p.
7. Shevchenko, A. A. Physics-chemistry and mechanics of composite materials: Textbook for HEE. A.A. Shevchenko. SPb: COP "Profession", 2010. 224p.
8. Method for determination of elastomers durability: patent 1791753 USSR, IPC G01N318//G01N1700. A. G. Alexeev, N. R. Prokopchuk, T. V. Starostina, L. O. Kisel (USSR). N4843144/08; appl.26.09.90; publ.30.01.93. Bull. N4.-8p.
9. Polymer articles for construction. Method of durability determination by activation energy of thermal-oxidative degradation of polymer materials. State Standard of the Republic of Belarus STB1333.0–2002.
10. Prokopchuk, N. R. Evaluation of durability of polymer articles. N. R. Prokopchuk. Standardization. 2008, No.1. pp. 41–45.

CHAPTER 24

A NOTE ON VISCOMETRY OF CHITOSAN IN ACETIC ACID SOLUTION

VALENTINA CHERNOVA, IRINA TUKTAROVA, ELENA KULISH,
GENNADY ZAIKOV, and ALFIYA GALINA

*Bashkir State University, 450074, Zaki Validi st., 32, Ufa, Russia;
E-mail: onlyalena@mail.ru*

CONTENTS

ABSTRACT

Some ways of estimating the values of the intrinsic viscosity of chitosan were analyzed. It is shown that the method of estimating the current value of the intrinsic viscosity of Irzhak and Baranov can adequately assess the conformational state of the macromolecular coil and its degree of swelling.

24.1   INTRODUCTION

Recently there has been a large number of studies related to the producing and studying of chitosan-based film materials for using them in different areas [1]. According to modern views, the supramolecular structure of the polymer films formed from the solution is determined by the conformational state of the polymer in the initial solution, which in turn, predetermines the complex physicochemical properties of the material. Viscosimetry is one of the most accessible and informative methods of studying the conformational state of polymers in solution. However, in the case of using viscosimetry to study polyelectrolyte solutions, reliable determination of intrinsic viscosity to assess their conformational state, faces with certain difficulties. In this paper we attempted to analyze some of the ways to estimate the values of the intrinsic viscosity of chitosan, which when dissolved in acidic aqueous media (e.g., acetic acid) acquires the properties of polyelectrolyte.

24.2   EXPERIMENTAL PART

As objects of study used a sample of chitosan (manufactured by "Bioprogress" Schyolkovo) obtained by alkaline deacetylation of crab chitin (deacetylation degree of ~ 84%). The molecular weigh of the original chitosan (CHT) was determined by combination of sedimentation velocity method and viscosimetry according to the formula [2]:

$$M_{s.} = \left(\frac{S_0 \eta_0 [\eta]^{1/3} N_A}{A_{hi}(1 - \bar{v}\rho_0)} \right)^{3/2}$$

where S_0 – sedimentation constant; η_0 – dynamic viscosity of the solvent, which is equal to $1{,}2269 \times 10^{-2}$ PP; $[\eta]$ – intrinsic viscosity, dl/g; N_A – Avogadro's number, equal to $6{,}023 \times 10^{23}$ mol^{-1}; $(1-\bar{v}\rho_0)$ – Archimedes factor or buoyancy factor, v – the partial specific volume, cm^3/g, ρ_0 – density of the solvent g/cm^3; A_{hi} – hydrodynamic invariant equal to 2.71×10^6.

The experiments were carried out in a two-sector 12 mm aluminum cell at 25°C. Thermostating was carried out with an accuracy of 0.1°C. Sedimentation constant was calculated by measuring the boundary position and its offset in time by means of optical schemes according to the formula [2]:

$$S_0 = \frac{1}{\omega^2 X_{max}} \frac{dX_{max}}{dt}$$

Archimedes factor was determined pycnometrically by standard methods [2] and was calculated using the following formulas:

$$v = v_0 \left[\frac{100}{m_0 \rho} \left(\frac{1}{m_0} - \frac{1}{m} \right) \right]$$

where v_0 – volume of the pycnometer; m_0 – weight of the solvent in the pycnometer; m – weight of the solution in the pycnometer; ρ=100 g /m_0; (g – solution concentration in g/mL).

$M_{s\eta}$ value for the CHT sample used in was 113000 Da.

In order to determine the constants K and α in Mark-Houwink-Kuhn equation the sample of initial chitosan was fractionated to 10 fractions, for each of them has been determined $M_{s\eta}$ value and an intrinsic viscosity value. Fractionation of chitosan was performed by sequential deposition [3]. As the precipitant was used acetone. Viscosimetric studies were performed according to standard procedure on an Ubbelohde viscometer at 25°C [4].

As the solvent was used acetic acid at a concentration of 1, 10 and 70 g/dL. Value of intrinsic viscosity of the initial CHT sample and its fractions in solutions of acetic acid was determined in two ways – by the method of Fuoss [5] and the method of Irzhak and Baranov [6, 7]. Dilution of initial solution of the polymer in the calculation of intrinsic viscosity by Irzhak and Baranov carried out by the solvent, while using the method Fuossa by water.

24.3   RESULTS AND DISCUSSION

Most often the values for determining the intrinsic viscosity [η] of Polymer solutions are described by Huggins equation [8]:

$$\frac{\eta_{sp}}{c} = [\eta] + k_1 [\eta]^2 c \qquad (1)$$

Where is c – concentration of the polymer in the solution (g/dl); k_1-constant reflecting the reaction of the polymer with a solvent; η_{sp} – specific viscosity of polymer solution equal to: $\eta_{sp}=\eta_{rel}-1$. Relative viscosity η_{rel} is the ratio viscosity of the polymer solution (η_p) to the viscosity of the solvent (η_0). The main condition for the applicability of the Huggins equation is linear dependence of the reduced viscosity (η_{sp}/c) of the polymer concentration in the solution and the absence of strong intermolecular interactions between the components macrochains. Value of intrinsic viscosity is calculated by extrapolation to zero concentration dependence of the reduced viscosity of the polymer concentration in the solution.

However, in the case of studies of solutions of polyelectrolytes in dilute solutions the macromolecular coil swells considerably, because there is a well-known effect of polyelectrolyte swelling. As a consequence there is a deviation from the linear dependence of the viscosity of the polymer concentration in the solution [9, 10]. For example, when studying solution polyelectrolyte HTZ in acetic acid, the experimental data in the viscosity of dilute solutions are also not described by the Huggins equation.

Fuoss and Strauss [8] proposed an empirical equation to determine the intrinsic viscosity of polyelectrolytes:

$$\frac{\eta_{sp}}{c} = \frac{[\eta]}{1+B\sqrt{c}} \tag{2}$$

where B – coefficient characterizing the electrostatic interaction of the polyion with simple ions.

When using Fuoss and Strauss approach dilution of the original concentration of the polymer solution is made by water instead of solvent. In the case where the solvent is a strong electrolyte (e.g., hydrochloric acid), the dilution water does not change the degree of dissociation of the solvent and does not change the ionic strength of the solution. Hence the value of the intrinsic viscosity, determined by extrapolation to zero concentration will indeed reflect the size of the coil in the solvent being. In our case, acetic acid is used as the solvent, which is weak and which dilution water is accompanied by an increase in its degree of dissociation. Increasing the degree of dissociation of acetic acid leads to additional protonation CHT macromolecules in solution and causes unfolding of the macromolecular coil. As a result, the value of intrinsic viscosity of CHT determined by Fuoss equation will not reflect the conformation of macromolecules in a solution of acetic acid concentrations investigated, but will reflect CHT conformation in solution infinitely dilute acetic acid.

More reasonable method for determining the intrinsic viscosity of polyelectrolytes in order to assess conformational state of the coil is considered the reception of isoionic dilution [11]. At the same time, dilution of the polyelectrolyte solution is made by low-molecular salt. This method is very time consuming, because it requires optimization. Meanwhile, in Refs. [9, 10], it was suggested that adequate alternative to method isoionic dilution is the method of determining the intrinsic viscosity of Irzhak and Baranov (method of assessing the current intrinsic viscosity), by the equation:

$$[\eta] = \frac{\partial \ln \eta_{rel}}{\partial c} \tag{3}$$

Using Eq. (3) the intrinsic viscosity can be determined by the initial slope of the $\ln \eta_{rel}$ from c. In this case, we can estimate $[\eta]$ of polyelectrolyte excluding inevitable, in our case, "overstatement" values $[\eta]$, concomitant determination of the intrinsic viscosity by the Eq. (2). Comparison of values of the intrinsic viscosity of the initial sample CHT defined by both analyzed methods (Fuoss and Irzhak-Baranov methods) in acetic acid solution of 1, 10 and 70% indicates that the value $[\eta]$, defined by Eq. (3) in all cases less than values defined by Eq. (2) (Table 24.1).

TABLE 24.1 Some Characteristics of Chitosan Calculated Using Methods Fuoss and Irzhak-Baranov

Concentration of acetic acid, %	Fuoss method				Irzhak-Baranov method			
	α	$K \times 10^5$	$[\eta]$, g/dl	M^*, kDa	α	$K \times 10^5$	$[\eta]$, g/dl	M^*, kDa
1	1.15	2.89	12.20	77.9	1.02	5.53	7.79	111.6
10	1.03	10.11	11.30	79.7	0.93	13.91	6.89	113.5
70	0.90	37.41	10.09	83.8	0.81	41.39	5.25	116.3

* M calculated by the Mark-Houwink-Kuhn equation.

After fractionation of the initial sample of CHT and determining the molecular weight of each fraction the values of intrinsic viscosity of fractions in 1, 10, and 70% acetic acid were determined by two analyzed methods – by Fuoss and by Irzhak and Baranov. This made it possible to estimate the parameters K and α in Mark-Houwink-Kuhn equation, testifying about the conformational state of macromolecules in CHT solution. Comparison of the values α in Mark-Houwink-Kuhn equation defined by two analyzed methods shows that for all the studied concentrations of acetic acid the macromolecular coil "by Fuoss" has more comprehensive conformation than the coil "by Irzhak." The reason for the discrepancy in the values of K and α, defined using two methods of determining the viscosity obviously is related to the fact that when using the Fuoss method the coefficients in the Mark-Kuhn-Houwink equation doesn't reflect CHT conformation in solution of acetic acid concentrations studied, but reflects CHT conformation in a dilute solution of acetic acid.

24.4 CONCLUSIONS

Value of the molecular weight calculated from the values [η], α and K according to the Irzhak and Baranov equation satisfactorily coincide CHT molecular weight obtained by the absolute method in acetate buffer, $M_{s\eta}$ =113000 Da. Thus, in the study of polyelectrolyte solutions in weak electrolytes, when calculating the values of the intrinsic viscosity is more correct to use the method of Irzhak and Baranov, rather than the more common method of Fuoss.

KEYWORDS

- chitosan
- polyelectrolytes
- viscosimetry
- conformation

REFERENCES

1. Vikhoreva, G. A., Galbraikh, L. S. Plenki i volokna na osnove hitozana in:. Skryabin, K. G., Vikhoreva, G. A., Varlamov, V. P. (Ed.), Hitin i hitozan: poluchenie, svoistva i primenenie. Nauka, Moskow 2002, 254–279.
2. Rafikov, S. R., Budtov, V. P., Monakov Yu.B. Vvedenie v fiziko-khimiyu rastvorov polimerov. Khimiya, Moskow 1978.
3. Nudga, L. A., Plisko, E. A., Danilov, S. N. Jurnal obshei himii. 1971. 41, 2555.
4. Tverdokhlebova, I. I. Konformatziya macromolecul (viskozimetricheskiy method otzenki). Khimiya, Moskow 1981.
5. Fuoss, R. M. Viscosity function of polyelectrolytes: Journal of *Polymer Science.* 1948. 3, 603.
6. Baranov, V. G., Brestkin Yu.V., Agranova, S. A. et al. Vysokomoleculyarnie soedineniya. 1986. 28B, 841.
7. Pavlov, G. M., Gubarev, A. S., Zaitseva, I. I. et al. Jurnal prikladnoy khimii. 2007. 79, 1423.
8. Huggins, M. L. Journal of the American Chemical Society. 1942. 64, 2716.
9. Tsvetkov, V. N., Eskin, V. E., Frenkel, C.Ya. Struktura macromolecul v rastvorah. Nauka, Moskow, 1964.
10. Tiger, A. A. Fiziko-khimia polymerov. Nuchniy mir, Moskow, 2007.
11. Tanford, Ch. Fizicheskaya khimiya polimerov. Khimiya, Moskow, 1964.

CHAPTER 25

MAGNETIC PROPERTIES OF ORGANIC PARAMAGNETS

M. D. GOLDFEIN, E. G. ROZANTSEV, and N. V. KOZHEVNIKOV

Saratov State University Named After N.G. Chernyshevsky, Russia, E-mail: goldfeinmd@mail.ru

CONTENTS

ABSTRACT

Communication between the phenomenon of a magnetism and paramagnetism, which stable radicals of different type possess is probed. It is shown that under the influence of an outside magnetic field there is a change of physical and chemical properties of some organic free radicals to the localized and nonlocalized unpaired electrons. Changes of properties of low-molecular and high-molecular organic paramagnets are caused by magnetic interactions occurring in them. Some iminoxyl polyradicals, possessing high value of a magnetic susceptibility, can weaken or increase strength of the enclosed magnetic field. One of the most important applications of stable paramagnets is their use as components to polarized proton targets in high energy physics. It allowed to create nuclear precession magnetometers for geophysics and astronautics.

25.1   INTRODUCTION

When a magnetic field H is imposed, all substances show a macroscopic magnetic moment M. The value M relates to the imposed field H with a coefficient of proportionality χ (the magnetic susceptibility):

$$M = \chi H$$

In diamagnetic substances with completely filled orbitals, the induced moment is oriented *against* the external field; their magnetic susceptibility is negative and temperature-independent. In paramagnetic substances with half-filled orbitals, the induced moment vector under the influence of the imposed magnetic field is directed *parallel* to the latter. For noninteracting (independent) spins, the value of the magnetic moment is inversely proportional to temperature and their susceptibility can be approximated by Curie's expression:

$$\chi = C/T$$

where C is Curie's constant, T is absolute temperature.

The value of magnetic susceptibility is usually recalculated to the effective magnetic moment μ_{eff} defined as

$$\mu_{eff} = [(3k/Na)\chi T]^{0.5} = \mu_s g[S(S+1)]^{0.5}$$

where k is Boltzmann's constant, N_a Avogadro's number, μ_b Bohr's magmeton, and S is spin.

For the case of interacting spins, numerous deviations from Curie's law are known. As a first approximation, such behavior is described by the Curie–Weiss law (Figure 25.1):

$$\chi = C/(T - \theta)$$

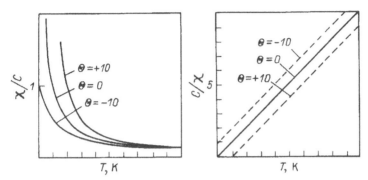

FIGURE 25.1 Temperature dependence of magnetic susceptibility (left) and that of inverse magnetic susceptibility (right) for noninteracting ferromagnetically coupled spins and antiferromagnetically coupled spins, respectively.

Here, the "characteristic temperature" θ is determined by the crystal field and may be either positive, corresponding to ferromagnetic interactions (with parallel spin orientation), or negative, corresponding to antiferromagnetic interactions (with antiparallel spin orientation).

The interradical interactions of uncoupled electrons are classified into two types, namely: dipole–dipole ones and exchange ones; the latter is determined by the overlap of the wave functions of uncoupled electrons and quickly decreases with increasing distance. The exchange interaction averages both the dipolar interaction between uncoupled electrons and the intraradical superfine interaction of uncoupled electrons with atomic nuclei. When there are a couple of electrons on neighboring centers with pronounced overlapping wave functions, some interaction between the spins S_1 and S_2 arises. It leads to the formation of singlet and triplet states. According to Heitler–London's description of chemical bonds this interaction is expressed by the Hamiltonian:

$$\hat{H} = -2J\hat{S}_1\hat{S}_2 \tag{1}$$

Extension of Eq (1) to a multielectron system is described by Heisenberg's exchange Hamiltonian as:

$$\hat{H} = -\sum_{i,j} J_{i,j}\hat{S}_i\hat{S}_j \tag{2}$$

where J_i, j is the exchange integral between atoms i and j having total spins S_i and S_j.

The exchange integral J characterizes the exchange interaction degree and is expressed in energy units. Negative J values correspond to interactions of the antiferromagnetic type (the state of the lowest energy with the antiparallel spin orientation, the ground state being a spin singlet). A positive exchange integral is associated with the ferromagnetic interaction (the ground state is a spin triplet) [1].

In 1963 McConnell [2] formulated an idea of the possible presence of particles with high positive and negative atomic π spin densities. In a crystal, such compounds can be packed in parallel to each other into stacks to form conditions for strong exchange interactions between the atoms with a positive spin density and the atoms with a negative spin density in the neighboring radicals. Ferromagnetic exchange interaction expressed by Heisenberg's exchange integral

$$H^{AB} = -\sum_{i,j} J_{ij}^{AB} S_i^A S_j^B = -S^A S^B \sum_{i,j} J_{ij}^{AB} \rho_i^A \rho_j^B \tag{3}$$

is a consequence of the incomplete compensation of the antiferromagnetic coupled spins, where S^A and S^B are the total spins of radicals A and B; ρ_i^A and ρ_j^B are the π-spin densities on atoms i and j in radicals A and B, J_{ij}^{AB} the interradical exchange integral for i and j.

Buchachenko [3] stated that "it would be almost impossible to realize this way since it is impossible to construct such a crystal lattice of the radical "to turn-on" the intermolecular exchange interaction among the atoms with opposite spin densities only and to "turn-out" it among the atoms with spin densities of an identical sign." McConnell's model, nevertheless, was involved to interpret complex interradical interactions found in the crystals of stable organic paramagnets. Rather recently [4], direct experimental evidence on *bis*-phenylmethylenyl-[2,2]-*p*-cyclophanes has been obtained that ferromagnetic exchange can be reached within McConnell's model.

Pseudo-*o*-, pseudo-*m*-, and pseudo-*p*-*bis*-phenylmethylenyl-[2.2]-*p*-cyclophanes

$S = 2$ $S = 0$ $S = 2$

respectively, were obtained through photolysis in a vitrified matrix at low temperatures. The spin-spin interaction between the two triplet diphenyl-carben fragments built into the [2,2]-p-cyclophane frame, was explored by the EPR technique.

For the pseudo-o-dicarben a quintet state has been revealed, and within a temperature range of 11–50 K the EPR signal intensity obeys the Curie law. When $T > 20$ K, another signal caused by a changed population of the triplet level was observed. Therefore, the pseudo-o-isomer is in its ground quintet state with $D = 0.0624$ and $E = 0.0190$ cm^{-1}, and the triplet state lies 63 cm^{-1} higher by energy.

Pseudo-m-bis-phenylmethylenyl-[2.2]-n-cyclophane gives no reso-nance signal at 11 K. But a triplet state with $D = 0.1840$ and $E = 0.0023$ cm^{-1} was recorded with increasing temperature. The pseudo-m-isomer is in its ground singlet state, and the value of singlet–triplet splitting is 98 cm^{-1}. At 15 K for the pseudo-n-isomer the quintet nature of the ground state with $D = 0.1215$ and $E = 0.085$ cm^{-1} has been established, but it is not stable chemically

25.2 MAGNETIC INTERACTIONS IN STABLE ORGANIC PARAMAGNETS

As our work deals with stable radicals only, it seems expedient to analyze liter-ature data, having limited ourselves to stable organic paramagnets. According to our goal, it is worthwhile focusing attention mainly on measurements of magnetic susceptibility and, in particular, on clarification of the dependence of the magnetic properties of substances on their chemical structure. One of the most studied stable aroxyls, the so-called galvinoxyl, possesses a highly delocalized uncoupled electron. The formula of galvinoxyl is

The crystals of galvinoxyl have monoclinic symmetry with the elementary cell parameters $a = 23.78$, $b = 10.87$, $c = 10.69$ nm and the angle of non-orthogonality $b = 106.6°$; a second-order symmetry axis; a $12°$ deviation from coplanarity, and a $134°$ angle formed by the C—C bonds at the central carbon atom [5]. The crystal structure of galvinoxyl allows the possibility of the formation of a magnetic linear chain structure extended along the c axis.

The temperature dependence of the paramagnetic susceptibility of galvinoxyl obeys the Curie–Weiss law with a positive Weiss constant $\theta = +19$ K above 85 K, which allows one to assume ferromagnetic interactions between neighboring particles. However, at 85 K a phase transition is observed, upon which the paramagnetic susceptibility sharply decreases, and at 55 K its value corresponds to the content of free radicals 1.1% [6].

It is interesting that galvinoxyl radicals form couples in a diluted crystal, which have a ground triplet state; and a thermally achievable excited singlet state lies 2 J higher [7]. Therefore, a ferromagnetic interradical exchange interaction with $2J_F = 1.5 \pm 0.7$ meV is realized in every radical couple. In other words, within a temperature range 10–100 K a diluted galvinoxyl crystal shows no phase transition since it retains ferromagnetic interactions. On the contrary, antiferromagnetic-type interactions with $2J_{AF} = -45 \pm 2$ meV prevail in chemically pure galvinoxyl below 85 K. Apparently, the phase transition in this case is caused by radical dimerization.

This is also confirmed by data on the temperature dependence of the magnetic susceptibility of mixed galvinoxyl crystals. From magnetization curves it follows that the spin multiplicity is almost proportional to the radical concentration in a mixed crystal. As calculations show [8], the ferromagnetic intermolecular interactions in galvinoxyl can be explained by superposition of the effects of intraradical spin polarization and charge transfer between free radicals.

Hydrazyl and hydrazidyl radicals are inclined to the formation of various complexes with solvents. This circumstance slightly influences the value of the g factor but strongly changes the EPR line width. The discordance in the magnetic data of different researchers is probably caused by the presence of impurities in the samples studied, owing to experimental difficulties in purification of organic paramagnets.

The magnetic susceptibility of 1,3,5-triphenyl verdazyl [8]

was measured in a temperature range of 1.6–300 K.

In the high-temperature range the magnetic susceptibility obeys the Curie–Weiss law with a negative Weiss constant $\theta = -8$ K. The susceptibility deviates from the Curie–Weiss law at lower temperatures and shows a wide maximum near 6.9 K.

The usage of Heisenberg's linear model with isotropic exchange interaction with $J/k = -5.4$ K above 6 K provides satisfactory agreement with experiment. The distant order of interactions caused by ferromagnetic-type interchain interactions arises at 1.7 K. The crystals of 1,3,5-triphenylverdazine have orthorhombic symmetry with the elementary cell parameters: $a = 18.467$, $b = 9.854$, $c = 8.965$ nm. All the four nitrogen atoms and the substituent at position 3 are almost coplanar, the two other phenyl groups turned relative to the C—N bond by 23 and 13°, respectively. The radicals in a possible magnetic chain are shown [9] to be bound with each other by a second-order screw axis parallel to the c axis so that interchain ferromagnetic exchange interactions are formed between these antiferromagnetically ordered chains.

In this regard, verdazyl biradicals with strongly delocalized uncoupled electrons are of interest, namely: n-di-1,5-diphenyl-3-verdazyl benzene and m-di-1,5-diphenyl-3-verdazyl benzene:

The susceptibility of the *n*-isomer obeys the Curie–Weiss law above 100 K with a Weiss constant $\theta = -100 \pm 20$ K and a Curie constant $C = 1.0 \pm 0.01$ K·emu/mol, and the χ *vs.* μ curve passes through a maximum at 19 ± 1 K when temperature reduces.

In the case of the *m*-isomer the susceptibility follows the Curie–Weiss law over the whole temperature range studied 1.8–300 K ($C = 0.90 \pm 0.05$ K·emu/mol and $\theta = -12 \pm 3$ K). Both biradicals are supposed to exist in a ground triplet state (J/k > 300 K). The J'/k value of the exchange interaction between the triplets in *n-bis*-verdazyl was estimated from the location of the maximum, it was negative (−7 K).

Classical aromatic hydrocarbonic radicals are often classified as so-called π-electronic radicals wherein an uncoupled electron is delocalized over the whole aromatic bond system. In their majority, arylmethyl radicals in solution exist in thermodynamic equilibrium with their dimer.

Ballester's perchloro-triphenylmethyl radicals sharply differ from classical hydrocarbonic ones by properties: they are rather stable in the absence of light, completely monomeric in both solution and their solid state.

The perchloro-triphenylmethyls studied in Ref. [10] within the range 293–77 K obey the Curie–Weiss law (Table 25.1).

The antiferromagnetic-type interactions found in the stable para-magnets of the trichloro-triphenylmethyl series, are well described by McConnell's above model, being in agreement with the crystal structure and spin density values.

Unlike classical aromatic radicals, the NO group in the iminoxyl radicals takes no part in the formation of a conjugated bond system; the uncoupled electron in such radicals is therefore mainly localized on the nitrogen–oxygen bond. The rather reliable steric shielding of the uncoupled electron (due to the effects of the voluminous methyl groups and the σ-bond system interfering uncoupled electron delocalization) provides conditions for nonradical reactions to proceed in the row of functionalized

TABLE 25.1 Characteristic Temperature π (K) of Some Perchloro-Triphenylmethyl Radicals: Ar, Ar^1, Ar^2 C$^{\cdot}$

Ar	Ar^1	Ar^2	θ, K	μ_{eff}
4H-C$_6$HCl$_4$	C$_6$Cl$_5$	C$_6$Cl$_5$	−4.8	1.76
4H-C$_6$HCl$_4$	4H-C$_6$HCl$_4$	4H-C$_6$HCl$_4$	+1.9	1.73
3H,5H-C$_6$H$_2$Cl$_3$	C$_6$Cl$_5$	C$_6$Cl$_5$	−10.4	1.76
3H,5H-C$_6$H$_2$Cl$_3$	3H,5H-C$_6$H$_2$C$_3$	3H,5H-C$_6$H$_2$Cl$_3$	−10.1	1.74
2H-C$_6$HCl$_4$	C$_6$Cl$_5$	C$_6$Cl$_5$	−12.0	1.71
2H-C$_6$HCl$_4$	2H-C$_6$HCl$_4$	C$_6$Cl$_5$	−3.3	1.69

radicals of this class. This allows synthesizing many chemically pure para-magnets of various chemical structures (Figure 25.2) [11].

One can easily see that the majority of works is devoted to 2,2,6, 6-tetramethyl-4-hydroxypiperidino-1-oxyl (TEMPOL) derivatives obtained by Rozantsev [12]. TEMPOL crystallizes in a monoclinic cell with the

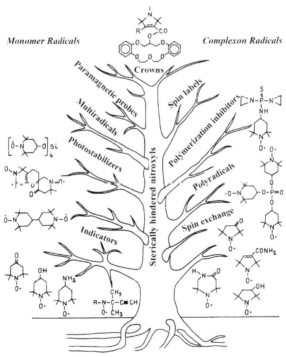

FIGURE 25.2 Genealogic tree of stable iminoxyl (nitroxyl) radicals, whose synthesis and application were promoted by the discovery of nonradical reactions of radicals.

axis parameters $a = 0.705$; $b = 1.408$; $c = 0.578$ nm; b=118°40 and belongs to the spatial group C (Figure 25.3).

Chains of the radicals bound to each other by hydrogen bonds are formed in a TEMPOL crystal (Figure 25.3). It is supposed that the strongest exchange interactions of radicals are oriented along the Z axis through the oxygen atoms. The direction along the a axis through the hydrogen bond could probably be the interaction next in contribution. Proceeding from structural reasons, weaker magnetic interactions can be expected between the ac planes.

Rozantsev and Karimov [13] investigated the magnetic susceptibility of chemically pure TEMPOL by the EPR method for the first time in 1966. They showed the EPR signal strength to deviate from Curie's law and to exhibit a wide and smooth maximum near 6 K. A wide maximum on the thermal capacity curve was found at 5 K. In the high-temperature range the susceptibility obeys the Curie–Weiss law with a negative Weiss constant $\theta = -6$ K. At lower temperatures, the χ vs. T curve deviates from the Curie–Weiss law and has a flat maximum at 6 K.

Such behavior of paramagnets is well described by Heisenberg's one-dimensional model with isotropic antiferromagnetic interactions. For Heisenberg's linear system with $S = 1/2$ the magnetic susceptibility should have a flat maximum determined by $\chi_{max}/(N_a g^2 \mu_B^2/J) \approx 0.07346$ at $kT_{max}/J \approx 1.282$.

The value of the exchange J/k parameter is estimated as -5 K. Therefore, independent studies of the magnetic susceptibility of TEMPOL evidence strong exchange interactions experienced by radicals in one direction, which results in the near order of interactions and the formation of linear antiferromagnetic chains near 6 K.

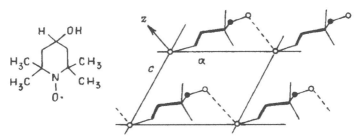

FIGURE 25.3   Projection of the 2,2,6,6-tetramethyl-4-hydroxypiperidyl-1-oxyl structure onto the ac plane.

As nonzero interaction always exists between the chains in one-dimensional magnetic systems, it could be expected that below some critical temperature it would get rather expressed to cause transition to a distant order of interactions. In the case of TEMPOL the distant order caused by interchain interactions arises at $T_N = 0.34$ K, the interchain to intrachain interaction ratio (J/J') estimated as 0.003, $J/k = 0.013$ K.

An alternating linear chain arises if $\gamma < 1$. The case $\gamma = 0$ corresponds to a simple dimer where paired interactions act only.

The stable di-2,2,6,6-tetramethyl-1-oxyl-4-piperidyl sulfite biradical exemplifies the alternating chain (Figure 25.4). In the high-temperature range its magnetic susceptibility is described by the Curie–Weiss law with $\theta = -9$ K. When temperature falls, the susceptibility of this biradical deviates from the Curie–Weiss law near 25 K, and then sharply drops down to 2 K.

As is seen from Figure 25.4, the susceptibility maximum of this biradical is less flat than it would be for a regular spin chain with $\gamma = 1$. The use of the model of an alternating spin chain with $J/k = 9.6$ K and $\gamma = 0.55$ provides satisfactory agreement with experiment, and the maximum exchange between interacting spins $J/k=9.6$ K is a result of structural exchange interactions in the crystal.

The adduct of copper hexafluoro acetylacetonate with the iminoxyl radical of 2,2,6,6-tetramethyl-4-hydroxypiperidyl-1-oxyl is another example of the alternating linear chain [14]

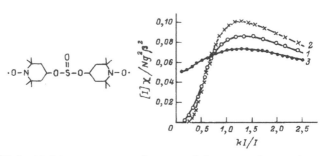

FIGURE 25.4   (a) A low-temperature fragment of the curve of magnetic susceptibility of the biradical; (b) calculated curve for Heisenberg's paired interaction with J/k = 9.6 K; (c) calculated curve for a regular spin chain with J/k = 9.6 K.

When its studying, strong (19 K) ferromagnetic interaction between the copper ion and iminoxyl was revealed, which followed from data obtained at temperatures above 4.2 K. At temperatures below 1 K, the magnetic susceptibility sharply increased, and a flat maximum, characteristic of antiferromagnetic linear chains, was found at ~ 80 meV. Analysis of these data was carried out in the assumption that the substance consists of chains with spins $S = 1$ and weak (2J = −78 meV) antiferromagnetic interaction between spins. Therefore, the alternation in this case arises because of alternation of the strong ferromagnetic and weak antiferromagnetic interactions.

For the silicon-organic iminoxyl polyradicals

$$R^• \diagdown \qquad\qquad \diagup R^•$$
$$R^• - \overset{R^•}{\underset{R^•}{Si}} - CH_2 - CH_2 - \overset{R^•}{\underset{R^•}{Si}} - R^•$$

I

$$R^• \diagdown \qquad\qquad \diagup R^•$$
$$R^• - \overset{R^•}{\underset{R^•}{Si}} - CH_2 - \overset{R^•}{\underset{R^•}{Si}} - R$$

II

$$R^• \diagdown \qquad\qquad \diagup R^•$$
$$CH_3 - \overset{}{\underset{R^•}{Si}} - CH_2 - CH_2 - \overset{}{\underset{R^•}{Si}} - CH_3$$

III

$$R^• \diagdown \qquad\qquad \diagup R^•$$
$$CH_3 - \overset{}{\underset{R^•}{Si}} - CH=CH-Si - CH_3$$

IV

$$R^• \diagdown \qquad\qquad R^•$$
$$C_6H_5 - \overset{}{\underset{R^•}{Si}} - CH_2 - CH_2 - \overset{}{\underset{R^•}{Si}} - C_6H_5$$

V

(R' ≡ 2,2,6,6-tetramethyl-1-oxyl-4-piperidyl fragment) paired spin-spin interactions are characteristic. All the studied paramagnets (I–V) exhibit low-temperature deviations of the course of their magnetic susceptibility from Curie's law $\chi = \text{const}/T$ (Figure 25.5a), which are due to the existence of correlation between uncoupled electrons. E.g., the susceptibility of tetraradical V passes through a maximum at 8 K and decreases by 10 times when temperature falls down to 2 K (Figure 25.5). Such a course of susceptibility is well described by a model offered for paired exchange interactions of uncoupled electrons

$$\chi = CT^{-1}[3 + \exp(J/kT)]^{-1}$$

If the ground state of such a couple is a singlet and the thermally excited state is a triplet (or a triplet magnetic exciton), it mainly contribute to the magnetic susceptibility. Excitons get energy for their excitation from thermal energy. Therefore, when kT becomes less than the exchange value J between electrons, the number of triplet states sharply falls and, hence, the susceptibility sharply decreases. From analysis of the course of susceptibility the exchange parameters of strongly bound spins (see Table 25.2) were estimated.

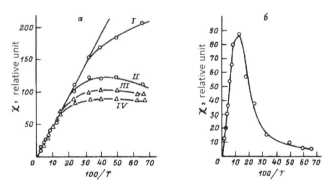

FIGURE 25.5 Magnetic susceptibility of polyradicals I–IV and V, respectively, graphs *a* and *b*.

TABLE 25.2 Exchange Interaction Parameters of Silicon-Organic Polyradicals

radical	I	II	III	IV	V
J/k,	K	2.2±0.5	3.2±0.5	4.6±0.5	5.2±0.5	14.6±0.5

It is interesting that for tetraradical V the exchange interaction parameter J/k between spin couples is 0.1 K.

The crystal structure of organic radicals, because of the asymmetry of the majority of chemical particles, as a rule, allows one to resolve topological linear chains of most strongly interacting spins. A study of the structure of, e.g., radical V (Figure 25.6) has shown that the nitrogen atoms of one radical heterocycle form a chain of paramagnetic centers with a link length about 6 nm parallel to the axis, and the nitrogen atoms of the other heterocycle form another spin chain parallel to the first one with the length of an elementary link of 6.6 nm. Besides, each paramagnetic center in the chain thus has two neighboring spins from other chains at distances of 6.4 and 6.6 nm, respectively.

In Ref. [15], the paired intermolecular interaction of the basic spin system in 1,4-*bis*-2,2,6,6-tetramethyl-1-oxyl-4-piperidyl-butane crystals was reported. This interaction is distinctly seen on the temperature dependence curve within a range of 10–300 K as the presence of a characteristic maximum near 40 K:

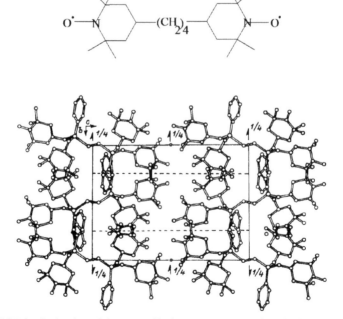

FIGURE 25.6   Projection of the tetraradical V structure onto plane (010).

The monoclinic crystals of this biradical have the elementary crystal cell parameters $a = 11.754$, $b = 10.980$, $c = 8.693$ nm with the P_12/b spatial group (Figure 25.7).

This structure features the existence of two systems of pairs of $=N^{o\cdot}$ radical fragments which are mirror symmetric about the ab plane. For the mirror symmetric couples the angle between the lines connecting the centers of the iminoxyl fragments is 50°. Inside each couple, the oxygen atoms are at a distance of 0.351 nm, and the nitrogen atoms are at a distance of 0.485 nm. The short distance between the NO fragments in a couple and the relative location of the C–N–C planes promote direct electronic exchange in these couples (Figure 25.8) [16, 17].

Really, the temperature course of the paramagnetic susceptibility in the crystals is well described within the model of antiferromagnetic paired exchange with a constant $J = -33.5$ K. The intramolecular exchange

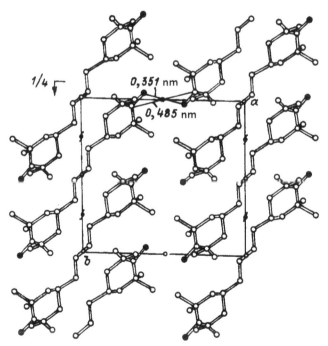

FIGURE 25.7 Projection of the 1,4-*bis*-2,2.6,6-tetramethyl-1-oxyl-4-piperidyl butane biradical structure.

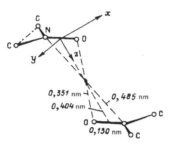

FIGURE 25.8 A scheme of the mutual arrangement of radical fragment =NO· pairs bound with strong exchange.

interactions J' transferred through the $-(CH_2)_4-$ bonds appear less than the hyperfine coupling constant, which corresponds to $J'J'<2\cdot10^{-3}$ K, that is, $J/J' > 10^4$.

The tanolic ester of octanoic acid

$$O^{\cdot}-N \overset{\displaystyle }{\bigcirc} -O-\underset{\underset{O}{\|}}{C}-(CH_2)_6-\underset{\underset{O}{\|}}{C}-O-\bigcirc N-O^{\cdot}$$

obeys the Curie–Weiss law with a positive constant $\theta = +1$ K in a temperature range 1.9–300 K. All the magnetic interactions of interest are rather weak and manifest themselves at temperatures below 1 K only. Apparently, a magnetic transition at $T = 0.38 \pm 0.01$ K proceeds in the system due to ferromagnetic ordering. Neutronography has established that the crystals of this paramagnetic are layered: the neighboring particles inside each layer are bound ferromagnetically with $J_1 = +1.1$ K and $J_2 = 0.07$ K, but the layers are connected among themselves by weaker antiferromagnetic interactions with $J' = -0.015$ K. It is believed [18] that the substance behaves as a metamagnetic with 2D ferromagnetic ordering. For 2,2,6,6-tetramethyl-4-oxo piperidyl-1-oxyl azine (TEMPAD), a maximum at 16.5 K is found on the curve of the temperature course of paramagnetic susceptibility, and its change under the Curie–Weiss law ($\theta = -15$ K) is observed within 77–273 K. In the case of diluted TEMPAD crystals, two values of the Weiss constant have been found: about -10 K in the high-temperature range and about -1 K in the low-temperature one (Figure 25.9) [19].

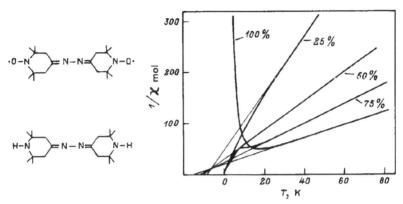

FIGURE 25.9 Temperature dependence of the inverse paramagnetic susceptibility of 2,2,6,6-tetramethyl-4-oxopiperidine-1-oxyl azine crystals in a matrix of triacetonamine azine.

The magnetic behavior of TEMPAD was interpreted within the theory of magnetic triplet paired transitions [46–48]. Nobody can exclude the existence of strong intermolecular exchange interaction along the a axis (J_1; $J_1/k = -12.8$ K), weak intramolecular interaction (J_2; $J_2/k \sim 2 \times 10^{-2}$ K), and interlayer interaction with $J_1/k \sim 1$ K [20].

In 2,2,6,6-tetramethyl-4-oxypiperidyl-1-oxyl phosphite (TEMPOP)

no near order of interactions was found, though at very low temperatures the course of inverse paramagnetic susceptibility deviated from the Curie–Weiss law (Figure 25.10). Theoretically, the effective magnetic moment of three noninteracting spins with $g = 2.00$ should be equal to $3.00 \, \mu_B$; in experiment (at high temperatures), a value of $3.01 \mu_B$ was obtained.

Interesting studies on nitronyl nitroxyls (or nitroxide nitroxyls) NIT(R) 2R-4,4,5.5-tetramethyl-4,5-dihydro-1H-imidazolyl-1-oxyl-3-oxides

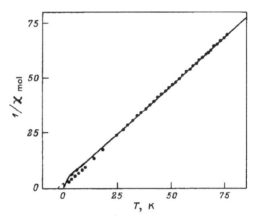

FIGURE 25.10 Temperature dependence of the inverse paramagnetic susceptibility of 2,2,6,6-tetramethyl-4-hydroxypiperidyl-1-oxyl phosphite ($\theta = -3.5$ K). The solid line is calculated for a three-spin cluster with J = −2.7 K.

with metal ions are presented in Refs [21–32]. Owing to the conjugation of a nitroxyl group with a nitroxide one, the exchange interaction between one oxygen atom and a metal ion can be transferred to another oxygen atom without attenuation. Therefore, these radicals with a delocalized uncoupled electron are capable of forming not only mononuclear complexes with metals [21–23] but also magnetic chains of various natures [24–27].

There exist a metal-containing compound Cu(hfac)$_2$(NIT)Me, where hfac is a hexaftoracetylacetonate ion [CF$_3$C(0)CHC(0)CF$_3$]$^-$, behaves as a one-dimensional ferromagnetic with an interaction constant of 25.7 cm^{-1}. The effective magnetic moment equal to 2.8 θ_B at 300 K monotonously increases with decreasing temperature and gets 4.9 θ_B at 4 K. This means that the effective spin of the system increases almost up to 3.

Replacement of a copper ion by Ni (II) or Mn (II) in the Ni(hfac)$_2$ (NIT)R and Mn(hfac)$_2$(NIT)R complexes with R = Me, Et, i-Pr, n-Pr, Ph leads to stronger antiferromagnetic-type interactions with J = −424 cm^{-1} for nickel and J = −230–330 cm^{-1} for manganese derivatives. When exploring Mn(hfac)$_2$(NIT)iPr monocrystals, noticeable anisotropy was discovered. As follows from the temperature dependence of susceptibility along the easy magnetization axis (this direction in the crystal coincides with the spin direction orientation), phase transition to a ferromagnetic state occurs at 7.6 K.

In Mn(hfac)$_2$(NIT)Et, Mn(hfac)$_2$(NIT)nPr and Ni(hfac)$_2$(NIT) Me ferromagnetic ordering occurs at 8.1, 8.6, and 5.3 K, respectively. Calculations confirm the dipole-dipole nature of the magnetic interactions in these compounds. Higher temperatures of magnetic phase transition were found for [Mn(F$_5$benz)$_2$]$_2$NITEt and [Me(F$_5$benz)$_2$]$_2$NITMe (where F$_5$benz is pentafluorobenzoate): 20.5 and 24 K, respectively. But there is no unambiguous confidence which type (ferrimagnetic or weak ferromagnetic) this ordering belongs.

The binuclear complex [CuCl$_2$(NITpPy)$_2$I$_2$], where NITpPy ≡ 2-(4-piridyl)-4,4,5,5-tetramethylimidizolinyl-3-oxide-1-oxyl, was studied. Potentially, the radical NITpFy could be a tridentate ligand. The copper ions are shown to be coordinated with two nitrogen atoms in the pyridine rings and three chlorine atoms, two of which are bridge ones. The NO groups of radicals belonging to different mononuclear fragments are rather close to each other. On the basis of magnetic susceptibility data and EPR spectra recorded at 4.2 K, it is supposed that six spins $S = 1/2$ are bound by antiferromagnetic exchange interaction. The interaction between copper and the radical through the pyridine cycle's nitrogen is preferable, which allows the authors to consider NITpFy as promising ligands in the synthesis of metal-containing magnetic materials.

25.3 ORGANIC LOW-MOLECULAR-WEIGHT AND HIGH-MOLECULAR-WEIGHT MAGNETS

The interest to low-molecular magnets [33–40] and high-spin compounds [41–57] is associated with the hope of obtaining compounds possessing spontaneous magnetization below their critical temperature. Though the critical temperatures reached are quite low, it is possible to state with confidence that the understanding of the necessary conditions for the design of a high-temperature organic magnetic material has become clearer than several years ago.

Wide-range studies on ion-radical salts like D$^+$A$^-$D$^+$A$^-$, where D is a cation (donor) and A is an anion (acceptor), were carried out by Miller et al. [33–40]. Decamethylferrocene Fe(II)(C$_5$Me$_5$)$_2$ is often used as a donor and flat 7,7,8,8-tetracyan-p-quinodimethane (TCNQ) and tetracyanoethylene (TCNE) serve as acceptors

TCNQ   TCNE

The complex $[Fe(C_5Me_5)_2]^+[TCNE]^+$ is characterized by a positive Curie–Weiss constant, $\theta = + 30$ K, Curie's temperature (T_c) equal to 4.3 K; in a zero magnetic field, spontaneous magnetization is observed for a polycrystalline sample ($M \sim 2000$ emu Gs/mol) [35]. The saturation magnetization is 16300 emu Gs/mol in oriented monocrystals. This result is in good agreement with the theoretical magnetic saturation moment at ferromagnetic spin alignment of the donor and acceptor and is by 36% higher than for metal iron (per one gram atom). At 2 K, hysteresis with a coercive force of 1000 Gs is observed, corresponding to the values for magnetically hard materials. Above 16 K, the magnetic properties are described by Heisenberg's one-dimensional model with ferromagnetic interaction ($J = + 27.4$ K). At temperatures near T_c, three-dimensional ordering prevails [36].

The compound $[Fe(C_5Me_5)_2]^+[TCNQ]^-$ shows metamagnetic signs with a Néel temperature $T_N = 2.55$ K and a critical field of ~ 1600 Gs. As a rule, metamagnetics are substances with strong anisotropy and, in the presence of concurrent interactions therein, first-order transition to a phase with a total magnetic moment can be observed [1]. E.g., for the salt $[TCNQ]^-$ the magnetization in fields with $H < 1600$ Gs is characteristic of a antiferromagnetic, while when $H > 1,600$ Gs, a hump-like increase of magnetization occurs up to the saturation value, which is characteristic of the ferromagnetic state [37].

Of the 2,5-disubstituted TCNQ salts of decamethylferripinium $[Fe(C_5Me_5)_2]^+[TCNQR_2]^+$ (R = Cl, Br, J, Me, OMe, OPh) [39], $[Fe(C_5Me_5)_2]^+[TCNQI_2]^-$ possesses the highest effective moment $\mu_{eff} = 3.96\ \mu_B$.

Above 60 K the magnetic susceptibility obeys the Curie–Weiss law ($\theta = + 9.5$ K) and the substance is a one-dimensional ferromagnetic. This feature, in combination with that $[TCNQI_2]^-$ exhibits stronger interchain antiferromagnetic interactions in comparison with $[Fe(C_5Me_5)_2]^+[TCNQ]^-$, provides no 3D ferromagnetic ground state at temperatures above 2.5 K.

The complex [40]

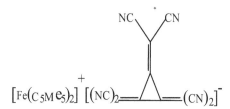

can exist in two polymorphic modifications, namely, monoclinic and triclinic, both obeying the Curie–Weiss law with $\theta = -3.4$ K, $\mu_{eff} = 2.98$ μ_B, and $\theta = -3.4$ K, $\mu_{eff} = 3.10$ μ_B, respectively Below 40 K, in a magnetic field of 30 Gs the monoclinic compound shows the Bonner–Fisher type of one-dimensional antiferromagnetic interaction, that is, has a typical flat maximum about 4 K. This is attributed to antiferromagnetic interaction along the cation chains with an exchange parameter of $J/k = -2.75$ K. To explain the magnetism of ion radical salts, the model of configuration interaction of the virtual triplet excited state with the ground state offered by McConnell [57] is applied. E.g., in the case of donor-acceptor pair D^+A^- it is supposed that the wave function of the ground state has the maximum "impurity" to the wave function of the lower virtual excited state with charge transfer. This state can arise due to direct virtual charge transfer $(D^+ + A^-) \rightarrow (D^{2+} + A^{2-})$, reverse charge transfer $(D^+ + A^- \rightarrow D^° + A^°)$ or disproportionation $(2D^+ \rightarrow D^{2+} + D^°)$.

If any of the states with charge transfer (either donor D or acceptor A, but not both) is triplet, the ground ferromagnetic state of the D^+A^- pair will be stabilized. Therefore, for ferromagnetism manifestation, an organic radical should possess a degenerated and partially filled valent orbital. An essential contribution of the lower virtual excited state with charge transfer to the ground state of the system is necessary; and the structure of the radical ion should be high-symmetric, without any structural or electronic dislocations breaking the symmetry and eliminating the degeneration [33,38].

A whole series of high-spin polycarbenes has been so far synthesized by means of photolysis of the corresponding polydiazo compounds [41–47]. Attempts to get high-spin macromolecules by iodine oxidation of 1,3,5-triaminobenzene have failed. Breslow et al. [57–59] have succeeded to synthesize stable organic triplet systems with C_3 and higher symmetry on the basis of hexaminotriphenylene and hexaaminobenzene derivatives.

where R = C_2H_5, $C_6H_5CH_2$, CF_3CH_2.

Studies on the material obtained by spontaneous polymerization of diacetylene monomer containing stable iminoxyl fragments of butadiyn-*bis*-2,2,6,6-tetramethyl-1-oxyl-4-oxi-4-piperidyl (BIPO) is a highly mysterious story…

The magnetic susceptibility of BIPO [60] obeys the Curie–Weiss law with $\theta = -1.8$ K. The effective magnetic moment equal to 2.45 μ_B at high temperatures corresponds to two independent spins of $S = 1/2$ per monomer unit. The exchange constant derived from analysis of the EPR line has appeared to be $J \sim 0.165$ K (0.115 cm^{-1}), and an estimation in the approach of molecular field has given a value of $J \sim 0.155$ K (0.108 cm^{-1}).

The thermal or photochemical polymerization of BIPO leads to the formation of black powder whose insignificant fraction (0.1%) shows ferromagnetic properties, its magnetization reaches above 1 Gs·g^{-1}. It is noted that ferromagnetism holds up to abnormally high temperatures (up to 200–300°C) and the paramagnetic centers thus die during polymerization (in some cases no more than 10% of their initial quantity remains).

Contrary to the earlier published analysis of these intriguing data, in a subsequent work [60] it was noted that the products of thermal decomposition of BIPO showed neither signs of 3D ferromagnetism nor magnetic interaction. Detailed static and dynamic magnetic data indicate the existence of weak intradimeric ferromagnetic (triplet) interaction with J ~ 10 K only.

Obviously, while solving this problem, the degree of reliability of obtained results will strongly depend on the chemical purity of the materials studied. It is possible to state without exaggeration that natural sciences progress is associated with obtaining and studying chemically pure materials.

Unfortunately, even superficial analysis of the available publications convinces us that the majority of experimental works in this field is associated with studying of structurally disordered "dirty" systems like spin glasses [61] with no coordinated magnetic interactions between chemical particles. The relative simplicity of obtaining "dirty systems" provokes the avalanche-like spreading of "impressive results," various fantastic models and theories, having nothing in common with true science.

It would be thoughtless to consider that chemical purity is sufficient to achieve success in the basic research of high-spin nonmetallic systems. Precision measuring equipment and a methodology including automated X-ray diffraction analysis and modern magnetometry with the usage of superconducting quantum interferometers (squids), without being limited to EPR equipment and high-temperature magnetic measurements, are, undoubtedly, other necessary conditions.

Only the successful development of the basic research of the magnetic properties of pure systems and their constituent chemical particles can provide real breakthrough in the technology of the design of materials of a new generation suitable for manufacturing competitive organic ferromagnets, antiferromagnets, and ferrimagnets, including metamagnets and speromagnets.

In 1990 Emsley [62] published a paper under an intriguing title where he reported about the synthesis of a stable iminoxyl radical (nitroxide nitroxyl triradical) with its properties of a "molecular" organic magnet. In other words, the discovery of a metal-free "organic magneton" with cooperatively ordered electronic interactions at the level of a discrete chemical particle was claimed:

Dulog and Kim [63] have found that some blue powder obtained by them possesses a high value of magnetic susceptibility and can strengthen or weaken the intensity of the applied magnetic field like metal magnets. Provided that the remarkable properties of blue trinitroxyl are not a trivial consequence of metallic pollution, the new material will be able to find applications when designing magnetic registering devices of a new generation, magnetoplanes, and other equipment.

Using the principle of orienting effect of the intraradical electrostatic field of nitroxide groups, it could be possible to design high-molecular-weight magnetic materials with magneto-ordered organic domains (magnetons) like blue nitroxide nitroxyl triradical of Stuttgart's chemists as monomeric links therein.

Stable paramagnets have found practical applications as additives to polarized proton targets in the experimental physics of high energy. The method of reaching ultralow temperatures by ^{3}He dissolution in ^{4}He opens new opportunities in the technology of polarized targets. For example, the high polarization obtained by a usual dynamic method in a strong and uniform magnetic field (25 kOe) can be kept for a long time after the termination of the dynamic polarization "pumping" if the working substance of the target is cooled rather quickly down to a temperature about 0.1÷0.01 K. Then, the intensity of magnetic field can be lowered down to ~5 kOe. This opens new prospects of the use of such targets in physical experiments.

In the existing polarized proton targets operating at temperatures as low as 0.5 K, the main working substances are butyl alcohol and ethylene glycol as frozen balls. However, these substances are not technological for their usage in cryostats with ^{3}He dissolution in ^{4}He.

A substance, solid at room temperature, rather rich with protons, and containing radicals stable at room temperature as paramagnetic additives, would be most convenient. Therefore, polyethylene used as either a 200 mμ film or powder with ~200 mμ grains was selected as the working substance on the basis of recommendations from Refs. Stable iminoxyl radicals were taken as a paramagnetic additive. To introduce such a radical into polyethylene, the necessary amount of the radical and polyethylene was placed into a tight glass ampoule, heated up to 80°C, and maintained at this temperature for 8–10 h. To study proton polarization in polyethylene, preliminary experiments were conducted at a temperature of 1.3 K in magnetic fields of 13 and 27 kOe. The stable volatile radical 2,2,6,6-tetra-methyl-piperidine-1-oxyl appeared most promising.

At an optimum concentration of the radical in a magnetic field of 13 kOe in a polyethylene film, a polarization of 5–7% has been reached, which corresponds to an amplification polarization factor of $E = 50 \div 70$. The period T_1 of proton spin-lattice relaxation was 2.5 min. Transfer into a magnetic field of 27 kOe led to no increase in the polarization amplification factor, only T_1 increased. However, it was revealed that the use of polyethylene of lower density and higher purity as powder with 200 mμ grains led to an increase in E up to $70 \div 100$. Thus, a 14–20% polarization was achieved in the field of 27 kOe at $T = 1.3$ K.

In experiments at ultralow temperatures, polyethylene powder was exposed to preliminary annealing in vacuum (10^{-5} mm Hg) at 80°C during 5–6 days. Then, the ampoule with this powder was filled with pure gaseous helium and the powder was saturated by the vapors of the radical in a helium atmosphere. Such an annealing procedure does not influence the rate of spin-lattice relaxation considerably, but increases the final polarization almost twice. The sample prepared in this amount (150 mg) was introduced, at room temperature, into a glass camera of ^{3}He dissolution in ^{4}He located in a running-wave microwave cell. The cell was placed into a superconducting solenoid and cooled down to a temperature of 1.3 K. The sample was in direct contact with the solution whose minimum temperature was 0.04–0.05 K.

For polarization pumping, a microwave generator of the OV-13 type ($\lambda = 4$ mm) with a power about 70 mW was used; bringing 1.5–2.0 mW into the cryostat was enough to increase the solution temperature up to 0.1 K.

Proton polarization up to 50% was reached in polyethylene samples as powder with an optimum concentration of the radical of 10^{-19} spins per polymer gram. The period of polarization pumping was about 3.5 h. After switching the microwave field off, the temperature of the ^{3}He–^{4}He system decreased down to 0.05 K. At this temperature in a magnetic field of 27 kOe almost no polarization decay was observed. In a magnetic field lowered down to 5 kOe, the relaxation time was not less than 30 h and only with the magnetic field decreased down to 1.5 kOe it reduced to 1.5 h.

The EPR spectrum of tetramethylpiperidine-1-oxyl introduced into polyethylene in the above quantity at room temperature represents a well-resolved triplet with a distance between the hyperfine coupling components of 15 kOe. At temperatures below 1 K the EPR spectrum is transformed to a line with its half-width of 80 Oe with an ill-defined superfine structure. To simplify the EPR structure, an attempt to replace the radical with ^{14}N by that with ^{15}N in the same concentration was made. However, such replacement gave no increase in the maximum polarization.

The experiments conducted have shown that now there is a real possibility to create a "frozen" polarized proton target in a magnetic field about 5 kOe.

Now, in geophysics and astronautics, nuclear precession magnetometers possessing a number of essential advantages are widely adopted. Their high sensitivity (to 0.01 gamma) and the accuracy of measurements, the absoluteness of indications, and the independence of temperature, pressure and sensor orientation are advantages of nuclear magnetometers.

Cyclic operation is a feature of precession magnetometers. The process of measurement consists of two consecutive processes, namely: polarization of the working substance of the magnetometer sensor during which nuclear magnetization is established, and measurements of the signal frequency of the nuclear induction determining the absolute value of the field measured.

The use of the phenomenon of dynamic polarization of atomic nuclei allows one to overlap the stages of polarization and measurement, and to essentially increase the speed of the magnetometer. Nuclear generators based on the phenomenon of dynamic polarization of atomic nuclei, allowing continuous monitoring of changes in the magnetic field were designed.

Earlier, only Fremy's diamagnetic salt dissociating into paramagnetic anions in aqueous solution was used as the working substance of the sensors of nuclear magnetometers based on the phenomenon of dynamic nuclei polarization.

Saturation of any hyperfine coupling line in the EPR spectrum of Fremy's salt solution leads to a significant increase in the nuclear magnetization of the solvent. This is the effect of dynamic polarization. The hydrolytic instability is a demerit of Fremy's salt. Even in distilled water, the radical anion of this salt is hydrolyzed to diamagnetic products during several dozen minutes, the process of degradation having autocatalytic character. A paramagnetic solution stabilized with an additive of potassium carbonate preserves about a month provided that its temperature would not exceed 40°C. Stable paramagnets of the iminoxyl class have indisputable advantages over Fremy's salt.

Rozantsev and Stepanov [64] proposed 2,2,6,6-tetramethyl-4-oxipiperidine-1-oxyl and 2,2,6,6-tetramethyl-4-oxopiperidine-1-oxyl well soluble in many proton-containing solvents and possessing a resolved hyperfine coupling in their EPR spectra within a wide range of magnetic fields as working substances for nuclear precession magnetometers in 1965.

In weak magnetic fields, the bond between the electronic and nuclear spins of nitrogen is not broken off, and the set of energy levels is characterized by a total spin number S taking on two values (1/2 and 3/2) and a magnetic quantum number ms taking on values $2S + 1$. The set of energy levels of iminoxyl radicals in weak magnetic fields is presented in Figure 25.11.

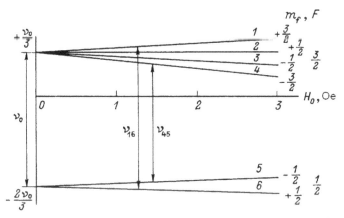

FIGURE 25.11　A scheme of the energy levels of iminoxyl radicals in weak magnetic fields.

The rule of selection $\Delta S = 0\pm1$ and $\Delta m_s = \pm1$ exists for π-electronic transitions. Saturation of one of such transitions by a strong radio-frequency field oriented perpendicular to the constant field significantly changes the electronic magnetization M_z of the solution, which, in turn, results in an increase in the proton magnetization m_z of solvent molecules according to the expression

$$m_z - m_o = fg(M_o - M_z)$$

where m_z, m_o, M_o, M_z are the nuclear and electronic magnetizations of the solution in a stationary mode and thermal equilibrium at saturation of the transition in the system of electronic energy levels, respectively, f is a coefficient determining the contribution of electronic–nuclear interaction into the mechanism of nuclear relaxation of the solvent, g a coefficient depending on the nature of the electronic–nuclear interaction.

Studying of iminoxyl radicals has shown that only saturation of the 1–6 and 4–5 transitions (Figure 25.11) leads to a considerable dynamic polarization of the nuclei of the solvent. By the value of proton dynamic polarization in solutions of iminoxyl paramagnets the latter ones are on a par with Fremy's salt applied earlier. They were more stable in water and organic solvents and did not change their initial characteristics within half a year. In the course of research, solutions of iminoxyl paramagnets were repeatedly heated up to 90°C. It is obvious that organic paramagnets represented essentially new working substances for nuclear precession magnetometers of a new generation. The main advantage of these new working substances consists in the possibility to choose a solvent for them to operate in any climatic conditions, with a high content of protons and a long period of proton relaxation. Little wider electronic transitions in comparison with an aqueous solution of Fremy's salt are easily eliminated by the usage of deuterated samples of iminoxyl radicals [65]. Numerous attempts of the authors to register the mentioned idea as a USSR author's invention certificate were steadily refused.

After these materials had been published, a group of French engineers designed a precession nuclear magnetometer with a deuterated solution of 2,2,6,6-tetramethyl-4-oxopiperidine-1-oxyl in dimethoxyethane as its working substance.

KEYWORDS

- **free radical**
- **interaction**
- **magnetic susceptibility**
- **magnetism**
- **magnetometer**
- **paramagnetic**
- **properties**

REFERENCES

1. Karlin, R. Magneto-chemistry. Moscow: World, 1989. [Russ].
2. McConnell, H. M. J. Chem. Phys. 1963. Vol. 39. P. 1910.
3. Buchachenko, A. L. Reports of USSR Science Academy. 1979. Vol. 244. P. 1146. [Russ].
4. Izuoka, A., Murata, S., Sugavara, T., Iwamura, H. J. Amer. Chem. Soc. 1987. Vol. 109. P. 2631.
5. Williams, D. Vol. Phys. 1969. Vol. 16. P. 1451.
6. Mukai, K., Mishina, T., Ishisu, K., Deguchi, Y. Bull. Chem. Soc. Jpn. 1977. Vol. 50. P. 641.
7. Agawa, K., Sumano, T., Kinoshita, M. Chem. Phys. Lett. 1986. Vol. 128. P. 587.
8. Agava, K., Sumano, T., Kinoshita, M. Chem. Phys. Lett. 1987. Vol. 141. P. 540.
9. Azuma, N., Yamauchi, J., Mukai, K., Ohya-Nisciguchi, H., Deguchi, Y. Bull. Chem. Soc. Jpn. 1973. Vol. 46. P. 2728.
10. Ballester, M., Riera, J., Castaner, J., Vonso, J. J. Amer. Chem. Soc. 1971. Vol. 93. P. 2215.
11. Rozantsev, E. G., Goldfein, M. D., Pulin, V. F. The Organic Paramagnets. Saratov State University. Russia. 2000. [Russ].
12. Rozantsev, E. G., Sholle, V. D. The Organic Chemistry of Free Radicals. Moscow. Chemistry. 1979. [Russ].
13. Karimov, U. S., Rozantsev, E. G. Physics of Solid State. 1966. Vol. 8. P. 2787. [Russ].
14. Nakajima, A., Ohya-Nisciguchi, H., Deguchi, Y. Bull. Chem. Soc. Jpn. 1985. Vol. 107. P. 2560.
15. Benelli, C., Gatteschi, D., Carnegie, J., Carlin, R. J. Amer. Chem. Soc. 1983. Vol. 105. P. 2760.
16. Zvarykina, A. V., Stryukov, V. B., Fedutin, D. N., Shapiro, A. B. Letters in JETPh. 1974. Vol. 19. P 3. [Russ].

17. Zvarykina, A. V., Stryukov, V. B., Umansky, S.Ya., Fedutin, D. N., Shibaeva, R. P., Rozantsev, E. G. Reports in USSR Academy. 1974. Vol. 216. P. 1091. [Russ].
18. Stryukov, V. B., Fedutin, D. N. Physics of Solid State. 1974. Vol. 16. P. 2942. [Russ].
19. Rassat, A. Pure and Appl. Chem. 1990. Vol. 62. P. 223.
20. Nakajima, A. Bull. Chem. Soc. Jpn. 1973. Vol. 46. P. 779.
21. Nakajima, A., Yamauchi, J., Ohea-Nisciguchi, H., Deguchi, Y. Bull. Chem. Soc. Jpn. 1976. Vol. 49. P. 886.
22. Caneschi, A., Gatteschi, D., Grand, A., Laugier, J., Rey, P., Pardi, L. Inorg. Chem. 1088. Vol. 27. P. 1031.
23. Caneschi, A., Laugier, J., Rey, P., Zanchini, C. Inorg. Chem. 1987. Vol. 26. P. 938.
24. Caneschi, A., Gatteschi, D., Laugier, J., Pardi, L., Rey, P., Zanchini, C. Inorg. Chem. 1988. Vol. 27. P. 2027.
25. Laugier, J., Rey, P., Bennelli, C., Gatteschi, D., Zanchini, C. J/ Amer. Chem. Soc. 1986. Vol. 108. P. 5763.
26. Caneschi, A., Gatteschi, D., Laugier, J., Rey, P. J. Amer. Chem. Soc. 1987. Vol. 109. P. 2191.
27. Caneschi, A., Gatteschi, D., Rey, P., Sessoli, R. Inorg. Chem. 1988. Vol. 27. P. 1756.
28. Caneschi, A., Gatteschi, D., Laugier, J., Rey, P., Sessoli, R. Inorg. Chem. 1987. Vol. 27. P 1553.
29. Caneschi, A., Gatteschi, D., Renard, J., Sessoli, R. Inorg. Chem. 1989. Vol. 28. P. 2940.
30. Caneschi, A., Gatteschi, D., Renard, J., Rey, P., Sessoli, R. Inorg. Chem. 1989. Vol. 28. P. 1976.
31. Caneschi, A., Gatteschi, D., Renard, J., Rey, P., Sessoli, R. Inorg. Chem. 1989. Vol. 28. P. 3314.
32. Caneschi, A., Gatteschi, D., Renard, J., Rey, P., Sessoli, R. J. Amer. Chem. Soc. 1989. Vol. 111. P. 785.
33. Caneschi, A., Ferraro, F., Gatteschi, D., Rey, P., Sessoli, R. Injrg. Chem. 1990. Vol. 29. P. 1756.
34. Miller, J., Epstein, A., Reiff, W. Chem Rev. 1988. Vol. 88. P. 201.
35. Miller, J., Epstein, A., Reiff, W. Acc. Chem. Res. 1988. Vol. 21. P. 114.
36. Miller, J., Calabrese, J., Rommelmann, Y., Chittapeddi, S., Zhang, J., Reiff, W., Epstein, A. J. Amer. Chem. Soc. 1987. Vol. 109. P. 769.
37. Chittapeddi, S., Cromack, K., Miller, J., Epstein, A. Phys. Rev. Lett. 1987. Vol. 22. P. 2695.
38. Candela, G., Swartzendruber, L., Miller, J., Rice, M. J/ Amer. Chem. Soc. 1979. Vol. 101. P. 2755.
39. Miller, J., Epstein, A. J. Amer. Chem. Soc. 1987. Vol. 109. P. 3850.
40. Miller, J., Calabrese, J., Harlow, R., Dixon, D., Zhang, J., Reiff, W., Chittapeddi, S. Selover, M., Epstein, A. J. Amer. Chem. Soc. 1990. Vol. 112. P. 5496.
41. Miller, J., Ward, M., Zhang, J., Reiff, W. Inorg. Chem. 1990. Vol. 29. P. 4063.
42. Sugamara, T., Bandow, S., Kimura, K., Itamura, H., Itoh, K. J. Amer. Chem. Soc. 1986. Vol. 108. P. 368.
43. Sugamara, T., Bandow, S., Kivura, K., H., Itamura, H., Itoh, K. J. Amer. Chem. Soc. 1984. Vol. 106. P. 6449.
44. Itamura, H. Pure and Appl. Chem. 1986. Vol. 58. P. 187.

45. Teki, Y., Takui, T., Itoh, K., Iwamura, H., Kobayaschi, K. J. Amer. Chem. Soc. 1983. Vol. 105. P. 3722.
46. Teki, Y., Takui, T., Yagi, H., Itoh, K., Imamura, H. J. Chem. Phys. 1985. Vol. 83. P. 539.
47. Sugamara, T., Murata, S., Kimura, K., Itamura, H. J. Amer. Chem. Soc. 1985. Vol. 107. P. 5293.
48. Sugamara, T., Tukada, H., Izuoka, A. J. Amer. Chem. Soc. 1986. Vol. 108. P. 4272.
49. Magata, N. Theor. Chim. Acta. 1968. Vol. 10. P. 372.
50. Itoh, K. Bussei. 1971. Vol. 12. P. 635.
51. Torrance, J., Oostra, S., Nazzal, A. Synth. Met. 1987. Vol. 19. P. 709.
52. Torrance, J., Bugus, P. J/ Appl. Phys. 1988. Vol. 63. P. 2962.
53. Breslow, R. Pure and Appl. Chem. 1982. Vol. 54. P. 927.
54. Breslow, R., Jaun, B., Kluttz, R. Tetrahedron. 1982. Vol. 38. P. 863.
55. Breslow, R. Pure and Appl. Chem. 1982. Vol. 54. P. 927.
56. Breslow, R., Maslak, P. J. Amer. Chem Soc. 1984. Vol. 106. P. 6453.
57. Le Pade, T., Breslow, R. J. Amer. Chem. Soc. 1987. Vol. 109. P. 6412.
58. Breslow, R. Mol. Cryst. Liquid cryst. 1985. Vol. 125. P. 261.
59. McConnell, H. Proc. R. A. Welch Found. Chem. Res. 1967. Vol. 11. P. 144.
60. Miller, J., Glatzhofer, D. et al. Chemistry of materials. 1990. Vol. 2. P. 60.
61. Steyn, D. L. In a world of science. 1989. Vol. 9. P. 26. [Russ].
62. Emsley, J. New Scient. 1990. Vol. 127. No 1727. P. 29.
63. Dulog, L., Kim, J. S. Angev. Chem. 1990. Bd. 29. S. 415.
64. Rozantsev, E. G., Stepanov, A. P. The geophysics apparatus. Sankt-Petersburg: Nedra. 1966. Vol. 29. P. 35. [Russ].
65. Rozantsev, E. G. Free iminoxyl radicals. Moscow: Chemistry. 1970. [Russ].

CHAPTER 26

HYPERBRANCHED 1,4-CIS+1,2-POLYBUTADIENE SYNTHESIS USING NOVEL CATALYTIC DITHIOSYSTEMS

SHAHAB HASAN OGLU AKHYARI,[1]
FUZULI AKBER OGLU NASIROV,[1,2] EROL ERBAY,[2] and
NAZIL FAZIL OGLU JANIBAYOV[1]

[1]*Institute of Petrochemical Processes of National Academy of Sciences of Azerbaijan, Baku, Azerbaijan, E-mail: j.nazil@yahoo.com*

[2]*Petkim Petrokimya Holding, Izmir, Turkey, E-mail: fnasirov@petkim.com.tr*

CONTENTS

26.1   INTRODUCTION

Branching degree of polydienes (polybutadiene, polyisoprene, etc.) is a very important parameter with the necessary specifications such as stereoregularity, molecular mass and molecular mass distribution. Branches in polydienes structure strongly reduce solution viscosity by improving its processability, increasing operating ability of vulcanizates. On the other hand, the branched polymers easily decompose *via* light and biological factors, that solves elimination problem of waste polymers from the environment.

By using known catalytic systems, we can achieve high molecular mass, linear 1,4-*cis*-polybutadienes of high solution viscosity. These prevent easily polymer processing and limits using them in preparing of impact resistant polystyrenes. At the same time, as it is known [1], because of their high crystallization temperature, high stereo, regular 1,4-*cis*-polybutadienes can not be used in the production of tires and technological rubbers, which are employed in cold climatic conditions. Therefore it is of critical importance to obtain cold resistant, hyperbranched high molecular 1,4-cis+1,2-polybutadiene with 20–40% of 1,2-vinyl bonds [1].

Earlier in the Institute of Petrochemical Processes of Azerbaijan National Academy of Sciences (IPCP of ANAS) we have developed the new bifunctional nickel- and cobalt-containing catalytic dithiosystems, which have shown high catalytic activity and stereo selectivity in butadiene polymerization process. Particularly, by using the cobalt-dithiocarbamate catalytic system (diethylditiocarbamate cobalt + alkylaluminummonochloride) it was possible to obtain linear high molecular 1,4-cis+1,2-polybutadiene of high solution viscosity [1].

For the synthesis of hyperbranched 1,4-cis+1,2-polybutadienes the new dialkyldithiocarbamate cobalt + alkylaluminumdichloride (or alkylaluminumsesquichloride) catalytic dithiosystems have been prepared and used in the butadiene polymerization process. It has been shown that by using the dialkyldithiocarbamate cobalt+ alkylaluminumdichloride (or alkylaluminumsesquichloride) catalytic dithiosystems it is possible to synthesize hyperbranched, low solution viscosity, high molecular weight 1,4-*cis*-polybutadiene with 20–40% 1,2-vinyl bond content.

In this paper, the main results of butadiene polymerization to high molecular weight hyperbranched 1,4-cis+1,2-polybutadienes using new dialkyldithiocarbamate cobalt+ alkylaluminumdichloride (or alkylaluminumsesquichloride) catalytic dithiosystems are given.

26.2 EXPERIMENTAL PART

Butadiene (99.8%, wt.), aluminum organic compounds (85.0%, wt. – in benzene) were obtained from Aldrich. Organic dithioderivatives – dithio-carbamates, dithiophosphates and xanthogenates of cobalt were synthe-sized according to relevant references [1–3].

After drying over metallic sodium for 24 h, polymerization was con-ducted in toluene, which was distilled and preserved over sodium. Where necessary, manipulations were carried out in 50–200 mL Schlenk-type glass reactors under dry, oxygen-free argon or nitrogen with appropri-ate techniques and gas tight syringes. The usual order of addition was: solvent, cobalt component, aluminum organic compound (at −78°C), and finally the monomer. Polymerizations were conducted at 25–60°C and then the polymerizate was poured to ethanol (or methanol) followed by the termination. Received polymer was washed several times with ethanol (or methanol) and was dried at 40°C in a vacuum to constant weight and stored under argon (or nitrogen).

Solution viscosities of polybutadienes were measured with an Ubbelohde viscometer in toluene solution at 30°C at a concentration of 0.2 g/dL [4]. The molecular weight of 1.4-cis+1.2-polybutadienes were determined by viscosimetric method [1, 4] using the following formula:

$$[\eta]_{30(toluene)} = 15.6 \times 10^{-5} \times M^{0.75}$$

The molecular weights (M_w and M_n) and molecular weight distribu-tion (M_w/M_n) of polybutadienes were measured by a Gel Permeation Chromatograph (GPC), constructed in the Czech Republic with a 6,000 A pump, original injector, R-400 differential refractive index detector, and styragel columns with nominal exclusions of 500, 10^3, 10^4, 10^5, and 10^6. The GPC instrument was calibrated according to the universal calibration method by using narrow molecular weight polystyrene standards [5].

The microstructure of the polybutadiene was determined using FTIR-spectrometry ("Nicholet NEXUS 670" with spectral diapason from 400 cm^{-1} till 4000 cm^{-1}, as a film on KBr, received from toluene solution) [6, 7].

Mooney viscosity was determined by the means of a SMV-201 Mooney viscometer (Shimadzu Co., Ltd.) in accordance with ASTM D 1646 [8, 9].

Branching degree was calculated based on the viscosity of polybutadiene at 25°C in 5% toluene solution. Solution viscosity (SV) in proportion to Mooney viscosity (MV) at 100°C (SV/MV), and its branching index (g_M) was calculated from proportional literature sources [8, 9]. Branching index for linear polybutadiene is $g_M = 1.0$ and for branched polybutadienes is $g_M < 1.0$.

26.3   RESULTS AND DISCUSSION

In this study, we have used dialkyldithiocarbamate cobalt as the main catalyst component with the structure as shown:

$$\begin{array}{c} R^1 \\ \diagdown \\ R \diagup \end{array} N-C \begin{array}{c} \diagup S \diagdown \\ \diagdown S \diagup \end{array} Co \begin{array}{c} \diagup S \diagdown \\ \diagdown S \diagup \end{array} C-N \begin{array}{c} \diagup R^1 \\ \diagdown R \end{array}$$

where, R and R^1 are alkyl, aryl, alkylaryl radicals and cocatalysts are the aluminum organic compounds (dialkylaluminumchlorides, alkylaluminumdichlorides, alkylaluminumsesquichlorides and aluminumoxanes) with formulas: AlR^2Cl_2 and/or $Al_2R^2_3Cl_3$, AlR^2_2Cl (where, R^2 is methyl, ethyl, isobutyl radicals, etc.).

We have investigated the (peculiarities?) of dialkyldithiocarbamate cobalt+ alkylaluminumdichloride (or alkylaluminumsesquichloride) catalytic dithiosystems in the polymerization of butadiene to highly branched and high molecular weight 1,4-cis+1,2-polybutadiene. These catalysts were studied in comparison with a known cobalt-containing catalytic dithiosystems for butadiene polymerization dependence of process outcomes, catalyst activity and selectivity on the nature of ligands in cobalt compounds, catalyst component concentration and ratios, as well as the influence of temperature and polymerization time. As may be, in polymerization the ligand nature of the transition metal is also crucial for catalyst activity and stereoselectivity. Ligands used in this work were mainly dithiocarbamates, dithiophosphates and xhanthogenates. As can be seen in Table 26.1, Exp. 28–31, that cobalt-dithiophosphates and cobalt-dialkylxhantogenates in combination with diethylaluminumchloride (DEAC), ethylaluminumdichloride (EADC) and methylaluminoxane (MAO) allow

TABLE 26.1 Results of Butadiene Polymerization to Hyper Branched and Linear Polybutadienes with Bifunctional Cobalt-containing Catalytic Dithiosystems

№	Cobalt Compound	Aluminum Organic Compounds (AOC)	[Co]·10^4, mol/L	Al:Co	[M], mol/L	T, °C	Reaction time, min.	Yield of PBD, %	Catalyst activity, kg of PBD/g Co·hour	[η], dL/g	Branching		M$_w$/M$_n$	Microstructure of PBD, %		
											Index (g$_M$)	Degree (SV/MV)		1,4-cis	1,4-trans	1,2-
1	2	3	5	6	7	8	9	10	11	12	13	14	15	16	17	18
1	DEDTC	EADC	1.0	100	5.0	25	60	96.0	44.0	2.5	0.35	2.5	2.3	70	5	25
2	DMDTC	EADC	1.0	100	5.0	25	60	92.0	42.0	3.5	0.55	2.7	2.5	60	5	35
3	DBenzDTC	EADC	1.0	100	5.0	25	60	97.0	45.0	3.3	0.40	2.6	2.6	60	10	30
4	DPhDTC	EADC	1.0	100	5.0	25	60	98.0	55.0	3.0	0.52	2.7	2.3	60	7	33
5	DEDTC	EASC	1.0	100	5.0	25	60	94.0	33.0	2.9	0.60	3.0	2.6	72	3	25
6	DBDTK	EASC	1.0	100	5.0	25	60	92.0	22.0	2.8	0.70	3.4	2.7	71	7	22
7	DEDTK	EADC	1.0	100	5.0	25	60	92.0	42.0	2.5	0.60	3.1	2.6	70	5	25
8	DMDTK	EADC	1.0	100	5.0	25	60	97.0	44.0	2.7	0.65	3.25	2.3	72	6	22
9	DBDTK	EADC	1.0	100	5.0	25	60	88.0	40.0	2.3	0.55	2.7	2.8	75	5	20
10	DEDTC	EADC	0.5	100	5.0	25	90	90.0	50.0	3.5	0.65	3.5	2.5	75	5	20
11	DEDTC	EADC	2.5	100	5.0	25	60	95.0	44.9	2.9	0.55	2.7	2.3	68	7	25
12	DEDTC	EADC	5.0	100	5.0	25	30	98.0	43.0	2.5	0.49	2.5	2.1	65	5	30

TABLE 26.1   (Continued)

№	Cobalt Compound	Aluminum Organic Compounds (AOC)	[Co]·10⁴, mol/L	Al:Co	[M], mol/L	T, °C	Reaction time, min	Yield of PBD, %	Catalyst activity, kg of PBD/g Co·hour	[η], dL/g	Branching Index (g_M)	Branching Degree (SV/MV)	M_w/M_n	Microstructure of PBD, % 1,4-cis	1,4-trans	1,2-
13	DEDTC	EADC	10.0	100	5.0	25	10	99.0	40.0	2.0	0.42	2.4	2.0	55	5	40
14	DEDTC	EADC	1.0	10	5.0	25	60	87.0	39.8	3.0	0.75	3.7	2.4	72	2	25
15	DEDTC	EADC	1.0	50	5.0	25	60	95.0	43.5	2.8	0.55	2.8	2.5	72	1	27
16	DEDTC	EADC	1.0	100	1.0	25	60	98.0	22.0	2.0	0.22	2.5	2.2	60	10	30
17	DEDTC	EADC	1.0	100	3.0	25	60	97.0	26.6	2.8	0.28	3.0	2.3	65	7	28
18	DEDTC	EADC	1.0	100	8.0	25	60	90.0	50.0	3.5	0.60	3.6	2.8	70	10	20
19	DEDTC	EADC	1.0	100	5.0	45	60	95.0	43.5	2.9	0.40	2.3	2.4	70	3	27
20	DEDTC	EADC	1.0	100	5.0	60	60	90.0	41.2	2.7	0.35	2.3	2.3	71	2	27
21	DEDTC	EADC	1.0	100	5.0	25	10	86.0	36.1	2.8	0.70	3.6	2.2	62	3	35
22	DEDTC	EADC	1.0	100	5.0	25	90	99.0	30.2	3.5	0.45	2.4	2.5	68	4	28
23	DEDTC	EADC	1.0	100	5.0	25	120	90.0	20.0	2.5	0.95	9.2	3.2	62	13	25
24	DMDTC	DIBAC	1.0	100	5.0	25	120	85.0	19.5	2.3	0.96	10.0	3.2	65	15	20
25	DBDTC	DEAC	1.0	100	5.0	25	120	92.0	22.0	2.7	0.94	8.8	3.5	66	12	22
26	DBenzDTC	MAO	1.0	100	5.0	25	120	88.0	20.0	2.7	0.95	9.0	3.6	65	13	22

TABLE 26.1 (Continued)

№	Cobalt Compound	Aluminum Organic Compounds (AOC)	[Co]·10⁴, mol/L	Al:Co	[M], mol/L	T, °C	Reaction time, min.	Yield of PBD, %	Catalyst activity, kg of PBD/g Co·hour	[η], dL/g	Branching Index (g_M)	Branching Degree (SV/MV)	M_w/M_n	Microstructure of PBD, % 1,4-cis	1,4-trans	1,2-
27	DPhDTC	DIBAC	1.0	100	5.0	25	120	90.0	21.0	2.8	0.96	9.5	3.5	63	12	25
28	DCDTPh	DEAC	1.0	100	3.0	25	60	98.0	12.0	1.8	0.92	8.6	1.8	90	8	2
29	DEDTPh	MAO	1.0	100	3.0	25	60	97.0	30.0	2.2	0.93	8.7	2.6	97	2	1
30	DPhDTPh	EADC	1.0	100	3.0	25	60	94.0	34.0	2.3	0.98	8.5	3.2	95	3	2
31	BuXh	EASC	2.0	100	3.0	25	60	80.0	28.0	2.5	1.0	10.0	2.6	95	2	3
32	EtXh	TEA	2.0	100	3.0	25	60	93.0	30.0	2.5	1.0	-	-	2	2	96
33	i-PrXh	TEA	2.0	100	3.0	25	60	98.0	35.0	2.3	1.0	-	-	1	1	98

Notes: *In experiments 7–9 benzene, chlorobenzene and hexane were used as solvent respectively;

**In experiments 24, 25, 27 alkylaluminum monochlorides (such as DEAC, DIBAC) were used as cocatalysts, respectively.

***In experiments 28–33 O,O-dialkyl substituted dithiophosphates and alkylxhantogenates were used as cobalt compounds, respectively.

Abbreviations: DCDTPh-Co – Cobalt-O,O′-di-4-methylphenyl dithiophosphate; DEDTPh-Co – Cobalt-Diethyldithiophosphate; DMDTC-Co – Cobalt-Dimethyldithiocarbamate; DBDTC-Co – Cobalt-Dibutyldithiocarbamate; DBenzDTC-Co – Cobalt-Dibenzyldithiocarbamate; DPhDTC-Co – Cobalt-Diphenyldithiocarbamate; DPhDTPh-Co – Cobalt-Diphenyldithiophosphate; EtXh-Co – Cobalt-Ethylxhantogenate; DCDTPh-Co – Cobalt-O,O′-di-4-methylphenyl dithiophosphate; EASC – Ethylaluminumsesquichloride; BuXh-Co – Cobalt-Butylxhantogenate; AlkXh-Co – Cobalt-Alkylxhantogenate; TEA – Triethylaluminum; iPrXh-Co – Cobalt-iso-Propylxhantogenate.

yield of linear high molecular weight 1,4-cis polybutadiene with poly-mer yield of 80.0–98.0%, catalyst activity of 12.0–34.0 kg PBD/g Co•h, intrinsic viscosity [η] of 1.8–2.5 dL/g and 1,4-cis content of 90.0–97.0%. Cobalt-dialkyldithiocarbamate catalytic system – cobalt-diethyldithio-carbamate (DEDTC-Co) + DEAC (or MAO) gives linear high molecu-lar weight 1,4-cis+1,2-polybutadiene with polymer yield of 85.0–90.0%, catalyst activity of 20.0–22.0 kg polybutadiene (PBD/g Co•h, intrinsic vis-cosity [η] of 2.3–2.8 dL/g, 1,4-cis content of 63.0–66.0% and 1,2-content of 20.0–25.0% (Table 26.1, Exp.24–27).

As seen from Table 26.1, Exp. 28–31, that cobalt-dithiophosphates and cobalt-dialkylxhantogenates in combination with DEAC, EADC and MAO allow yield of linear high molecular weight 1,4-cis polybutadiene with a polymer yield of 80.0–98.0%, catalyst activity of 12.0–34.0 kg PBD/g Co·h, intrinsic viscosity [η] of =1.8–2.5 dL/g and 1,4-cis content of 90.0–97.0%.

The cobalt-dialkyldithiocarbamate catalytic system – DEDTC-Co + DEAC (or MAO) gives linear high molecular weight 1,4-cis+1,2-polybuta-diene with polymer yields of 85.0–90.0%, catalyst activity of 20.0–22.0 kg PBD/g Co·h, intrinsic viscosity [η] of 2.3–2.8 dL/g, 1,4-cis content of 63.0–66.0% and 1,2-vinyl content of 20.0–25.0% (Table 26.1, Exp.24–27).

Cobalt-dialkylxhantogenates + TEA catalytic dithiosystems give high molecular and high crystalline syndiotactic 1,2-PBD (1,2-SPBD). In their presence, polymer yield was 93.0–98.0%, catalyst activity was 30.0–35.0 kg PBD/g Co·h, and the synthesized 1,2-SPBD had an intrinsic viscosity (135°C, tetralin) of 2.3–2.5 dL/g, 1,2-vinyl content of 96–98% (Table 26.1, Exp. 32, 33).

Only using the dialkyldithiocarbamate cobalt + EADC (or EASC) as catalyst results in obtaining of hyperbranched high molecular weight 1,4-cis+1,2-polybutadiene with the yields of 92.0–98.0%, intrinsic vis-cosity [η]= 2.8–3.5 dL/g, 1,2-vinyl content of 22.0–35.0%, branching index of 0.35–0.70 and with the capability of regulation of polymers chain branching in wide intervals. Productivity of these catalysts is 22.0–55.0 kg PBD/g Co·hour (Table 26.1, Exp. 1–23).

The influence of the organic solvents (toluene, benzene, chlorobenzene, and hexane) on the activity and stereoselectivity of the diethyldithiocar-bamate cobalt + ethylaluminumdichloride (DEDTC-Co+EADC) catalytic dithiosystem was studied at [Co] = 1.0×10^{-4} mol/L [M] = 5.0 mol/L,

Al:Co= 100:1. The butadiene polymerization reactions were conducted at 25°C for an hour.

From the results shown in Table 26.1 Exp.1 and 7, it seems that when toluene and benzene were used as the solvent the polymer yield was 96.0 and 92.0%, and the catalyst activity was 44.0 and 42.0 kg PBD/g Co·hour, respectively. The obtained polybutadiene had 1,2-content – 30.0 and 25.0%, branching index of – 0.35 and 0.60, respectively. The used catalyst showed the same results in terms of activity and stereoselectivity when chlorobenzene and hexane were used as the process solvent (Table 26.1, Exp. 8 and 9).

For future investigations, toluene, a largely used industrial polymerization process solvent, has been chosen as an optimal solvent in the polymerization of butadiene to hyperbranched polybutadiene. In these experiments, polymerization of butadiene was carried out at 25°C and the cobalt compound concentration was varied within the limits – $[Co]=(0.5–10) \times 10^{-4}$ mol/L at Al:Co= 100:1. The results are shown in Table 26.1, Exp.1, 10–13. An increase in cobalt-compound concentration from 0.5×10^{-4} mol/L up to approximately 10.0×10^{-4} mol/L resulted in an increase in the yield of polymer from 90.0 to 99.0%. Increasing the concentration of cobalt-compound resulted in a decrease in the branching index of polymer from 0.65 to 0.42 and intrinsic viscosity – $[\eta]$ from 3.5 to 2.0 dL/g. Content of the 1,2-vinyl chain increased in the range of 20.0–44.0%.

The influence of Al:Co ratio in the interval of (10–100):1 on the activity, stereoselectivity and productivity of catalyst was investigated at $[Co]=1.0 \times 10^{-4}$ mol/L and t = 25°C. From the results of Table 26.1, Exp.1, 14, 15, it seems that an increase of Al:Co ratio from 10:1 to 100:1 resulted in an increase of polybutadiene yield of 87.0–95.0% and 1,2-vinyl content of 25.0–30.0%. It was also observed that the intrinsic viscosity decreased to 3.0–2.5 dL/g and branching index of polymer decreased to 0.75–0.35.

The influence of monomer concentration in a range of 1.0 to 8.0 mol/L was studied at $[Co]=1.0 \times 10^{-4}$ mol/L, t = 25°C (Table 26.1, Exp.1, 16–18). Increasing the monomer concentration resulted in a decrease in the polymer yield of 98.0–90.0% and 1,2-vinyl content of 30.0–20.0 with an increase in the catalyst activity to 22.0–50.0 kg PBD/g Co•h and branching index to 0.22–0.60.

The influence of temperature on the yield of the butadiene polymerization reaction was investigated between 25°C and 60°C. As it seems from

Table 26.1, Exp. 1, 19, 20, an increase in temperature resulted in a decrease in polybutadiene yield of 96.0–90.0% and intrinsic viscosity [η]= 2.9–2.5 dL/g. But this has no influence on the 1,2-vinyl content (25.0–27.0%) and branching index (0.35–0.40) of the product.

Experimental results show that only the new cobalt alkyldithiocarbamate + alkylaluminumdichloride (or alkylaluminumsesquichloride) catalytic dithiosystems allow us to obtain high molecular weight 1,4-cis+1,2-polybutadienes with high activity and stereoselectivity, high cold resistance and various branching index.

In the presence of known catalytic systems, such as cobalt-dithiophosphate+AOC, cobalt-xhantogenate + AOC and cobalt-dithiocarbamate + dialkylaluminumchloride catalytic systems did not yield branched polybutadienes. Obtaining of branching polymers in the presence of alkyldithiocarbamate cobalt + alkylaluminumdichloride (or alkyl-aluminumsesquichloride) could be explained by the cationic nature of an alkylaluminumdichloride and alkylaluminumsesquichloride cocatalysts. These cocatalysts act as catalysts in the polymerization process of butadiene (I. Direction) and in the copolymerization of received polymer with a new molecule of monomer (II. Direction) or with another molecule of polymer (III. Direction) as resulting in the formation of branching with various lengths in the polybutadiene chain as shown in scheme 26.1.

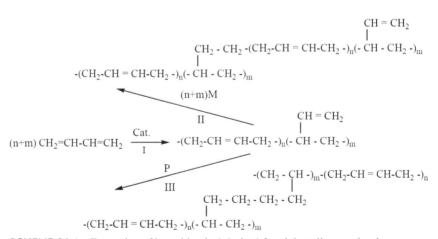

SCHEME 26.1 Formation of branching in 1,4-cis+1,2-polybutadiene molecule.

26.4 CONCLUSIONS

High activity and stereo selectivity cobalt-dithiocarbamate catalytic dithiosystems for butadiene polymerization process have been developed. Based on alkyldithiocarbamate cobalt as the main catalyst compound in combination with alkylaluminumdichloride (or alkylaluminumsesquichloride) as the cocatalyst they provide for the preparation of high molecular hyperbranched 1,4-cis+1,2-polybutadienes with 20–40% 1,2-vinyl bond content.

Productivity of these catalysts is 20.0–40.0 kg PBD/g Co•h with the yields of polybutadiene 85.0–99.0%. The obtained hyperbranched polybutadienes have a branching index of 0.70–0.20, versus the branching index of linear polybutadienes, which is between 0.96–0.94. Known catalytic systems based on cobalt-dithiophosphates or cobalt-xhantogenates as the main cobalt compound and aluminum organic compounds (such as, dialkylaluminumchlorides, alkylaluminumdichloride, alkylaluminumsesquichloride and trialkylaluminum) and cobalt-dithiocarbamate + dialkylaluminumchloride do not lead to formation of hyperbranched polybutadienes.

Synthesized hyper-branched high molecular 1,4-cis+1,2-polybutadiene can be used in the production of tires, technological rubbers and impact resistant polystyrenes.

KEYWORDS

- branching degree of polydienes
- molecular mass
- polydienes structure
- biological factors
- climatic conditions
- solution viscosity

REFERENCES

1. Nasirov F. A. Bifunctional Nickel- or Cobalt containing catalyst-stabilizers for poly-butadiene production and stabilization (Part I): Kinetic study and molecular mass stereoregularity correlation, Iranian Polymer Journal, 12, 4, 217–235, 2003.
2. Nasirov F. A. Organic Dithioderivatives of Metals – Components and Modificators of Petrochemical Processes, Petrochemistry, 6, p. 403–416, 2001 (in Russian).
3. Byrko V. M. Dithiocarbamates. Moscow, Nauka, 342 p., 1984 (in Russian)
4. Rafikov S. P., Pavlova S. A., Tvyordokhlebova I.I. Methods of determination molecular weight and polydispersity of high molecular materials. Moscow, Academy of Sciences of USSR, 336 p., 1963 (in Russian).
5. Deyl Z., Macek K., Janak J. (Eds.). Liquid Column Chromatography, Elsevier, Amsterdam Scientific V.1, 2, 1975.
6. Haslam J., Willis H. A. Identification and analysis of Plastics, London, Iliffe Books, Princeton, New Jersey: D. Van Nostrand Co., p.172–174, 1965.
7. Bellami L. J. The Infra-red Spectra of Complex Molecules, London, Methuen and Co, New York: J. Wiley, 592 p., 1957.
8. Grechanovsky V. A. Branching in polymer chains. Uspekhy Khimii, T. 38, 12, c.2194–2219, 1969 (Russian).
9. Jang Y.-C., Kim P.-S., Kwag G.-H., Kim A.-J., Lee S.-H. Process for controlling degree of branch of high 1,4-cis polybutadiene. US Patent 20020016423 A1, 2002.

CHAPTER 27

COBALT ALKYLXHANTHOGENATE +TRIALKYLALUMINUM CATALYTIC DITHIOSYSTEMS FOR SYNTHESIS OF SYNDIOTACTIC 1,2-POLYBUTADIENE

NEMAT AKIF OGLU GULIYEV,[1] FUZULI AKBER OGLU NASIROV,[1,2] and NAZIL FAZIL OGLU JANIBAYOV[1]

[1]*Institute of Petrochemical Processes of National Academy of Sciences of Azerbaijan, Baku, Azerbaijan, E-mail: j.nazil@yahoo.com*

[2]*Petkim Petrokimya Holding, Izmir, Turkiye, E-mail: fnasirov@petkim.com.tr*

CONTENTS

27.1 INTRODUCTION

Syndiotactic 1,2-polybutadiene is a photodegradable polymer with molecular irregularities-double bonds in its polymeric backbone. Studies have shown a 95% property loss after photodegradable materials have been exposed to direct sunlight over a year.

It is used in the manufacturing of tires, packing polymer-film materials, microcapsules for the medical purposes, ceramics, semi permeability membranes, adhesives, synthetic leather, oil resistant tubes, coatings for semiconductor devices, nonwoven materials, and carbon fibers, etc. [1, 2].

Syndiotactic 1,2-polybutadienes have been synthesized using various catalysts based on compounds of Ti, Cr, Pd, Co, V, Fe, and Mo. The results, relating to the catalysts and processes of syndicotactic 1,2-polybutadiene synthesis, have been described in many papers [3–11] and patents [12].

Ashitaka H. et al., in a series of articles [3–6], have described the $Co(acac)_3 + AIR_3 + CS_2$ catalyst for polymerization of butadiene to high stereoregularity and high melting point syndiotactic 1,2-polybutadiene. This catalyst had very low catalytic activity that polymer yield is only 10–20% in toluene solution (the mostly used industrial polymerization processes as solvent).

Chinese researchers have recently focused on the synthesis and characterization of syndiotactic 1,2-polybutadiene [7–11], but catalytic systems have shown very low catalytic activity and stereoregularity.

Earlier we have developed the new bifunctional nickel- and cobalt-containing catalytic dithiosystems for the polymerization of butadiene [13, 14]. In this article, the results of the polymerization of butadiene to the photodegradable highly crystalline syndiotactic 1,2-polybutadiene in the presence of a new cobalt alkylxhantogenate (AlkXh-Co) + trialkylaluminum (TAA) catalytic dithiosystems will be shown.

27.2 EXPERIMENTAL PART

Butadiene (99.8%, wt) and aluminum organic compounds (85.0%, wt in benzene) were used as obtained from Aldrich.

Organic dithioderivatives (dithiophosphates, dithiocarbamates, and xhantogenates) of cobalt were synthesized according to [13–16].

Polymerization was conducted in toluene, which, after predrying over metallic Na for 24 h, was distilled and preserved Na. Where necessary, manipulations were carried out under dry, oxygen-free argon or nitrogen in 50–200 mL-glass reactors. The desired volumes of toluene, monomer, TEA (or DEAC, MAO), and cobalt compound solutions from the calibrated glass reservoir were added to the reactor under stirring at controlled temperature. The usual order of addition was: solvent, cobalt component, aluminum organic compound (at −78°C), and finally the monomer. All polymerizations were conducted at a temperature range of 0–100°C. After polymerization, the polymerizate was poured to ethanol or methanol and the polymerization reactions were terminated. The precipitated polymer was washed several times with ethanol (or methanol). Polybutadiene was dried at 40°C in a vacuum to constant weight and stored under argon or nitrogen.

The viscosity of dilute solutions of syndiotactic 1,2-polybutadienes were measured with a Ubbelohde viscometer in tetralin at 135°C or in o-dichlorobenzene (o-DCB) at 140°C at a concentration of 0.2 g/dL [2, 3]. The molecular mass of high molecular weight and high crystalline syndiotactic polybutadiene was determined by viscosimetric method [17] using the relationship:

$$[\eta]_{135(\text{tetralin})} = [\eta]_{140(\text{o-DCB})} = 9.41 \times 10^{-5} \times M^{0.854}$$

The molecular masses (Mw and Mn) and molecular mass distribution (M_w/M_n) were measured by a Gel Permeation Chromatograph (GPC), which was constructed in Czech Republic, consisting of a 6,000 A pump, original injector, R-400 differential refractive index detector, and styragel columns with nominal exclusion of 500, 10^3, 10^4, 10^5, and 10^6. The GPC was operated at a flow rate of 0.8 mL/min with o-dichlorobenzene as solvent. The sample concentration was kept at about 0.3–0.6% with a sample volume of 100–200 mL. The GPC instrument was calibrated according to the universal calibration method by using narrow molecular weight polystyrene standards [18].

The microstructure of the polybutadiene was determined by a FTIR spectrometer (Nicholet NEXUS 670, with spectral diapason from 400 cm^{-1} to 4000 cm^{-1}, as a film on KBr, received from THF or o-dichlorobenzene solution) [19,20]. Tactisity and crystallinity of polymer were determined accordingly [21]. Melting point (mp) was determined under nitrogen by differential scanning calorimeter (DSC Q20 V23.4) [21, 22].

27.3 RESULTS AND DISCUSSION

We have investigated the peculiarities of cobalt alkylxhantogenate + tri-alkylaluminum catalytic dithiosystems in polymerization of butadiene to syndiotactic 1,2-polybutadiene [23–26]. These catalysts were studied in comparison with known cobalt-containing catalytic dithiosystems for butadiene polymerization dependence of process outcomes on the nature of ligands of cobalt-compounds, catalyst components concentration and ratio, as well the influence of temperature on catalyst activity and selectivity.

Apart from the metal chosen for polymerization, the ligand nature is also of high importance. Ligands that have been used for polymerization of butadiene were mainly dithiophosphates, dithiocarbamates and xhan-thogenates. The results are given in Table 27.1.

It can observed from Table 27.1 that codithiophosphates and coalkyl-xhantogenates in combination with diethylaluminumchloride (DEAC) allow obtaining of high molecular mass 1,4-cis polybutadiene with 1,4-cis contents of 90.0–96.0%. Cobalt alkyldithiocarbamate catalytic system (DEDTC-Co+DEAC) yields high molecular mass 1,4-cis+1,2-polybutadiene with 1,4-cis content of 58.0% and 1,2-content of 34.0%.

Only the cobalt alkylxhantogenate catalytic dithiosystems (Cobalt alkylxhantogenates + TEA) yield high molecular and high crystalline syndiotactic 1,2-PBD (1,2-SPBD). In their presence, polymer yields are 93.0–99.0% and the obtained 1,2-SPBD has intrinsic viscosity (135°C, in tetralin) of 2.2–3.5 dL/g, 1,2-content of 94.0–99.0%, crystallinity of 86.0–95.0%, and mp of 175–208°C.

The effect of the organic solvents (toluene, benzene, chlorobenzene, methylene chloride, and hexane) on the activity and selectivity of the iso-propylxhantogenate (iPrXh-Co) + TEA catalytic dithiosystem was studied at: [Co] = $1.0 \cdot 10^{-4}$ mol/L [M] = 3.0 mol/L, Al:Co= 100:1. For that the necessary amount of a particular solvent was mixed with iPrXh-Co followed by the addition of triethylaluminum (TEA) and butadiene into reactor at −78°C. The butadiene polymerization reactions were conducted at 40°C for 120 min.

From the results shown in Table 27.2 it appears that when toluene and benzene were used as the solvent the polymer yield was 96.0–99.0%.

TABLE 27.1 Comparison of Efficiency of Different Cobalt-Containing Catalytic Dithiosystems CoX$_2$+AOC in the Butadiene Polymerization Process

Item No	Cobalt-containing catalytic system (CoX$_2$+AOC)	Polymer yield, % (mass)	Crystallinity, %	Syndiotacticity, %	Melting point, °C	Intrinsic viscosity, [η], dL/g	Molecular mass		Microstructure, %			Yield of polymer, kg PBD/g Co·h
							M$_w$×10^{-3}	M$_w$/M$_n$	1,4-cis	1,4-trans	1,2-	
1	DCDTPh-Co+ DEAC	95	-	-	-	2.62	570	2.1	92	6	2	57
2	TBDTPh-Co+ DIBAC	98	-	-	-	2.6	610	2.3	90	7	3	109
3	4 m-6-TBPh-Co+ MAO	92	-	-	-	2.5	625	2.5	92	5	3	59
4	X-Co+DEAC	95	-	-	-	2.9	610	1.9	93	6	1	57
5	NGDTPh-Co+ DIBAC	88	-	-	-	3.1	630	1.85	93	5	2	46
6	DEDTC-Co+ DEAC	58	-	-	-	2.55	600	2.3	58	8	34	44
7	EtXh-Co+DEAC	95	-	-	-	2,8	540	1.85	96	2	2	127

TABLE 27.1 (Continued)

Item No	Cobalt-containing catalytic system (CoX₂+AOC)	Polymer yield, % (mass)	Crystallinity, %	Syndiotacticity, %	Melting point, °C	Intrinsic viscosity, [η], dL/g	Molecular mass $M_w \times 10^{-3}$	M_w/M_n	Microstructure, % 1,4 cis	1,4 trans	1,2-	Yield of polymer, kg PBD/g Co·h
8	BuXh-Co+ DIBAC	90	-	-	-	2,5	410	1.65	95	2	3	105
9	EtXh-Co+TEA	95	92	98.5	208	2.2	270	2.1	2	1	99	96
10	i-PrXh-Co+TEA	99	95	97.0	205	2.3	255	1.82	1	1	98	115
11	BuXh-Co+TEA	98	93	95.5	200	2.5	240	1.6	3	1	96	125
12	HeXh-Co+TEA	95	88	94.0	185	2.9	330	1.75	5	1	94	108
13	OcXh-Co+TEA	93	86	94.5	175	3.5	485	2.2	6	2	95	112

TABLE 27.2 The Influence of Organic Solvents Type on the Conversion of Butadiene, Selectivity and Productivity of iPrXh-Co+TEA Catalytic System $[Co] = 1.0 \times 10^{-4}$ mol/L $[M] = 3.0$ mol/L, Al:Co = 100:1, T = 40°C, $\tau = 120$ min

Solvent	Yield of PBD, % (mass)	$[\eta]_{135}$, dL/g	Crystallinity, %	Syndiotacticity, %	Melting point, °C	Microstructure, %		
						1,4-cis	1,4-trans	1,2-
Toluene	99.0	2.5	95	97.5	208	1	1	98
Benzene	96.0	2.3	93	95.0	203	3	1	96
Chlorobenzene	92.0	1.75	91	94.0	194	3	2	95
Methylene Chloride	90.0	1.65	90	93.8	185	3	2	95
Hexane	70.0	1.5	76	91.0	190	6	2	92
-	56.0	3.5	93	95.0	195	4	1	95

The obtained polybutadiene had 1,2-vinyl contents of 96.0–98.0%, mp of 203–208°C, crystallinity of 93.0–95.0% and syndiotacticity of 95.0–97.5%. The known catalytic systems for butadiene 1,2-polymerization showed very low catalytic activity and stereoregularity in toluene and benzene solutions. For future investigations, toluene, a commonly used industrial polymerization process solvent, has been chosen as an optimal solvent in the 1,2-polymerization of butadiene.

An increase in cobalt concentration from 0.2×10^{-4} mol/L up to approximately 10.0×10^{-4} mol/L resulted in an increase in the initial reaction rate. As the concentration of cobalt compound was increased, the polybutadiene yield increased from 50.0 to 100.0%, 1,2-vinyl content decreased from 99.0 to 95.0%, and melting point decreased from 212 to 187°C as shown in the results in Table 27.3.

The increase of Al:Co ratio from 10:1 to 200:1 resulted in an increase of polybutadiene yield to 65.0–99.0% and a decrease in the intrinsic viscosity to 2.9–1.5 dL/g, 1,2-content of 99.0–94.0%, crystallinity of 96.0–76.0%, and mp of 212–175°C (Table 27.3).

With an increase in temperature there was an increase in polybutadiene yield to 26.0–100.0% as expected. This results in decreasing of intrinsic viscosity to 3.5–1.3 dL/g, 1,2-content of 99.0–94.0%, crystallinity of 98.0–63.0%, and melting point of 215–177°C.

Experimental results show that the high activity and stereoregularity of the new cobalt alkylxhantogenate + trialkylaluminum catalytic dithiosystems in toluene solution allow for the formation of syndiotactic 1,2-polybutadienes of varying crystallinity, syndiotacticity and molecular mass when experimental conditions are varied.

The obtained experimental results allow us to establish the optimal parameters for synthesis of high molecular mass and highly crystalline syndiotactic 1,2-polybutadiene [42–45]: [Co] = $(1.0–2.0) \times 10^{-4}$ mol/L; [M] = 3.0 mol/L; Al:Co = (50–100): 1; T = 40–80°C.

Reaction conditions: [Co] = 2.0×10^{-4} mol/L [M] = 3.0 mol/L; Al:Me = 100; T= 25°C; τ = 60 min, solvent – toluene.

Note: In experiments 9–13 an intrinsic viscosity was measured in tetraline at 135°C.

TABLE 27.3 The Influence of Butadiene Polymerization Parameters on the Conversion of Butadiene, Selectivity and Productivity of iPrXh-Co+TEA Catalytic System (Solvent-Toluene)

[Co]·10⁴, mol/L	[M], mol/L	Al: Co	Temperature, °C	Reaction duration, min	Yield of 1,2-SPBD, % (mass)	$[\eta]_{135}$, dL/g	Crystallinity, %	Syndiotacticity, %	Melting point, °C	Microstructure, %		
										1,4-cis	1,4-trans	1,2-
0.2	3.0	100	40	180	50	3.0	68	98	212	1	-	99
0.5	3.0	100	40	120	65	2.8	72	98	212	1	-	99
1.0	3.0	100	40	120	95	2.5	76	97	210	1	1	98
2.0	3.0	100	40	60	99	2.3	85	98	205	1	1	98
5.0	3.0	100	40	15	99	2.1	90	96	197	2	1	97
10.0	3.0	100	40	15	99	1.7	95	95	187	3	2	95
2.0	1.5	100	40	60	97	2.0	84	96	207	2	1	97
2.0	6.0	100	40	45	90	3.2	87	98	203	1	-	99
2.0	3.0	10	40	180	65	2.9	96	98	212	1	-	99
2.0	3.0	25	40	90	83	2.7	93	98	208	1	-	99
2.0	3.0	50	40	60	91	2.4	78	97	206	1	1	98
2.0	3.0	150	40	45	99	2.0	73	96	186	3	1	96
2.0	3.0	200	40	30	99	1.5	76	94	175	4	2	94
2.0	3.0	100	0	180	26	3.5	98	97	215	1	-	99
2.0	3.0	100	10	120	45	2.8	93	97	212	1	1	98
2.0	3.0	100	25	60	95	2.4	90	96	209	2	1	97

TABLE 27.3 (Continued)

[Co]·10⁴, mol/L	[M], mol/L	Al: Co	Temperature, °C	Reaction duration, min	Yield of 1,2-SPBD, % (mass)	[η]₁₃₅, dL/g	Crystallinity, %	Syndiotacticity, %	Melting point, °C	Microstructure, %		
										1.4-cis	1.4-trans	1.2-
2.0	3.0	100	55	45	99	2.0	81	95	194	4	1	95
2.0	3.0	100	80	45	99	1.5	68	93	180	4	2	94
2.0	3.0	100	100	45	99	1.3	63	93	177	5	1	94

27.4 CONCLUSIONS

In this work, novel highly active and stereo-regular cobalt alkylxhanto-genate + trialkylaluminum catalytic dithiosystems have been developed. Activity and stereo regularity of these catalysts were studied in comparison with known cobalt-containing catalytic dithiosystems of butadiene polym-erization. These experiments were conducted under varied conditions including catalyst concentration, ratio, and temperature. The developed catalytic systems allowed for the synthesis of high molecular weight and high crystalline syndiotactic 1,2-polybutadiene with yields of 50.0–99.0% and 1,2-contents of 94.0–99.0%, intrinsic viscosity between 1.3–3.5 dL/g, crystallinity between 63.0–98.0%, and melting points between 175–212°C in toluene solution.

The optimal conditions for the synthesis of high-molecular mass and highly crystalline syndiotactic 1,2-polybutadiene were established with the following parameters: $[Co] = (1.0–2.0) \times 10^{-4}$ mol/L; $[M] = 3.0$ mol/L; Al:Co = (50–100): 1; T = 40–80°C.

KEYWORDS

- photodegradable polymer
- catalytic activity
- polymerization processes
- highly crystalline syndiotactic
- Butadiene
- aluminum organic compounds

REFERENCES

1. Obata, Y., Ikeyama, M. Bulk properties of syndiotactic 1,2-polybutadiene. I. Thermal and viscoelastic properties, Polym. J., 7, 02, 207–216, 1975.
2. Junji, K., Shoko, S. Syndiotactic 1,2-polybutadiene rubber characteristics and applications., JETI, 46, 111–115, 1998.

3. Ashitaka, H., Ishikawa, H., Ueno, H., Nagasaka, A. Syndiotactic 1,2-polybutadiene with Co-CS2 catalyst system. I. Preparation, properties, and application of highly crystalline syndiotactic 1,2-polybutadiene, J. Polym. Sci., Polym. Chem. Ed., 21, 1853–1860, 1983.

4. Ashitaka, H., Ishikawa, H., Ueno, H., Nagasaka, A. Syndiotactic 1,2-polybutadiene with Co-CS2 catalyst system. II. Catalysts for stereospecific polymerization of butadiene to syndiotactic 1,2-polybutadiene, J. Polym. Sci.: Polym. Chem. Ed., 21, 1951–1972, 1983.

5. Ashitaka, H., Ishikawa, H., Ueno, H., Nagasaka, A. Syndiotactic 1,2-polybutadiene with Co-CS2 catalyst system. III. 1H-and 13C-NMR Study of highly syndiotactic 1,2-polybutadiene, J. Polym. Sci.: Polym. Chem. Ed., 21, 1973–1988, 1983.

6. Ashitaka, H., Ishikawa, H., Ueno, H., Nagasaka, A. Syndiotactic 1,2-polybutadiene with Co-CS2 catalyst system. IV. Mechanism of syndiotactic polymerization of Butadiene with Cobalt compounds-organoaluminum-CS2, J. Polym. Sci.: Polym. Chem. Ed., 21, 1989–1995, 1983.

7. Cheng-zhong, Z. Synthesis of syndiotactic 1,2-polybutadiene with silica gel supported CoCl2-Al(I-Bu)$_3$-CS$_2$ Catalyst, China Synthetic Rubber Industry, 22, 04, 243, 1999.

8. Cheng-zhong, Z. Synthesis and Morphological Structure of Crystalline Syndiotactic 1,2-Polybutadiene, Chemical Journal of Chinese Universities, 11, 2003.

9. Cheng-zhong, Z. Investigation on Synthesis of High Vinyl Polybutadiene with Iron-Based Catalysts. I. Effect of Triphenyl Phosphate, Chinese Journal of Catalysis, 08, 1219, 2004.

10. Cheng-zhong, Z., Zhen, D., Li-hong, N. Research Progress of Syndiotactic 1,2-Polybutadiene, Chemical Propellants & Polymeric Materials, 04, 2005.

11. Lan-guo, D., Weijian, H., Cheng-zhong, Z. Preparation and characterization of high 1,2-syndiotactic polybutadiene/polystyrene in situ blends, China Synth. Rubber Industry, 03, 2005.

12. Pat. USA 4751275 A, 1988; Pat. USA 5239023 A, 1993; Pat. USA 5356997 A, 1994; Pat. USA 5677405 A, 1997; Pat. USA 5891963 A, 1999; Pat. USA 6720397 B2, 2004; Pat. USA 6956093 B1, 2005; Pat. USA7186785 B2, 2007.

13. Nasirov, F. A. Dissertation Prof. Doctor (Chemistry). Baku: IPCP, Azerbaijan National Academy of Sciences, 376 p., 2003 (in Russian).

14. Nasirov, F. A. Bifunctional Nickel- or Cobalt containing catalyst-stabilizers for polybutadiene production and stabilization (Part I): Kinetic study and molecular mass stereo regularity correlation, Iranian Polymer Journal, 12, 217–235, 2003.

15. Djanibekov, N. F. Dissertation Prof. Doctor (Chemistry). Baku: IPCP, Azerbaijan Academy of Sciences, 374 p., 1987 (in Russian).

16. Nasirov, F. A. Organic Dithioderivatives of Metals – Components and Modificators of Petrochemical Processes, Petrochemistry, 6, 403–416, 2001 (in Russian).

17. Rafikov, S. P., Pavlova, S. A., Tvyordokhlebova, I. I. Methods of determination molecular weight and polydispersity of high molecular materials. M.: Academy of Sciences of USSR, 336 p., 1963 (in Russian).

18. Deyl, Z., Macek, K., Janak, J. Liquid Column Chromatography, Elsevier, Amsterdam Scientific, 1975.

19. Haslam, J., Willis, H.A. Identification and analysis of Plastics, London, Iliffe Books, Princeton, New Jersey: D. Van Nostrand Co., 1965.

20. Bellami, L. J. The Infra-red Spectra of Complex Molecules, London, Methuen and Co, New York: J. Wiley, 592 p., 1957.
21. Rabek, J. F. Experimental Methods in Polymer Chemistry: Physical Principles and Applications, John Wiley & Sons: New York, 1982.
22. Huhne, G. W. H., Hemminger, W. F., Flammersheim, H., -J. Differential Scanning Calorimetry, Springer, 2003.
23. Nasirov, F. A., Novruzova, F. M., Azizov, A. G., Djanibekov, N. F., Golberg, I. P, Guliev, N. A. Method of producing syndiotactic 1,2-polybutadiene, Pat. 20010128, Azerbaijan, 1999 (in Azerbaijanian).
24. Nasirov, F. A., Novruzova, F. M., Golberg, I. P., Aksenov, V. I. New Bifunctional Catalyst for Obtaining of Syndiotactic 1,2-Polybutadiene. In.: Proc. of III. Baku International Mamedaliev Petrochemistry Conference, Baku, 289, 1998.
25. Nasirov, F. A., Guliev, N. A., Novruzova, F. M., Azizov, A. G., Djanibekov, N. F. Peculiarities of Butadiene Polymerization to Syndiotactic 1,2-Polybutadiene in Toluene Solution. In.: Proc. of IV. Baku International Mamedaliev Petrochemistry Conference, Baku, 266, 2000.
26. Nasirov, F. A., Azizov, A. H., Novruzova, F. M., Guliyev, N. A. Polymerization of Butadiene to Syndiotactic 1,2-Polybutadiene in the Presence of Co-xantogenate + AlEt$_3$ Catalytic Systems. In.: Proc. of Polychar –10 World Forum on Polymer Applications and Theory, Denton, USA, 226, 2002.

CHAPTER 28

SILOXANE MATRIX WITH METHYLPROPIONATE SIDE GROUPS AND POLYMER ELECTROLYTE MEMBRANES ON THEIR BASIS

NATIA JALAGONIA,[1] IZABELA ESARTIA,[1] TAMAR TATRISHVILI,[1] ELIZA MARKARASHVILI,[1] DONARI OTIASHVILI,[2] JIMSHER ANELI,[2] and OMAR MUKBANIANI[2]

[1]Iv. Javakhishvili Tbilisi State University, Department of Chemistry, I. Chavchavadze Ave., 3, Tbilisi 0179, Georgia

[2]Iv. Javakhishvili Tbilisi State University, Faculty of Exact and Natural Sciences, Institute of Macromolecular Chemistry and Polymeric Materials, I. Chavchavadze Ave., 13, Tbilisi 0179, Georgia; E-mail: omarimu@yahoo.com

CONTENTS

28.1 INTRODUCTION

Solvent-free polymer electrolytes may be formed by the interaction of polar polymers with metal ions. Ion transport in polymer electrolytes extensively studied since Wright [1] discovered that polyethyleneoxide (PEO) can act as a host for sodium and potassium salts, thus producing a solid electrical conductor polymer/salt complex. The unique idea of employing these polymer electrolytes in battery applications belong to Armand et al. [2].

Transport mechanism models developed by Ratner et al. [3] indicated that polymers with low T_g have extremely high free volumes, which favors to ion transport. Better results are obtained for polymers with highly flexible backbones, bearing oligo(ethylene glycol) side chains.

Interest in polysiloxane-based polymer electrolytes arose early in the 1980 s. PEO – substituted polysiloxanes as ionically conductive polymer hosts have been previously investigated [4–6]. Their relatively high ionic conductivity was ascribed to the highly flexible inorganic backbone, which produced a totally amorphous polymer host. In recent years, improved battery performance has been observed for systems containing polymer electrolytes, with a Li+ transference number close to unity [7]. Efforts have also been made to design and synthesize siloxane-based single-ion conductors [8, 9]. Polysiloxanes are promising components for comb polyelectrolytes because they possess a flexible backbone that enhances the transports of ions. Their amorphous and highly flexible $[Si\text{-}O]_n$ backbone produces glass transition temperatures as low as $-100°C$ and yields little or no crystallinity at room temperature. In addition, each monomer unit has two sites for cross-links or functional side chains through bond formation with silicon. Simulations indicate that comb polyelectrolytes should display higher conductivity values than their analogs to local motion of the bound anions in comb systems [10–12].

28.2 EXPERIMENTAL PART

28.2.1 MATERIALS

D_4^H (Aldrich), platinum hydrochloric acid (Aldrich), Karstedt's catalyst $(Pt_2[(VinSiMe_2)_2O]_3)$ or platinum(0)-1,3-divinyl-1,1,3,3-tetramethyldisiloxane complex (2% solution in xylene) (Aldrich), vinyl triethoxysilane (Aldrich), methyl acrylate (Aldrich) were used as received. Lithium tri-fluoromethylsulfonate (triflate) and lithium bis(trifluoromethylsulfonyl) imide were purchased from (Aldrich). Toluene was dried over and distilled from sodium under an atmosphere of dry nitrogen. Tetrahydrofuran (THF) was dried over and distilled from K–Na alloy under an atmosphere of dry nitrogen.

28.2.2 CHARACTERIZATION

FTIR spectra were recorded on a Varian 660/670/680-IR series spectrometer. ^{1}H, ^{13}C NMR and ^{29}Si NMR spectra were recorded by a Varian Mercury 300 VX NMR spectrometer, using DMSO and CCl_4 as the solvent and as an internal standard. Differential Scanning Calorimetric (DSC) investigation was performed on a Netzsch DSC 200 F3 Maia apparatus. Thermal transitions including glass transition temperatures T_g were taken as the maxima of the peaks. The heating and cooling scanning rates were 10 K/min.

Size exclusion chromatographic study was carried out with the use of Waters Model 6000A chromatograph with an R 401 differential refractometer detector. The column set comprised 10^3 and 10^4 Å Ultrastyragel columns. Sample concentration was approximately 3% by weight in toluene and typical injection volume for the siloxane was 5 µL, flow rate – 1.0 mL/min. Standardization of the SEC was accomplished by the use of styrene or polydimethylsiloxane standards with the known molar mass.

Wide-angle X-ray analysis was performed on a Dron-2 (Burevestnik, Saint Petersburg, Russia) instrument. A-CuK$_\alpha$ was measured without a filter; the angular velocity of the motor was $\omega \approx 2°$/min.

Determination of $\equiv$Si-H content was calculated according to the method [13].

28.2.3 HYDROSILYLATION REACTION OF D$_4$^{H} WITH METHYL ACRYLATE AND VINYLTRIETHOXYSILANE

D$_4$^{H} (5.000 g, 0.0208 mol) were transferred into a 100 mL flask under nitrogen using standard Schlenk techniques. High vacuum was applied to the flask for half an hour before the addition of methyl acrylate (5.3636 g,

0.0624 mol) and vinyltriethoxy silane (3.9562 g, 0.0208 mol). The mixture was then dissolved in 7 mL of toluene and 0.1 M solution of platinum hydrochloric acid in tetrahydrofuran ($5/9 \times 10^{-5}$ g per 1.0 g of starting substance) was introduced. The homogeneous mixture was degassed and placed into an oil bath, which was previously set to 60°C and reaction continued at 60°C. The reaction was controlled by decrease of intensity of active ≡Si-H groups. Then 0.1wt. % activated carbon was added and refluxed for 12 h for deactivation of catalysts.

All volatile products were removed by rotary evaporation and the compound was precipitated at least three times into pentane to remove side products. Finally, all volatiles were removed under vacuum for 24 h to isolate 13.7 g (95.6%) of colorless viscous compound I – 2.4.6.8-tetramethtyl-2.4.6-tri(methyl propionate)-8-ethyltriethoxysilanecyclotetrasiloxane ($D_4^{R, R'}$).

28.2.4   RING-OPENING POLYMERIZATION REACTION OF $D_4^{R, R'}$

The 1.1365 g (1.4046 mmol) of compound $D_4^{R, R'}$ was transferred into 50 mL flask under nitrogen. High vacuum was applied to the flask for half an hour. Then the compound was dissolved in 1.8 mL dry toluene and 0.01% of total mass powder-like potassium hydroxide was added. The mixture was degassed and placed in an oil bath, which was previously set to 60°C was polymerized under nitrogen for 25 h. After the reaction, 7 mL of toluene was added to the mixture and the product was washed by water. The crude product was stirred with $MgSO_4$ for 6 h, filtered and evaporated and the oligomer was precipitated at least three times into pentane to remove side products. Finally, all volatiles were removed under vacuum to isolate 1.06 g (93%) colorless viscous oligomer (II).

Ring-opening polymerization reaction of compound I at various temperatures had been carried out by the same manner.

28.2.5   GENERAL PROCEDURE FOR PREPARATION OF CROSS-LINKED POLYMER ELECTROLYTES

In a typical preparation, 0.75 g of polymer II was dissolved in 4 mL of dry THF and thoroughly mixed for half an hour before the addition of catalytic amount of acid (one drop of 0.1 M HCl solution in ethyl alcohol) to initiate the cross-linking process. After stirring for another 3 h, required amount of lithium triflate from the previously prepared stock solution in THF was added to the mixture and stirring continued for another 1 h. The mixture was then poured into a Teflon mold with a diameter of 4 cm and solvent was allowed to evaporate slowly overnight. Finally, the membranes were dried in an oven at 70°C for 3 days and at 100°C for 1 h. Homogeneous and transparent films with average thickness of 200 μm were obtained in this way. These films were insoluble in all solvents, only swollen in THF.

28.2.6 *AC IMPEDANCE MEASUREMENTS*

The total ionic conductivity of samples was determined by locating an electrolyte disk between two 10 mm diameter brass electrodes. The electrode/electrolyte assembly was secured in a suitable constant volume support, which allowed extremely reproducible measurements of conductivity to be obtained between repeated heating–cooling cycles. The cell support was located in oven and the sample temperature was measured by thermocouple positioned close to the electrolyte disk. The bulk conductivities of electrolytes were obtained during a heating cycle using the impedance technique (Impedance meter BM 507 –TESLA for frequencies 50 Hz–500 kHz) over a temperature range between 20 and 100°C.

28.3 RESULTS AND DISCUSSION

Comb-type polymers for solvent free solid polymer electrolytes usually are obtained via hydrosilylation or dehydrocondensation reactions of industrial linear polymethylhydrosiloxanes (PMHS) with donor group containing vinyl-, allyl- or hydroxyl containing organic compounds. It should be noted that often these reactions proceeds incompletely with obtaining irregular defect structures.

The aim of our work is synthesis of organocyclotetrasiloxane with desired propionate donor side groups at silicon via hydrosilylation reaction of 2.4.6.8-tetrahydro-2.4.6.8-tetramethylcyclotetrasiloxane (D_4^H) with methylacrylate and vinylthiethoxsilane at 1:3:1 of initial compounds in the presence of platinum catalysts; polymerization reactions of organo-cyclotetrasiloxane in the presence of nucleophilic catalysts and obtaining comb-type polymers with regular arrangement of methyl propionate and ethoxyl group; obtaining of solid polymer electrolyte membranes by the incorporation of lithium salt into polymer matrices; investigation of ionic conductivity of membranes via impendence method.

Separately preliminary heating of initial compounds in the temperature range of 50–60°C in the presence of catalysts showed that under these conditions polymerization of D_4^H or allyl acetoacetate, and scission of siloxane backbone does not take place. No changes in the NMR and FTIR spectra of initial compounds were found. It was established that in melt condition, the hydrosilylation reaction proceeds vigorously with initiation of side reactions [14], therefore for obtaining fully addition product hydrosilylation reaction have been carried out in dilute solutions.

Hydrosilylation reaction of D_4^H methylacrylate and vinylthiethoxysilane at 1:3:1 of initial compounds in the presence of platinum catalysts was carried out in 50% solution of dry toluene or tetrahydrofuran at 60–70°C temperature. It was established that hydrosilylation in the presence of Karstadt's and platinum hydrochloride acid catalyst proceeds very slowly, so catalyst Pt/C (5%) in hydrosilylation reaction we did not use.

In order to increase the reaction rate the reaction temperature at the final stage the reaction mixture was heated up to 80° C temperature rained.

The reaction generally proceeds according to the following scheme 1:

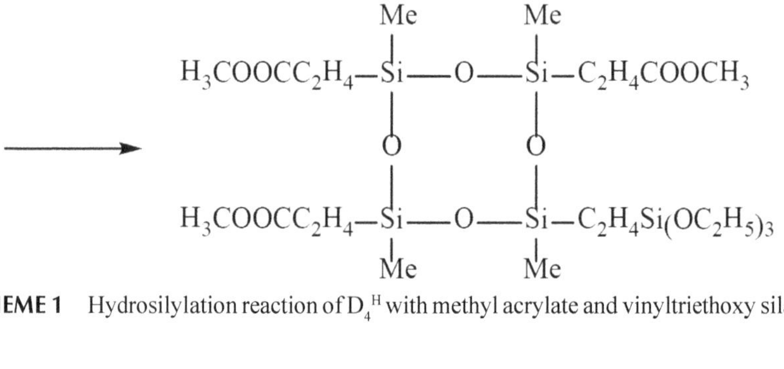

SCHEME 1   Hydrosilylation reaction of D_4^H with methyl acrylate and vinyltriethoxy silane.

From Literature it is known that hydrosilylation with unsaturated bonds may be proceeded according Markovnikov and anti-Markovnikov (Farmer) rules [11, 12].

In addition methyl acrylate is conjugated compound. As it is known from literature in conjugated systems hydrosilylation may be proceeded not only 1.2-direction, but also in the direction of 1.4 (Scheme 2):

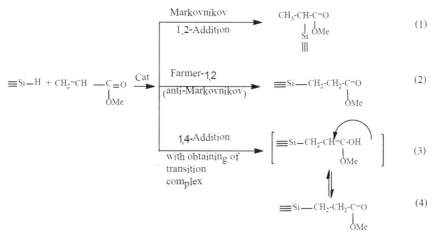

SCHEME 2   Possible addition of ≡Si-H bonds to methyl acrylate.

As it is seen from the possible Scheme 2.3, in the case of 1.4-hydride addition, reaction proceeds with obtaining of intermediate-transition complex. By regrouping of the intermediate product according to the Eltekov rule (2.4), the products of 1.2-addition by anti-Markovnikov rule are obtained.

In addition, in hydrosilylation reaction the mixture of isomeric cyclic compounds (*cis*- and trans – isomeric mixture) can be take place. The obtained substance is a transparent viscous liquid, which is well soluble in common organic solvents. Compound I was identified by IR, ^{29}Si, ^{1}H and ^{13}C NMR spectral data, determining the molecular mass and molecular refractivity: $n_D^{20} = 1.4389$; $d_4^{20} = 1.1185$; Calculated – $M_{RD} = 153.36$; found $M_{RD} = 152.67$.

In the FTIR spectra of compound I absorption bands at 1020 and 1197 cm^{-1}, characteristic for asymmetric valence oscillation of ≡Si-O-Si≡, Si-O-C and CO-O-C bonds. Also one can observe absorption bands at 1257 and 2860–2970 cm^{-1} region, characteristic for valence oscillation of ≡C-H bonds. In the FTIR spectra of compound I at 2167 cm^{-1}, there is no absorption bands characteristic for un-reacted ≡Si-H bonds. In the spectra absorption bands at 1736 cm⁻¹ characteristic for carbonyl groups is observed.

In ²⁹Si NMR spectra of compound I one can see signal with chemical shift δ≈−18.98 and δ≈−26.86 ppm corresponding to the presence of RR'SiO (D) units in cyclotetrasiloxane fragment [15, 16], on the other hand, two signals δ≈−46.09÷−47.43 and δ≈−58.28÷−61.03 ppm corresponding to the presence of $M^{(OR)}_2$ ꙇꙍ D^{OR} fragments and the signal with chemical shift δ≈−65.66 ppm T units at −55 and −65 ppm is observed [16], which is confirmed reaction direction, with formation of above structure. The direction of the reaction is confirmed by the structure formation.

In the ¹H NMR spectrum of compound I (Figure 28.2), one can observe singlet signal with a chemical shift δ=0.1–0.2 ppm characteristic for protons in ≡Si-Me groups in isomeric mixture of cyclotetrasiloxane fragments, triplet signal for methylene protons in ≡Si-CH₂- fragment with chemical shift δ=0.50 ppm (during addition to anti-Markovnikov rule), triplet signal for methyl protons in =CH-CH₃ group with chemical shift δ=1.1 ppm (during addition to Markovnikov rule). Also, one can observe multiplet signals with center of chemical shifts δ=2.3, 3.6–3.8 ppm characteristics for proton in CH₃CH=, -CH₂CO- and OCH₃ group accordingly.

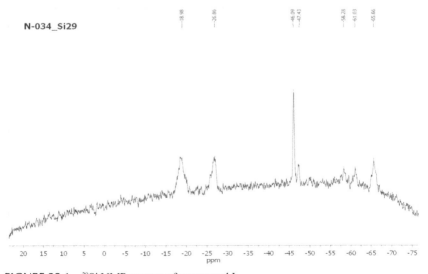

FIGURE 28.1   ²⁹Si NMR spectra of compound I.

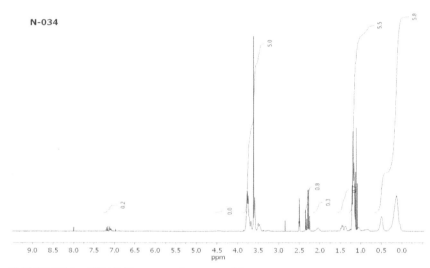

FIGURE 28.2   I ¹H NMR spectra of compound I.

In the ¹³C NMR spectra of isomeric mixture of compound I the resonance signals with chemical shifts $\delta\approx$-3.83, −2.45, 1.01, 17.75, 26.38, 30.13, 50.45 and 57.33 ppm characteristic for $\equiv$Si-Me, CH$_{3to}$CH=, $\equiv$Si-CH$_2$-, -CH$_2$CO-, -CH(CH$_3$)- and OCH$_3$, accordingly. ¹³C NMR spectra correspond to ¹H NMR spectra data.

The synthesized compound I was used in polymerization and copolymerization reactions. From literature [12] it is known that during polymerization reactions of ethoxyl group containing compound D$_4$$^{R,R'}$ proceeds with obtaining of cross-linking systems, which may be explained via intermolecular condensation reactions of $\equiv$Si-OH and -Si(OC$_2$H$_5$)$_3$ groups. Therefore we have investigated copolymerization reactions of D$_4$$^{R,R'}$ with hexamethyldisiloxane as a terminating agent at 8:1 ratio of initial compounds. In most cases, terminating agent helps to regulate the molecular masses [17].

The polymerization reactions of compound I in the presence of anhydrous powder like potassium hydroxide (0.01 mass %) as a catalyst in dilute solution of dry toluene (C = 0.8606 mol/L), at 60–80°C temperature have been investigated. It was established that polymerization reactions proceeds slowly during 80–100 h. The optimal condition of reaction temperature is 80–90°C. During polymerization reaction of D$_4$$^{R,R'}$ in toluene

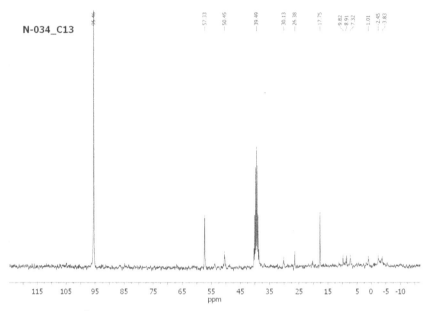

FIGURE 28.3 ^{13}C NMR spectra of compound I.

the polymer precipitated from the solvent, but during polymerization in THF the polymer remain in solution. The copolymerization reaction generally proceeds according to the Scheme 3:

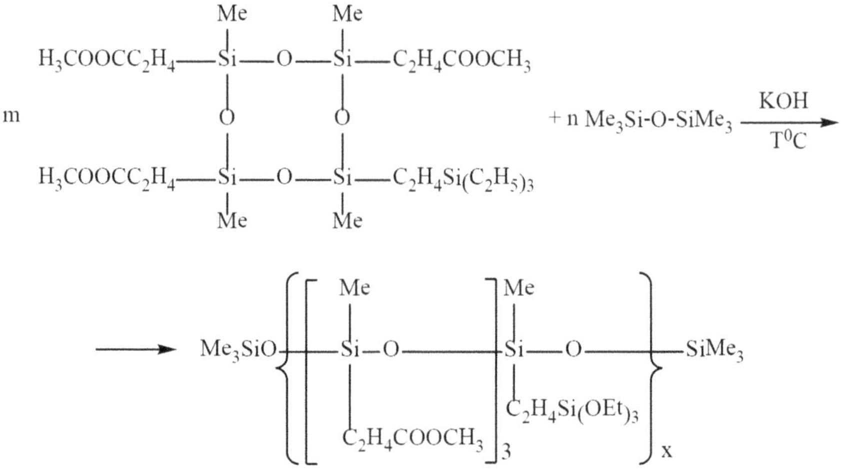

SCHEME 3 Copolymerization reaction of compound I.

where: m:n=8:1. II[1] (60°C), II (80°C).

The synthesized polymers are vitreous, viscous products which are well soluble in ordinary organic solvents with specific viscosity $\eta_{sp} \approx 0.14-0.27$. The structure and composition of polymers were determined by elemental analysis, molecular masses – by FTIR and NMR spectra data. Elemental composition yields and some physical-chemical properties of oligomers are presented in Table 28.1.

In FTIR spectra of polymers the absorption band at 1080 cm^{-1} is characteristic for asymmetric valence oscillation of $\equiv$Si-O-Si$\equiv$ bonds. In the spectrum are reserved all the absorption bands characteristic of the initial monomeric compound I.

In ^{29}Si NMR spectra of polymers (Figure 28.4) one can see resonance signal with chemical shifts $\delta\approx-19.10$ and $\delta\approx -26.97$ ppm corresponding to the presence D to M$^{(OR)}_2$ and D^{OR} units on the other hand the signal at $\delta\approx-65.73$ ppm, corresponds to T units.

On the ^{1}H NMR spectra of polymers singlet signal with a chemical shift $\delta\approx0.1-0.2$ ppm characteristic for protons in $\equiv$Si-Me groups, triplet

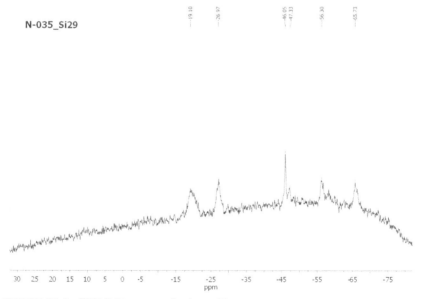

FIGURE 28.4 ^{29}Si NMR spectra of polymer II.

signal for methylene protons in $\equiv$Si-CH$_2$- fragment with chemical shift $\delta\approx$0.50 ppm (during addition to anti-Markovnikov rule), triplet signal for methyl protons in =CH-CH$_3$ group with chemical shift δ=1.1 ppm (during addition to Markovnikov rule). Also, one can observe multiplet signals with center of chemical shifts $\delta\approx$2.3, 3.6–3.8 ppm characteristics for proton in CH$_3$CH=, -CH$_2$CO- and OCH$_3$ group, accordingly.

On the Figure 28.6 ^{13}C NMR spectra of polymer II is presented. The signal with chemical shifts at $\delta\approx$ $\delta\approx$–3.83, –2.45, 1.01, 17.75, 26.38, 30.13, 50.45 and 57.33 ppm characteristic for $\equiv$Si-Me, CH$_3$CH=, $\equiv$Si-CH$_2$-, -CH$_2$CO-, -CH(CH$_3$)- and OCH$_3$ accordingly is preserved. ^{13}C NMR spectra of oligomer I is in accordance with ^{1}H NMR spectra.

For obtaining polymer II the molecular masses by ebuliometric methods has been determined M$_n$= 6100. As is known from the literature data, the molecular masses of obtained polymer, is sufficient for a solid polymer – membranes electrolytic production.

DSC calorimetric investigation of polymer II was carried out. As it is seen from Figure 28.7 the polymer is characterized only one glass transition temperature, which is equal to Tg=–81.7°C.

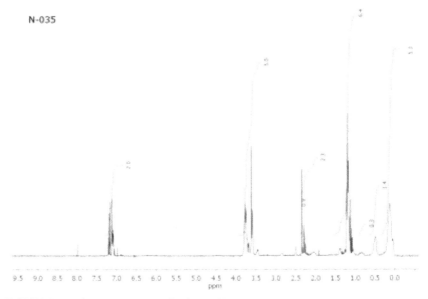

FIGURE 28.5 ^{1}H NMR spectra of polymer II.

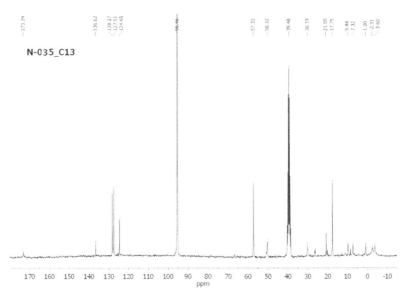

FIGURE 28.6   ¹³C NMR spectra of polymer II.

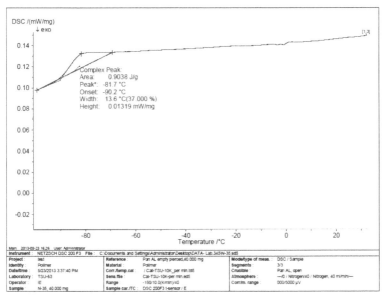

FIGURE 28.7   DSC curve of polymer II.

28.3.1 POLYMER ELECTROLYTE MEMBRANES

For obtaining of the membranes we prepared the polymer solutions of the Lithium salt striflate($LiSO_3CF_3$) and lithium-bis(triftormethylsulfonilimide) [$(CF_3SO_2)_2NLi^+$] in the tetrahydrofuran, where these salts were about 5–20 wt% of the polymer full mass. The solution contained 0.8 g salt in the 2 mL tetrahydrofuran.

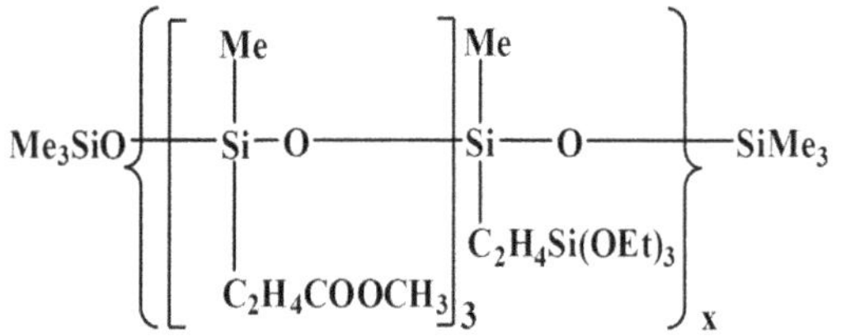

The solution of polymer in the tetrahydrofuran was prepared in the cylindrical form (diameter 4 cm) vessel made from Teflon. After the salt solution in the tetrahydrofuran and 1–2 drops of 0.1 N alcohol solutions were added to the initial mixture. Mixing elapsed during 30 min. The mixture was stayed in the inert atmosphere and after out gassed, in result of which the transparent-yellow films were formed. At adding of the hydrochloric acid alcohol solution the sol gel processes conducted, which is accompanied by crosslink reactions with following Scheme (4).

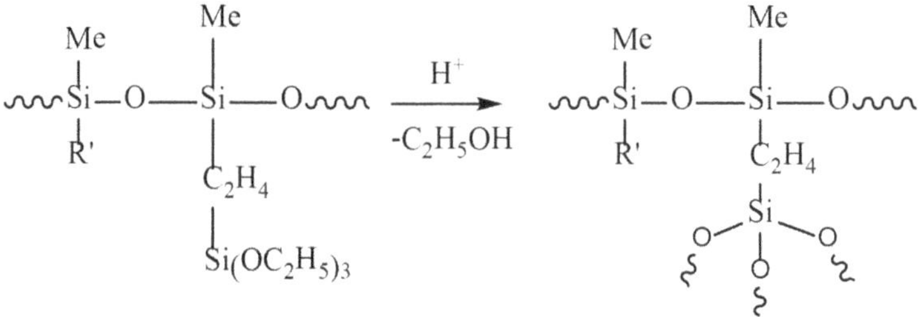

SCHEME 4   The sol-gel cross linking reaction of the polymer.

There were conducted the investigation of the electric-physical properties of synthesized polyelectrolytes. It was measured the conductivity and its dependence on the temperature. We prepared the solid polymerelectrolyte membranes with 5, 15 and 20 wt% of lithium salts. It was studied the dependence of the specific volumetric electric conductivity of the membranes on temperature. The curves on the Figure 28.8 show that their character corresponds to analogical dependences for polyelectrolytes based on silicon-organic polymers and some lithium salts and are well described by so called Vogel-Taman-Fulcher formula [8,9]:

$$C(T) = a/T \exp(-E/kT),$$

where a is the preexponential factor, E – the activation energy of charge transfer and k –Boltzmann constant.

On the basis of preliminary measures of the membranes described above it was established that the conductivity of membrane containing 5 wt % salt is so small (near to 10^{-9} S/cm) that we decided not investigate this material, as electrolytes with such level of conductivity practically are unusable in few of creation of the chemical current of sources.

Dependence of conductivity of membranes based on polymer II with 15 (1) and 20 (2) wt % content of lithium salt – triflate on the temperature is presented on the Figure 28.8.

The character of voltamogrames obtained for membranes based on polymer II and 15 and 20 wt % of triflate salt is in full accordance with the values of their conductivity (Figure 28.9).

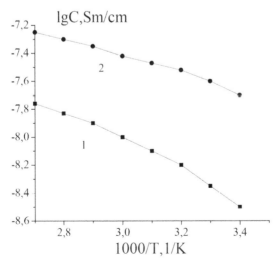

FIGURE 28.8 Dependence of conductivity of membranes based on polymer II with 15 (1) and 20 (2) wt % of lithium salt – triflate on the temperature.

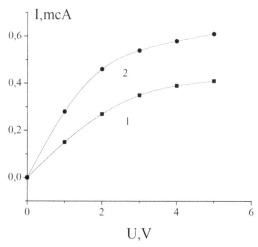

FIGURE 28.9   Voltamogrames of membranes based on the polymer II and 15 (1) and 20 wt % (2) of lithium triflate.

Thus, on the basis of the temperature dependences and corresponding voltamogrames for the membranes based on polymer II and lithium triflate salt it may be expressed an opinion that the ionic conducting character essentially is due to both structure of electrolytes polymer matrix and behavior of last under conditions of the change of external temperature. First of all it is expressed in the essential difference between initial and end values of the conductivity at increasing temperature. It may be proposed that microstructure of this type membranes is high permeable for the ions at heating, which is due to increasing of the concentration of microempties in the polymer matrix because of high level of heterogeneity of the macromolecules with different side groups and, respectively, charge transfer in the material grows.

28.4   CONCLUSION

Hydrosilylation reaction of tetrahydro tetramethylcyclotetrasiloxane with methyl acrylate and vinyltriethoxysilane at 1:3:1 ratio of initial compounds, in the presence of Karstadt's catalyst have been studied and corresponding addition product have been obtained. Via copolymerization reaction of organocyclotetrasiloxane and hexamethyldisiloxane as a terminated agent corresponding comb-type polymers have been obtained. Via sol-gel processes of doped with lithium trifluoromethylsulfonate (triflate) or lithium bis(trifluoromethylsulfonyl)imide oligomer systems solid polymer electrolyte membranes have been obtained. The dependence of ionic conductivity as a function of temperature and salt concentration was investigated. The electric conductivity of investigated polymer electrolyte membranes at room temperature changes in the range 3×10^{-9}–2×10^{-8} S/cm.

KEYWORDS

- **siloxane matrix**
- **methylpropionate side groups**
- **polymer electrolyte membranes**
- **transport mechanism models**
- **conductive polymer**
- **performance**

REFERENCES

1. Wright, P. V. Electrical conductivity in ionic complexes of poly(ethylene oxide), British Polymer Journal, 7, 319–327, 1975.
2. Armand, M.B., Chabagno, J. M., Duclot, M. J. Polyethers as solid electrolytes, Second International Meeting on Solid Electrolytes, St. Andrews, Scotland, 1–4, 1978.
3. Ratner, M. A., Shriver, D. F. Ion transport in solvent-free polymers, Chemical Review, 88, 109–124 (1988).
4. Nagaoka, K., Naruse, H., Shinohara, I., Watanabe, M. High ionic conductivity in poly(dimethyl siloxane-coethylene oxide) dissolving lithium perchlorate, Journal Polymer Science, Polymer Letter Edition, 22, 659–665, 1984.
5. Albinsson, I., Mellander, B. E., Stevens, J. R. Ionic conductivity in poly(ethylene oxide) modified poly (dimethylsiloxane) complexed with lithium salts, *Polymer*, 32, 2712–2715, 1991.

6. Fish, D., Khan, I. M., Smid, J. Conductivity of solid complexes of lithium perchlorate with poly{ω-methoxyhexa(oxyethylene)ethoxymethylsiloxane}, Macromolecular Chemistry, Rapid Communications, 7, 115–120, 1986.

7. Doyle, M., Fuller, T. F., Newman, J. The importance of the lithium ion transference number in lithium/polymer cells, Electrochimica Acta, 39, 2073–2081, 1994.

8. Karatas Yu., Kaskhedikar, N., Wiemhofer, H. D. Synthesis of cross-linked comb polysiloxane for polymer electrolyte membranes. Macromolecular Chemistry and Physics, 207, 419–425, 2006.

9. Zhang Zh., Lyons, L. J., Jin, J. J., Amine Kh., West, R. Synthesis and ionic conductivity of cyclosiloxanes with ethyleneoxy-containing substituent's, Chemistry of Materials, 17, 5646–5650, 2005.

10. Snyder, J. F., Ratner, M. A., Shriver, D. F. Polymer electrolytes and polyelectrolytes: Monte Carlo simulations of thermal effects on conduction, Solid State Ionics, 147, 249–257, 2002.

11. Mukbaniani, O., Koynov, K., Aneli, J., Tatrishvili, T., Markarashvili, E., Chigvinadze, M. Solid Polymer Electrolyte Membranes Based on Siliconorganic Backbone. Macromolec. Symposia, 328(1), 38–44, 2013.

12. Mukbaniani, O., Aneli, J., Esartia, I., Tatrishvili, T., Markarashvili, E., Jalagonia, N. Siloxane Oligomers with Epoxy Pendant Group. Macromolec. Symposia, 328(1), 25–37, 2013.

13. Iwahara, T., Kusakabe, M., Chiba, M., Yonezawa, K. Synthesis of novel organic oligomers containing Si-H bonds. J. Polym. Sci., A, 31, 2617–2631, 1993.

14. Mukbaniani, O., Tatrishvili, T., Markarashvili, E., Esartia, E. "Hydrosilylation reaction of tetramethylcyclotetrasiloxane with allyl butyrate and vinyltriethoxysilane. Georgian Chemical Journal, 2(11), 153–155 (2011).

15. Uhlig, F., Marsmann, H. Chr. ^{29}Si NMR Some Practical Aspects, Springer, 2008.

16. Khan, I. M., Yuan, Y., Fish, D., Wu, E., Smid, J. Comb-like polysiloxanes with oligo (oxyethylene) side chains. Synthesis and properties, Macromolecules, 21, 2684–2689, 1988.

17. Mukbaniani, O., Aneli, J., Tatrishvili, T., Markarashvili, E., Chigvinadze, M., Abadie, M. J. M. Synthesis of cross-linked comb-type polysiloxane for polymer electrolyte membranes. E-polymer #089, 1–14, 2012.

CHAPTER 29

COMPOSITES ON THE BASIS OF GLYCIDOXYGROUP CONTAINING PHENYLSILSESQUIOXANES

MARINA ISKAKOVA,[1] ELIZA MARKARASHVILI,[2,3] JIMSHER ANELI,[3] and OMAR MUKBANIANI[2,3]

[1]Ak. Tsereteli Kutaisi State University, Department of Chemical Technology,

[2]Iv. Javakhishvili Tbilisi State University, Department of Chemistry, I. Chavchavadze Ave. 1, 0179 Tbilisi, Georgia

[3]Institute of Macromolecular Chemistry and Polymeric Materials, I. Javakhishvili Tbilisi State University, I. Chavchavadze Ave. 13, 0179 Tbilisi, Georgia
E-mail: marinaiskakova@gmail.com

CONTENTS

29.1   INTRODUCTION

There is recent intense interest in the development of silsesquioxane based materials because of their 3D nature, their ability to offer a very high degree of functionalization, their ease of synthesis and typically high thermal stability. This is evidenced by the fact that now there are some 3 reviews on silsesquioxanes and the related silicates [1–3]. These references describe their potential application in a broad range of areas from biomedical, to organic light emitting diodes, to nanocomposites and etc. In the literature data a lot of information about depending of the substance properties on the polyorganosiloxane organic framing at the silicon atom. For example, polyorganosiloxanes containing aromatic radicals different from polydimethylsiloxanes are characterized by high thermostability and dielectric characteristics [4].

For improving of dielectric properties of the modified filler compounds the oligotetraepoxysiloxane oligomers with organosilsesquioxane fragments in the chain have been synthesized. As initial compound for the synthesis of tetraepoxyphenylsilsesquioxane during condensation reaction of epichlorohydrin the *cis*-2.4.6.8-tetrahydroxy-2.4.6.8-tetraphenylcyclotetrasiloxane and tetrahydroxyphenylsilsesquioxane oligomer (n=2÷10) with sodium hydroxide have been used and phenylethoxysilsesquioxanes PhES-80 (n–1) and PhES-50 (n=2) in the presence of catalysts iron chloride (III) 0.01% by weight have been investigated.

29.2   EXPERIMENTAL PART

29.2.1   *MATERIALS AND TECHNIQUES*

Initial *cis*-2.4.6.8-tetrahydroxy-2.4.6.8-tetraphenylcyclotetrasiloxane and tetrahydroxyphenylsilsesquioxane have been obtained according to the literature [5, 6].

FTIR spectra were recorded on a Varian 660/670/680-IR series spectrometer. Chromatographic analysis of the purity of the starting reactants and the reaction was conducted on a chromatograph grade LKhM −80,

Model 2 (column 3000 chromatografic-4 mm). Media-Chromosorb W or Chromaton. For analysis the following phases were used: Silicone SE-30 resin, REOPLEX-400 and carrier gas-helium. Match of the applied phase separating substances possessed the ability to OV-17 supported on Chromaton-NAW. For better separation of different monomers the temperature was selected in the range 130–200°C.

29.2.2 CONDENSATION REACTION OF EPICHLOROHYDRIN WITH CIS-2.4.6.8-TETRAHYDROXY-2.4.6.8-TETRAPHENYLCYCLOTETRASILOXANE

In a four-necked flask equipped with a thermometer, mechanical stirrer, reflux condenser and dropping funnel was charged 5.52 g (0.01 mol) of tetrahydroxytetraphenylcyclotetrasiloxane and 30 mL of diethyl ether and at stirring at 40°C 7.52 g (0.08 mol) of epichlorohydrin (100% excess) was added, and then slowly and gradually distilled of ether, the reaction temperature rising to 60–70°C, re stirred and heated for 1 h. After these procedures in three portions 25% sodium hydroxide solution 0.4 mol (16 g NaOH, 48 g H$_2$O) was added. After addition of the last portion of the alkaline solution mixture, stirring maintained for a further 1 h at a temperature of 70°C and 9.3 g of epichlorohydrin was added and the reaction mixture was diluted with 100 mL of ether. Ether was removed by suction. Unreacted part of epichlorohydrin was evacuated at 30°C for 3 h at a residual pressure of 1 mm Hg. In result of these manipulations about 6.8 g slow-moving viscous mass (compound I) with a yellowish color at 96% yield was obtained.

29.2.3 CONDENSATION REACTION OF EPICHLOROHYDRIN WITH OLIGOPHENYLSILSESQUIOXANE (N≈3)

Compounds (II) were prepared by Scheme 2 and the same procedure. The flask was charged with 15.84 g (0.01 mol) of tetrahydroxypoly-phenyloligotetrole in 60 mL of dry toluene and 7.52 g (0.08 mole) epichlorohydrin (100% excess). 16.74 g (96%) tetraepoxypoly-phenylsilsesquioxane (II) was obtained. Condensation reaction of epichlorohydrin with oligotetraphenysilsesquioxane (n≈10) have been carried out with the same manner.

29.2.4   CONDENSATION REACTION OF EPICHLOROHYDRIN WITH TETRAETHOXYTETRAPHENYLCYCLOTETRASILOXANE (PHES-80)

The reaction between an industrial product tetraethoxytetraphenylcyclotetrasiloxane PhES-50 (n=1) and PhES-80 (n=2) with epichlorohydrin was conducted in a four-necked flask equipped with a dropping funnel, a reflux condenser, mechanical stirrer and thermometer. This flask was charged with 118 g (0.1 mol) of PhES-80 (phenethylsilsesquioxane-80), 2.4 g of iron chloride (III) 0,01% by weight of the reaction mixture and 75.2 g (0.8 mol) epichlorohydrin (100% excess). The reaction mixture was heated up to 80–85°C. The reaction was carried out for 5–6 h and low molecular byproduct C_2H_5Cl was collected in a flask. The resulting mass was centrifuged to precipitate the catalyst. After evacuating of unreacted products, the 112.5 g of a dark brown color gummy product I' was obtained with a 95% yield.

Similarly, the condensation reaction of epichlorohydrin with PhES-50 (n=1) in the presence of catalysts, iron (I') and aluminum chloride (IV') have been carried out and dark brown color gummy products have been obtained.

29.3   RESULTS AND DISCUSSION

There are some ways of synthesis of epoxyorganosiloxanes. From these methods it is significant to denote oxidative epoxidation of unsaturated bond containing organosilanes and siloxanes [7, 8], hydrosilylation reactions of ≡Si-H bond containing silanes and siloxanes to allyl glycidyl ether in the presence of catalysts [9, 10] and the reactions of oganosilanoles or organosiloxanoles with epichlorohydrin [11, 12].

The condensation reaction of epichlorohydrin with cis-2.4.6.8-tetrahydroxy-2.4.6.8-tetraphenylcyclotetrasiloxane with the excess epichlorohydrin in the presence of 25% sodium hydroxide solution have been performed. The reaction proceeds according to Scheme 29.1:

SCHEME 29.1   The condensation reaction of epichlorohydrin with *cis*-2.4.6.8-tetrahydroxy -2.4.6.8-tetraphenylcyclotetrasiloxane.

In result of reaction the transparent yellow viscous compound I well soluble in ordinary organic solvents have been obtained. The composition and structure of the compound I obtained have been proved on the basis of elemental analysis, definition of number of epoxy groups, determination of molecular masses and FTIR spectra data. Some physicochemical data of the synthesized compounds are presented in Table 29.1.

In the FTIR spectra of obtained compounds in the range of asymmetric and symmetric valence oscillations $\equiv$Si-O-Si$\equiv$ bond the bifurcation of the strips with maximums v_{as} – 1045 cm^{-1}, 1145 cm^{-1} and v_{as} – 455, 480 cm^{-1} are observed. In the case of phenylcyclotetrasiloxane ring condensation at presence of nucleophilic sodium may be occur with the opening of cyclotetrasiloxane ring, which leads to formation of the structure different from *cis*-configuration (see Figures 29.1 and 29.2). This opinion is in accordance with known data [13, 14].

TABLE 29.1   Some Physicochemical Data of the Synthesized Compounds

#	n	Yield, %	Amount of epoxide** groups, %	M*$_{mass}$	Elemental analysis**, %		
					C	H	Si
I	1	96	24.16	712	60.67	5.62	15.73
			24.05	709	60.40	5.60	15.66
II	3	95	9.86	1744	57.80	4.59	19.27
			10.00	1769	58.63	4.66	19.55
III	10	96	3.23	5324	56.80	4.13	21.04
			3.18	5242	55.92	4.07	20.72

*Molecular masses have been determined via ebulliometric method.
**Over line calculated values, under line found values.

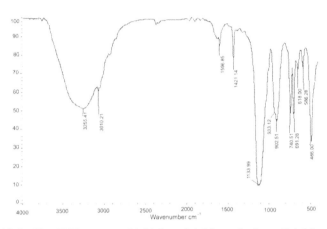

FIGURE 29.1 The FTIR spectra of initial *cis*-2.4.6.8-tetrahydroxy-2.4.6.8-tetraphenyl-cyclotetrasiloxane.

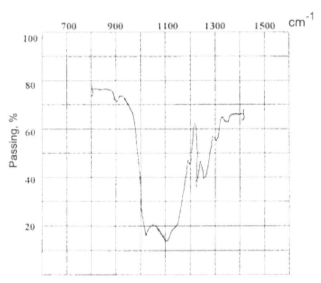

FIGURE 29.2 FTIR spectrum of 2.4.6.8-glycidoxy-2.4.6.8-tetraphenylcyclotetrasiloxane (I).

However, the realization of *cis*-isotactic structure is associated with several steric hindrances and characterized by a short length of the molecules. During the tetrole polycondensation without initiator of the basis type, which allows the obtaining of the polymer without break of siloxane bond in the organocyclosiloxane, the conditions of synthesis ensures the perfect cyclolinear ladder structure and the macromolecules with the Kuhn segment with length about 50 Å are formatted.

The spectroscopic investigations [14] give the fundament for propose, than the structure of macromolecules of PPSSO, obtained by condensation of tetrole T_4, and differs from structure of one, obtained from phenyltrichlorosilane by anion polymerization at present of the initiator of basis type. The conducted experimental investigations allow suggest that the macromolecules with Kuhn segment about 50 Å have the structure near for of *cis*-anti-*cis*-tactic one [13, 14]:

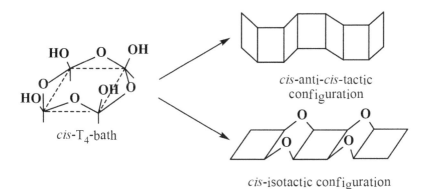

SCHEME 29.2   *Cis*-anti-*cis*-tactic and *cis*-isotactic configuration of PPSSO.

A combination of the T_4 fragments with trans-displacement of the functional-groups the *cis*-sindiotactic structure of chain may be formatted with *cis*-syndiotactic configuration (Scheme 29.3).

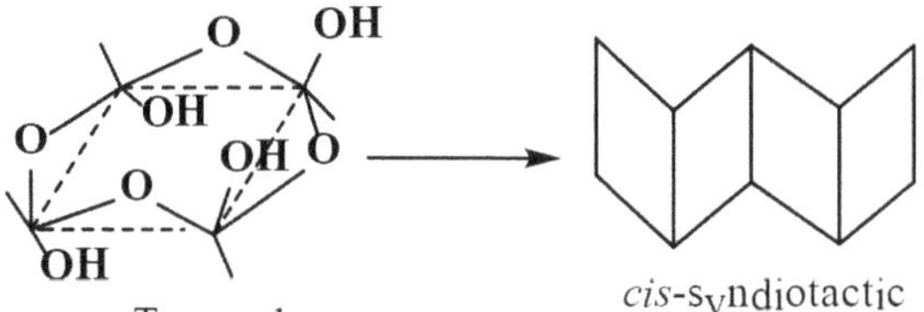

SCHEME 29.3   *Cis*-syndiotactic configuration of PPSSO.

Such structure of the chain is satirically more profitable and is characterized by high rigidity of the chain. Therefore, the dual structure of the molecules of ladder fragments lay under hydrolytic condensation organotriclorosilane and products of it partial hydrolysis and condensation.

So, it was established that in the presence of sodium hydroxide the rearrangement of polyphenylsilsesquioxane skeleton takes place and *cis*-anti-cis structure of silsesquixane structure turn to *cis*-sindiotactic structure, on which indicates the doublet character of absorption bands of ≡Si-O-Si≡ bonds.

With the aim of definition of the effect of organocyclosiloxane fragment length in the tetraepoxy compounds on the properties of a flood composition materials based on them we have investigated the reactions analogically reaction (scheme 29.6) the condensation reactions of tetrahydroxyoligo-organosilsesquioxanes with bifurcated epichlorohydrin at present of 25% solution of sodium hydroxide at room temperature were investigated.

The initial tetrahydroxyphenylsilsesquioxanes (n≈3, 10) with *cis*-anti-*cis*-tactic structure as initial compounds obtained by thermal condensation of *cis*-2.4.6.8-tetrahydroxy-2.4.6.8-tetraphenylcyclotetrasiloxane were used [5, 6]. At interaction of the fragments of T_4 with cis–location of functional groups polyphenylsilsesquioxanes with *cis*-isotactic configuration may be formed if the atoms of silicon in the tetrole molecules are in one and same plane and in *cis*-anti-*cis*-tactic configuration, if the T_4 cycles are in the parallel planes (Scheme 29.2).

However, the realization of *cis*-isotactic structure is associated with several steric hindrances and characterized by a short length of the molecules. During the tetrole polycondensation without initiator of the basis type, which allows the obtaining of the polymer without break of siloxane bond in the organocyclosiloxane, the conditions of synthesis ensures the perfect cyclolinear ladder structure and the macromolecules with the Kuhn segment with length about 50 Å are formatted (Scheme 29.2).

Performing spectroscopic research [13] give grounds to assume that the structure of macromolecules polyphenylsilsesquioxane obtained by thermal condensation tetrol T_4 toluene solution is different from the structure of the macromolecules polyphenylsilsesquioxane obtained from products pheeniltrichlorsilane anionic polymerization initiator in the presence of the basic type. The experimental studies have assumed that the macromolecule with Kuhn segment ~ 50 Å have a structure close to the *cis*-anti-*cis*-tactic configuration [14].

So the condensation reaction of tetrahydroxyoligoorganosilsesquioxanes with epichlorohydrin proceeds according to the following Scheme 29.4:

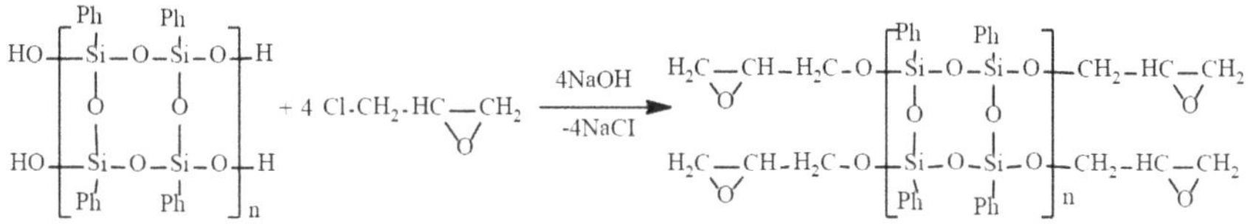

SCHEME 29.4 Condensation of tetraethoxypolyphenylsilsesquioxane (n=3, 10) with epichlorohydrin in the presence of sodium hydroxide.

where n=3 (II), 10 (III).

The obtained compounds II and III are amber color viscous products well soluble in the acetone, methylethylketone and ethyl acetate. The yield, number of epoxy-groups, molecular masses and elemental analysis of obtained silicon-organic oligomers are presented in the Table 29.1.

For obtained epoxy-phenylsilsesquioxsane the reaction between an industrial product tetraethoxyphenylsilsesquioxanes PhES-50 (n=1) and PhES-80 (n=2) with epichlorohydrin have been investigated. The reaction is carried out at a temperature of 80–100°C in an inert gas atmosphere in the presence of a catalytic amount of iron chloride (III) or aluminum chloride, in Scheme 29.5:

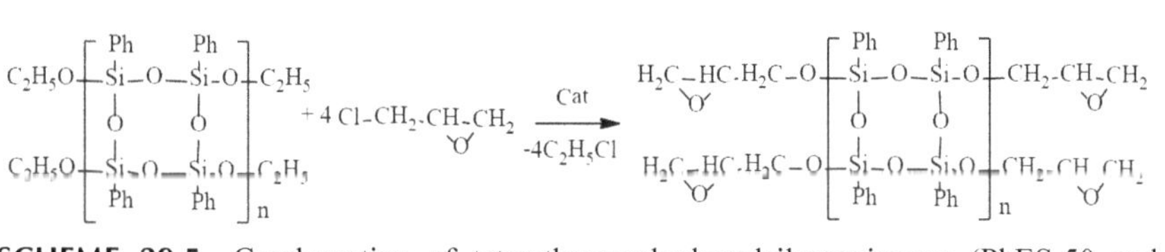

SCHEME 29.5 Condensation of tetraethoxypolyphenylsilsesquioxane (PhES-50 and PhES-80) with epichlorohydrin in the presence of catalyst.

where Cat-FeCl$_3$: n=1 – I, ' 2 (IV); Cat – AlCl$_3$, n=2 – IV'.

It is interesting to note, that iron (III) chloride paints the reactionary mix in dark brown color. Decolorizing of an obtained product neither processing on a centrifuge, nor adsorption of a solution of siliconorganic oligomers on the activated coal was not possible. Therefore the further researches were carried out at the presence of catalytic amounts of aluminum chloride.

The structure and composition of synthesized oligomers were determined by means of elementary and functional analyzes by finding of molecular masses, FTIR and ^{1}H NMR spectra data. In the FTIR spectra of compounds I' and IV the absorption bands both for ν_{as} asymmetric, and for symmetric valence oscillation of $\equiv$Si-O-Si$\equiv$ bonds are kept in the field of absorption at 1060–1010 cm^{-1}, and also there are new absorption bands at 820–840, 917 cm^{-1} characteristic for epoxy rings. In the spectra does not observe the absorption bands, characteristic for ethoxy groups that testifies to full replacement of ethoxy groups by epoxy one. In the ^{1}H NMR spectra of synthesized oligomers one can observe multiplet signals with center of chemical shift δ= 3.3 ppm characteristic for methylene protons in the -CH$_2$O- group; multiplet signal characteristic for methine group of oxirane cycle with center of chemical shift δ=3.01 ppm and also multiplet signal characteristic for methylene group of oxirane cycle with center of chemical shift δ=2.5 ppm. Yields, amount of epoxy groups and molecular masses $\overline{M}_n$, $\overline{M}_\omega$, $\overline{M}_z$ and polydispersity of synthesized epoxy group containing siliconorganic oligomers I", IV and IV' are presented in Table 29.2.

It was established that the synthesized tetraepoxypolyphenylsilsesquioxanes I' and IV do not change their characteristic properties during three months that specifies their long viability (Table 29.3).

Epoxy group functionalized polyphenylsilsesquioxane

TABLE 29.2   Yields, Amount of Epoxy Groups and Molecular Masses of Synthesized Tetraepoxypolyphenylsilsesquioxanes (I–II)

№	Yield, %	Epoxy group, %	M*	$\overline{M}n$	$\overline{M}\omega$	$\overline{M}z$	$\overline{M}\omega/\overline{M}n$
I'	94	4.74	3626	–	–	–	–
		4.80	3672				
IV	93	13.52	1272	1260	2460	6250	1.95
		13.58	1277				
IV'	95	13.52	1272	1200	2080	4100	1.83
		13.48	1268				

*In numerator the calculated values are presented and in denominator experimental values; Molecular masses were calculated from the values of epoxy groups.

TABLE 29.3 Change of Specific Viscosity, Amount of Epoxy Groups for Oligomers I' and IV Depending on Duration of Their çstorage

№	η_{sp} 50%-benzene solution at 20°C					Amount of epoxy groups, %				
	After reaction	After 10 days	After 1 month	After 2 months	After 3 months	After reaction	After 10 days	After 1 month	After 2 months	After 3 months
XVI	2.6	2.8	2.8	2.8	2.9	24.30	24.29	24.28	24.28	24.28
XVII	5.2	5.2	5.3	5.5	5.5	7.19	7.18	7.17	7.17	7.17

From obtaining epoxyorganosilicon the compounds I-III were used as modifiers for epoxy-dian resin ED-22.The compounds were prepared on the basis of ED-22 with 20–24% epoxy-groups (100 mass parts) at different ratio of modifier and hardener: D-1 – methylphenyldiamine (MPDA), D-2–4,4' – diaminodiphenylsulfone (4,4'-DADPS), D-3–4,4'diaminodiphenylmethane (4,4'-DADPM), D-4–4,4'-diaminotriphenyloxide (4,4'- DATPO) and D-5 – metatetrahydrophtalanhydride (MTHFA). The hardening of composites was conducted both at room and high temperatures. Composition and the hardening regime of obtained composites were conducted also both at room and high temperatures. The composition and hardening regime of obtained compounds are presented on the Table 29.4 and physical-mechanical and electric properties – on the Table 29.5. The obtained compounds have the high dielectric, mechanic characteristics and thermal-oxidative stability. The results of investigation of Dependence of physical-mechanical and dielectric properties on the number of the introduced modifier are presented on the Table 29.6. The study of change of compounds viscosity in the process of hardening on temperature shows that compound differ from analogous by high viability in the range 80–90°C, however at temperatures 150–160°C its viability sharply decreases, which is connected with quick destruction of the functional groups and its consequent hardening.

TABLE 29.4 Composition and Hardening Regime of the Composites Based on Tetraepoxypolyphenylsilsesquioxanes and Epoxy-Resin ED-22

Composite	Composite content	Mass part	Composite Color	Hardening tem-re, T°C	Hardening time, h
G-1	Epoxy pitch ED-22	100	Light yellow viscous mass	160	4
	Oligomer I	40–45			
	Binder D-1	13			
G-2	Epoxy pitch ED-22	100	Light yellow, transparent mass	155	4
	Oligomer III	40–45			
	Binder D-2	13			
G-3	Epoxy pitch ED-22	100	Dark-yellow viscous mass	155	5
	Oligomer IV	40–45			
	Binder D-4	10–15			

TABLE 29.5 Mechanical and Electric Properties of Compounds Based on Tetraepoxypolyphenylsilsesquioxanes and Epoxy Resin ED-22

№	Characteristic	Compound			Unit
		G-1	G-2	G-3	
1	Electrical strength	24.5	22	23.8	кB/mm
2	Specific surface electrical resistivity at 20°C	1×10^6	1.2×10^6	0.6×10^{17}	°m
3	Specific volumetric electric resistivity at 20°C	1.5×10^{13}	2×10^{14}	5×10^{13}	°m/cm
4	Strength at bending	111.0	113.5	120.0	MPA
5	Heat resistance according to Martens	77.8	77.5	87.5	°C

By study of thermal oxidation destruction of obtained composites it is established that the polymers are more high thermal proof systems in comparison with initial epoxy resin.

The Figure 29.3 shows that at increasing of the ladder fragment length of a chain increases mainly relative stability of the composites. Nearly same situation has place at study of stability of the composites with different hardeners as in preliminary case in accordance with the Figure 29.3.

All composites G–D based on epoxy resin ED-22 modified by synthetic tetraepoxypolyphenylsilsesquioxanes and hardened by amine type hardeners are divided on two stages. On the first stage composites destruction is followed by high loss of the mass. This process proceeds at temperatures from 180° up to 420°C. Probably on this stage the compounds organic groups burn out (Figure 29.4).

In search of polymers with different properties along with the research of new methods for the synthesis of oligomers and polymers special place choice of optimal curing conditions modified composites. In this regard, particular importance has an elaboration of rational designs temperature curing of composites. We have studied the hardening of composite G-D with different amine hardener D-1, D-2, D-3, D-4, D-5 at different temperatures from 80°C up to 160–180°C.

At obtaining of superposed composites of epoxyorganosilicon oligomers (I and III) and epoxy-dian resin ED-22 the chemical interaction between them and decreasing of the functional epoxy groups were not

TABLE 29.6 Physical-Mechanical Properties of Epoxy Resin ED-22, Modified By Epoxyorganosilicon Oligomer XII With Hardener MTGPA (D-5)

№	Containing of modifier, mass parts (at 100 mass parts of the resin)	Destruction stress at stretching, MPa	Relative Elongation at rupture, %	Hardness by Brinnel, MPa	Electric durability, kV/mm	Tangent of dielectric losses angle at 10^3Gz	1. Dielectric penetration at 10^3Gz	Specific volumetric electric resistance, $\times 10^{13}$ Ohm $\times$ cm
1	10	7.15	10	1.0	61	0.022	5.92	4.3
2	15	8.1	12	1.0	68	0.029	6.06	4.3
3	20	6.8	10	1.2	52	0.018	5.85	4.3
4	25	5.2	9	1.4	46	0.011	5.02	4.2
5	35	5.01	8	1.5	44	0.011	5.02	4.2
6	45	5.01	7	1.7	42	0.010	5.02	4.1
7	0	3.5	0.8	1.0	20–25	0.0045	3.90	–

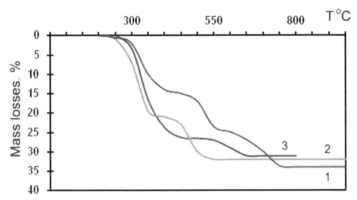

FIGURE 29.3 Thermogravimetric curves for the compounds G-1(1), G-2 (2), G-3 (3) hardened by hardener D-1.

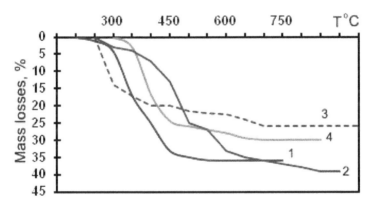

FIGURE 29.4 Thermogravimetric curves of thermal oxidative destruction for the compound G-2 hardened by hardeners D-1 (1), D-2 (2), D-3 (3) and D-4 (4).

observed. However, at hardening their behavior is another. The present of the functional groups with high reaction ability even at present of amine type hardeners leads to chemical interaction and step-by-step disappearing of the functional groups.

From Table 29.7 full curing of the composite G-1 with binder D-2 at a temperature of 150°C is reached within 4–5 h.

The monomers containing organosilicon fragments and epoxy groups in the molecule are cured in the presence of amine curing agents and give

TABLE 29.7   Yields of Gel-Fraction of the Composite K-1 With Hardener D-1 in Dependence of Hardening Temperature

№	Hardening time, min	Hardening temperature, °C	Yield of gel-fraction, %
1	20	18	0
2	60	45	0.1
3	120	60	0.2
4	160	80	2.8
5	180	110	10.0
6	240	120	40.0
7	270	130	60.0
8	300	140	80.0
9	360	150	100.0

a three-dimensional cross-linked polymers, deformation heat resistance, which is in the range 250–300°C. Combinations of different types behave differently when cured. During curing of composites by hardeners of cold cure, for example polyethylene polyamine (PEPA), increasing of the hardening time till 3 days is observed.

The hardening kinetics for composites G-D and for pure ED-22 at using of hardeners PEPA and MTGPA have been studied. It was shown that hardening rate for ED-22 is essentially higher than in case of other compounds. At heating of these composites the fullness of hardening increases, however, the hardening rate for epoxy resin is higher than for all organosilicon composites. This fact proves once more that the modified composites are characterized with lower reaction ability and more high viability than that for epoxy-resin. Therefore these composites are more suitable embedding materials.

At account of noted above it may be concluded that hardening of the epoxy resin modified by organosilicon oligomers (I-III) may be divided on two stages. On the first stages proceeds the process of the slow formatting of the linear polymers and increasing of polymer grid because of interaction between functional groups and hardener, the consumption on which reaches about 30%. The temperature of reaction mass on the first stage of hardening must be not exceed 60°C. To it ensures the application of MTGPA instead of PEPA.

The second stage of the process takes place by heating the composition up to 120–150°C during 3 h. Under such conditions, insoluble and infusible polymers are obtained by the interaction of the functional groups and the formation of new cross-links due to the reaction of the secondary hydroxyl groups.

In the threefold system (ED-22 + epoxyorganosilicon oligomer + hardener) interaction of hardener occurs with ED-22, and with epoxyorganosilicon modifier. And the second reaction is intense during heat treatment. The hardened material presents relatively rare grid of epoxy polymer filled with thermoplastic product epoxyorganosilicon modifier + hardener and therefore has more high than pure epoxy polymer strengthening and thermal stability.

Increasing of hardener amount in the threefold system leads to obtaining of the polymer with high density of the grid. Such polymer is characterized by more stress structure, which leads to decreasing of its mechanical strengthening and polymer becomes fragile. For hardening of composites of the threefold system the optimal amount of hardener is 10–15 mass. part (20–25 wt%). For the samples hardened during 5 h at 150°C decreasing of softening temperature and increasing of high elastic deformation are observed. This fact explained by full finishing of the chemical processes of hardening of thermal treated samples.

So, it seems that polymer materials based on epoxy-dian resin ED-22 modified by epoxyorganosilicon oligomers possess with the complex of the valuable exploitation properties: high thermal stability, improved dielectric, physical and mechanical properties. The complex of these properties is fully corresponds to modern technical demands and is perspective.

29.4 CONCLUSIONS

The condensation reaction of tetraethoxyphenylsilsesquioxane with the excess epiclorohydri in the presence of catalysts has been investigated and corresponding tetraepoxy derivatives has been obtained. These composite compounds may be used as a potting material.

KEYWORDS

- composites
- silsesquioxane based materials
- synthesis
- potential application

REFERENCES

1. Abeand, Y., Gunji, T. *Prog. Polym.* Sci. 29, 149, 2004.
2. Baney, R. H., Itoh, M., Sakakibara, A. Suzuki, T. *Chem. Rev.* 95, 1409, 1995.
3. Cordes, D. B., Lickissand, P. D., Rataboul, F. *Chem. Rev.* 110, 2081, 2010.
4. Choi, J., Yee, A. F., Laine, R. M. *Macromol.*, 36(15), 5666, 2003.
5. Mukbaniani, O. V., Khananasvili, L. M., Inaridze, I. A. Intern. J. Polymeric Materials, 24, 211, 1994.
6. Mukbaniani, O. V. and Zaikov, G. E. The book, New Concepts in Polymer Science, "Cyclolinear Organosilicon Copolymers: Synthesis, Properties, Application". Printed in The Netherlands, "VSP," Utrecht, Boston, 2003.
7. Prilezhaev, E. N. Prilezhaev Reaction. Electrophilic oxidation, Moscow, "Nauka", 332 (1974) (In Russian).
8. Huirong, Y., Richardson, D. E. J. Am. Chem. Soc. 122, 3220, 2000.
9. USA Patent 6124418, 2000.
10. Valetski, P. M. Polymer Chemistry and Technology 2, 64, 1966 (In Russian).
11. Mukudan, A. L., Balasubramian, K., Srinisavan, K. S. V. Polym. Commun., 29 (10), 310, 1988.
12. Iskakova, M. K., Markarashvili, E. G., Mindiashvili, G. S., Shvangiradze, G. M., Gvirgvliani, D. A., Mukbaniani, O. V. Abstract of Communications of X All Russian Conference "Organosilicon Compounds: Synthesis, Properties and Application", Moscow, Russian Federation, 25–30 May, 3C12, 2005.
13. Volchek, B. Z., Purkina A.B *Macromol.*, 28A(6), 1203, 1976.
14. Tverdochlebova, I. I., Mamaeva, I. I. *Macromol.*, 26A(9), 1971, 1984.

CHAPTER 30

THE COMPARATIVE STUDY OF THERMOSTABLE PROTEIN MACROMOLECULAR COMPLEXES (CELL PROTEOMICS) FROM DIFFERENT ORGANISMS

D. DZIDZIGIRI, M. RUKHADZE, I. MODEBADZE,
N. GIORGOBIANI, L. RUSISHVILI, G. MOSIDZE, E. TAVDISHVILI,
and E. BAKURADZE

*Department of Biology Faculty of Exact and Natural Sciences,
Iv. Javakhishvili Tbilisi State University, Tbilisi, Georgia, E-mail:
d_dzidziguri@yahoo.com*

CONTENTS

30.1   INTRODUCTION

Identification of the functions of proteins and other polymeric complexes or cell proteomes, in which the achievements of proteomics contributes greatly, is the subject of intensive research. [13]. Modern technology now allows us to investigate not only individual protein molecules in living cell, but also to understand their interaction with other macromolecules and reveal their previously unknown functions. Several facts are determined: participation of polyfunctional macromolecular protein complexes in the biosynthesis of fatty acids, involvement of erythrocyte membrane proteins macromolecular complexes in exchange of CO_2/O_2, biological effects of some growth factors (polyfunctional proteins), which sometimes is achieved by interactions of other protein complexes, etc. ([4, 5]; http://belki.com.ua/belki-struktura.html).

Earlier we have identified the protein complex with such properties in various cells of adult white rats [1, 4, 6]. The main feature of those complexes is the thermostability of the containing components. With gel electrophoreses and chromatography of hydrophobic interaction it was determined that components with high molecular weight (45–60kD) are hydrophobic whereas components with low molecular weight (11–12kD) are hydrophilic according to the column retention time. Through the inhibition of transcription, complex reduces the mitotic activity of homo- and heterotypic cells in growing animals [6, 10]. Components of complex are water-soluble and its not characterized by species specificity. Thus, we can assume that they are maintained in cells of phylogenetic distance organisms and in case of confirmation of this fact they may belong to the conservative family of proteins. In order to determine general regularities of effects of complex described by us, it is necessary to more detailed examination of components phylogenesis.

30.1.1   THE GOAL

Extraction and comparative characterization of thermostable protein complexes from phylogenetic distance organisms.

30.2   EXPERIMENTAL PART

30.2.1   MATERIAL AND METHODS

The thermostable protein complexes obtained by alcohol extraction from normal cells of various organisms (bacteria, snail, lizard, guinea pig, rat, etc., also human postsurgical material and cell culture) were used for research.

Thermostable protein fractions were obtained by the method of alcohol precipitation of Balazs and Blazsek [2], with modification. Animals were decapitated under diethyl ether. Organs were removed quickly, separated from capsules of connective tissues and vessels, rinsed with the physiological solution, and crushed. Aqueous homogenates were prepared in a tissue/distilled water ratio of 1:8. The homogenates were saturated step-wise with 96% ethanol to obtain 81% ethanol fraction, which was heated in a water bath (100°C) for 20 min, cooled and centrifuged (600 g, 15 min). The supernatant was frozen in liquid nitrogen and dried in an absorptive-condensate lyophilizer. As a result a residue was obtained of a thermostable protein complex (TSPC), a light-gray powder soluble in water. Samples were kept at 4°C. Protein concentration was determined by the method of Lowry et al. [9].

Hydrophobic interaction chromatography (HIC) was used for comparative analysis of TSPC [11]. A hydrophilic polymeric sorbent, HEMA BIO Phenyl-1000 (particle size 10 mm) modified by phenyl groups, served as the stable phase. The mobile phase was phosphatebuffer (pH 7.4) with ammonium sulfate. Elution was performed with the mobile phase in molar concentration range from 2.0 M to 0.0 M (pure buffer) with respect to $(NH4)_2SO_4$. For coelution of hydrophilic and hydrophobic components of protein fractions, Brij-35 polyoxyethylene dodecyl ether with increasing concentration from 0% to 3% was added to the mobile phase. UV detection was usually set at 230 nm.

30.3   RESULTS AND DISCUSSION

Dynamic Interaction between the protein molecules (protein – protein) determines vital activity of the cells. In the last few years as a result of intensive research in this field some knowledge about formation and function of protein complex has been accumulated. Dynamic interactions of proteins are studied not only in individual species but also in different types of cells and tissues. Therefore, in the first stage the aim of the research was to obtain and compare the thermostable protein complex from organisms of different classes.

It was established that all the protein complex samples contain qualitatively different two groups of proteins. I group is hydrophilic and II group – hydrophobic proteins with a column retention time of 5–6 min. and 20 min., respectively, (Figure 30.1a). The same subfractions were revealed in case of protein complexes obtained from different tissues of white rat by using this method (Figure 30.1b).

It is known that dysfunctions of protein-protein interactions can lead to development of various diseases, including cancer, neurodegeneration, autoimmune diseases and more. Therefore, the analysis of protein networks based on the protein-protein interaction may be used in developing various therapeutic approaches [14]. On the next stage of the research we have performed comparative analyzes of protein complexes obtained from rat kidney and normal and transformed renal tissue of human. The differences between normal and transformed cells were revealed in this experiment. In particular, the components with low molecular weight was not observed in the protein complex obtained from human kidney cancer, which indicates on the changes of the complexes composition and their function during of cancer development (Figure 30.2).

The low molecular weight subfraction of protein complex has mitosis-inhibitory properties, reduces number of active and moderately active

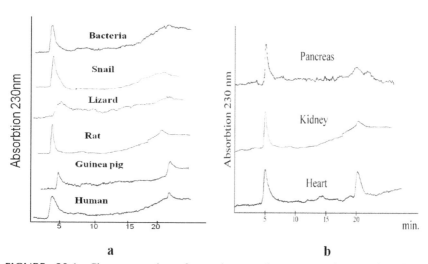

FIGURE 30.1 Chromatography of protein complex: a. protein complex from phylogenetically distance animals; b. protein complex form various tissues of white rat.

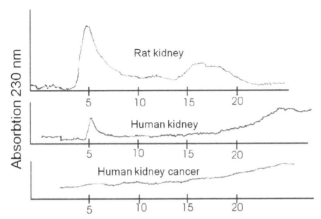

FIGURE 30.2 Chromatography of protein complexes obtained from rat kidney and normal and transformed renal tissue of human.

nucleoli, and decreases the activation of RNA synthesis in nuclei of cardiomiocytes of newborn rat [12]. Usually, this subfraction is essential component of any thermostable protein complex obtained from various organs of adult rat. It is always seen as a major subfraction in native gel-electrophoresis (PAAG electrophoresis) (Figure 30.3).

Consequently, on the next stage of experiment the low molecular weight components of protein complexes obtained from intact (rat pancreas, kidney, brain) and transformed cells (with different degrees) (human papillary carcinoma, CLL, hemangioma) were examined.

As it is shown at Figure 30.3 component with low molecular weight are well expressed in Protein complexes from various organs (kidney,

Rat: pancreas (I), kidney (II), brain (III).   Human: papillary carcinoma (IV), CLL (V), hemangioma (VI).

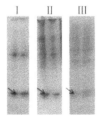
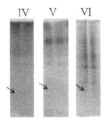

FIGURE 30.3 Electrophorogram of TPC from different organ of white rat and human (arrows indicate on low molecular weight subfraction of TPC).

pancreas and brain) of adult rat (Figure 30.3). Different picture is shown in complexes from transforming cells. In particular, electrophorogrames shows that protein components with low molecular weight are manifested in the samples obtained from benign as well as malignant cells. However, intensity of silver nitrate staining is much lower compared to the norm.

30.4 CONCLUSIONS

From these results it can conclude that pro- and eukaryotic cells contain a thermostable protein complex that inhibits cell proliferation. Quantitative content of protein components in the complex is changing with the growth of transformation degree of cells. Currently, development of the relative proteomics allows us to determine the identity of proteins within these complexes to reliably identify the set of proteins which is responsible for and participates in the regulation of proliferation are constantly presented in the cell. With the help of comparative proteomics it was identified Nilaparvata lugens proteins that are involved in the process of proliferation and their expression changes in response to insecticide treatment [8].

KEYWORDS

- Comparative Study
- Protein Macromolecular Complexes
- polymeric complexes
- cell proteomes
- biosynthesis
- gel electrophoreses

REFERENCES

1. Amano, O., Iseki, S. Expression and localization of cell growth factors in the salivary gland. Kaibogaku Zasshi, 76(2), 201–212, 2001.
2. Balazs, A., Blazsek, I. Control of cell proliferation by endogenous inhibitors. Akademia Kiado (Budapest), 302, 1979.

3. Dijke, P. T., Iwata, K. K. Growth factors for wound healing. Biotechnology, 7(8), 793–798, 1992.

4. Dzidziguri, D., M. Iobadze, T. Aslamazishvili, G. Tumanishvili, V. Bakhutashvili, T. Chigogidze, L. Managadze. The influence kidney protein factors on the proliferative activity of MDSK cells. Tsitologiya, 46(10), 913–914, 2004.

5. Ge, L. Q., Cheng, Y., Wu, J. C., Jahn, G. C. Proteomic analysis of insecticide triazophos-induced mating-responsive proteins of *Nilaparvata lugens* Stal (Hemiptera: Delphacidae), *J. Proteome Res.*, 10 (10), 4597–4612, 2011.

6. Giorgobiani, N., D. Dzidziguri, M. Rukhadze, L. Rusishvili, G. Tumanishvili. Possible role of endogenous growth inhibitors in regeneration of organs: Searching for new approaches. Cell Biology International, 29, 1047–1049, 2005.

7. http://belki.com.ua/belki-struktura.html

8. Lin-Quan Ge, Yao Cheng, Jin-Cai Wu, Gary, C. Jahn. Proteomic Analysis of Insecticide Triazophos-Induced Mating-Responsive Proteins of Nilaparvata lugens Stål (Hemiptera: Delphacidae). School of Plant Protection, Yangzhou University, Yangzhou 225009, P. R. China, J. Proteome Res., 10 (10), 4597–4612, 2011.

9. Lowry DH, Rosebrough, N. J., Farr, A. L., Randell, R. J.. Protein measurement with the folin phenol reagent. J Biol Chem, 193, 265–275, 1951.

10. Modebadze, I., M. Rukhadze, E. Bakuradze, D. Dzidziguri. Pancreatic Cell Proteome – Qualitative Characterization And Function. Georgian Medical News, 7–8(220–221), 71–77, 2013.

11. Queiroz, J. A., Tomaz, C. T., Cabral, J. M. S. Hydrophobic interaction chromatography of proteins. J Chromatogr, 87, 143–59, 2001.

12. Rusishvili, L. Giorgobiani, N., Dzidziguri, D., Tumanishvili, G. Comparative analysis of cardiomiocyte growth-ihibitory factor in animals of different classes. Proc. Georgian Acad. Sci., Biol. Ser. B, 1(1–2), 42–45, 2003.

13. Shaojun Dai, Taotao Chen, Kang Chong, Yongbiao Xue, Siqi Liu, Tai Wang. Proteomics Identification of Differentially Expressed Proteins Associated with Pollen Germination and Tube Growth Reveals Characteristics of Germinated, *Oryza sativa*. Pollen *Mol Cell Proteomics*, 6, 207–230, 2007.

14. Terentiev, A. A., Moldogazieva, N. T., Shaitan, K. V. The dynamic proteomics in the modeling of living cell. Protein-protein interactions. Success of biological chemistry (Advance in Biological Chemistry), 49, 429–480, 2009 (article in Russian).

INDEX